T0222622

Springer Undergraduate Mathematics Series

Advisory Editors

M. A. J. Chaplain, *St Andrews, UK*
Angus MacIntyre, *Edinburgh, UK*
Simon Scott, *London, UK*
Nicole Snashall, *Leicester, UK*
Endre Süli, *Oxford, UK*
Michael R. Tehranchi, *Cambridge, UK*
John F. Toland, *Bath, UK*

The Springer Undergraduate Mathematics Series (SUMS) is a series designed for undergraduates in mathematics and the sciences worldwide. From core foundational material to final year topics, SUMS books take a fresh and modern approach. Textual explanations are supported by a wealth of examples, problems and fully-worked solutions, with particular attention paid to universal areas of difficulty. These practical and concise texts are designed for a one- or two-semester course but the self-study approach makes them ideal for independent use.

More information about this series at http://www.springer.com/series/3423

Matej Brešar

Undergraduate Algebra

A Unified Approach

 Springer

Matej Brešar
Department of Mathematics
Faculty of Mathematics and Physics
University of Ljubljana
Ljubljana, Slovenia

Department of Mathematics and Computer Science
Faculty of Natural Sciences and Mathematics
University of Maribor
Maribor, Slovenia

ISSN 1615-2085   ISSN 2197-4144   (electronic)
Springer Undergraduate Mathematics Series
ISBN 978-3-030-14052-6   ISBN 978-3-030-14053-3   (eBook)
https://doi.org/10.1007/978-3-030-14053-3

Library of Congress Control Number: 2019933855

Mathematics Subject Classification (2010): 12-01, 13-01, 16-01, 20-01

© Springer Nature Switzerland AG 2019
This work is subject to copyright. All rights are reserved by the Publisher, whether the whole or part of the material is concerned, specifically the rights of translation, reprinting, reuse of illustrations, recitation, broadcasting, reproduction on microfilms or in any other physical way, and transmission or information storage and retrieval, electronic adaptation, computer software, or by similar or dissimilar methodology now known or hereafter developed.
The use of general descriptive names, registered names, trademarks, service marks, etc. in this publication does not imply, even in the absence of a specific statement, that such names are exempt from the relevant protective laws and regulations and therefore free for general use.
The publisher, the authors and the editors are safe to assume that the advice and information in this book are believed to be true and accurate at the date of publication. Neither the publisher nor the authors or the editors give a warranty, expressed or implied, with respect to the material contained herein or for any errors or omissions that may have been made. The publisher remains neutral with regard to jurisdictional claims in published maps and institutional affiliations.

This Springer imprint is published by the registered company Springer Nature Switzerland AG
The registered company address is: Gewerbestrasse 11, 6330 Cham, Switzerland

Preface

The term *algebra* is used in different contexts. We learn elementary algebra in high school, and linear algebra in one of the first mathematical courses at university. This book provides an introduction to the algebra that is more precisely called *abstract algebra*. And it really is abstract. We are able to illustrate abstract algebraic concepts through concrete examples and soon get used to handling them. But what is their purpose? Questions like this often distract beginners. The basic definitions are so general that it can be difficult to imagine how they can be useful. And yet, it is a fact that algebraic methods are an extremely effective tool for solving various kinds of mathematical problems. Before we can begin to use them, however, we need to truly understand them—and this takes time. Some examples of their applicability will be presented in this book, but only in the last chapters. We comprehend algebra when we grasp the whole picture: the introduction of abstract concepts and their elementary consideration gradually lead to solutions of difficult concrete problems. The proofs are often short and elegant. There is not much computing; instead, intertwining seemingly simple ideas is what yields results. Algebra is stunningly beautiful.

The book is based on lectures I gave to second year undergraduate students at the University of Ljubljana during the last four academic years. The original version is written in Slovenian. The present English version is not only a translation, but contains additional material and some modifications.

There is a myriad of books on basic abstract algebra, so why does the world need another one? The present book covers the usual topics one would expect to find. Its special feature, however, is that these topics are grouped together in a non-traditional way, thereby giving the reader a somewhat different perspective on the main concepts of algebra than one gets from reading standard textbooks.

In a typical abstract algebra course, basic group theory and basic ring theory are presented as different areas. Yet, as everyone working in algebra knows, only the frameworks are different, but the ideas on which both theories are based are the same. So why not teach algebra in the way that we algebraists understand it? That is, why not treat conceptually similar themes on different algebraic structures concurrently, in a unified way, rather than scatter them all over the place? This has been in my mind for many years while teaching a standard algebra course in Ljubljana. Four years ago, I decided to give it a try and present the material to the class in the way that is now presented in the book. After teaching this way for four years, I can say that it was a very good decision. There are more opportunities to point out essential ideas, tedious repetitions are avoided, and, most importantly, I feel that students understand algebra better than before. Especially now, when the Slovenian version of the book is available to them, I can really feel the difference. This has encouraged me to translate the text into English.

The book is divided into two parts. Part I, entitled *The Language of Algebra*, is the one in which different algebraic structures are treated concurrently. Part II, entitled *Algebra in Action*, is more traditional, studying groups, rings, and fields in separate chapters.

Most scientists do not use deeper mathematical theorems, but need the language of mathematics to present their discoveries in an exact manner. Similarly, mathematicians working in other areas do not use profound algebraic theorems everyday, but need the language of algebra to communicate with other mathematicians. Familiarity with the content of Part I is thus a necessity for every mathematician as well as for every serious user of mathematics. This includes the understanding of numerous abstract concepts which may be demanding for some students. Yet in many ways Part I is easy, containing mostly simple results with straightforward proofs. Some of the readers will surely be able to find these proofs themselves and only after that compare them with those in the book, which is certainly the most instructive and entertaining way of studying. The wealth of examples and some challenging exercises make Part I more colorful, but its main purpose is just learning the language of algebra and giving a solid foundation for tackling serious algebraic topics. It consists of four chapters. Chapter 1 introduces basic algebraic structures and some other fundamental notions. It includes some examples, but only simple ones which should not be entirely new to the reader. Chapter 2, on the other hand, is devoted exclusively to examples. Its main message is that algebraic structures crop up all over mathematics. We are encountering them all the time, but have to learn to recognize them. Chapter 3 considers homomorphisms. Besides basic properties and examples, it also includes results on various natural embeddings such as Cayley's Theorem, the embedding of an integral domain into its field of fractions, etc. Chapter 4 discusses quotient structures. A particular emphasis is given to their connection with homomorphisms.

Part II is much more exciting from the mathematical point of view. The problems are intriguing and many of the proofs are striking. The strength and beauty of the abstract way of thinking become apparent. The reader can now see the point of building the algebraic machinery in Part I, which is used throughout as a fundamental tool. The table of contents says it all: the main themes are commutative rings (Chap. 5), finite groups (Chap. 6), and field extensions (Chap. 7). When discussing advanced topics such as modules over principal ideal domains and Galois theory, I focused only on the essential matter, having in mind that presenting too many different aspects could confuse the reader who is encountering abstract algebra for the first time. Also, I have tried to make the text accessible to readers who are only seeking information on a particular topic, and have no intention to study the whole book systematically. For this reason, I have introduced as little new terminology and notation as possible.

The final sections of each of the three chapters of Part II may be considered optional. The basic material is always covered in the first sections. The contents of Sects. 5.1, 6.1, 6.2, 7.2, and 7.3 should probably be included in any undergraduate algebra course, while the rest depends on available time and the level of course difficulty. Parts of some sections in Part I (for example, Subsects. 3.5.2, 3.5.3, 4.4.4, and 4.5.2) may also be skipped, and of course it is not absolutely necessary to cover all of the examples.

Each section ends with exercises. They tend to be arranged progressively, from easier to more difficult. However, this is not a strict rule since they are also sorted by topics. In any case, it is advisable to do the exercises in the given order. Some of them may be used in subsequent exercises.

The material treated in this book is so standard that I did not find it necessary to include references to the literature. Yet it goes without saying that I have consulted various sources and was influenced by them. In particular, some of the exercises can be found in many other textbooks. On the other hand, the concept of this book made it possible to include some more original exercises that consider parallels and connections between different algebraic structures. Another special feature of the exercises is that quite a few of them require finding concrete examples and counterexamples. The solutions are often rather straightforward, but this type of question may nevertheless be difficult for some students since it demands an insightful understanding rather than following a pattern.

The book is written on the assumption that the reader has previously taken a linear algebra course. Nevertheless, all details, from definitions to proofs, concerning vector spaces, matrices, linear maps, etc., are presented in a succinct manner. Vector spaces (as well as algebras over fields) are actually treated as one of the algebraic structures, perhaps slightly overshadowed by groups, rings, and fields, but still important.

A short chapter preceding the main text surveys the basic set theory and other elementary topics needed to understand the book. A few exercises and examples may require more prerequisites, but can be skipped.

Finally, I would like to express my gratitude to Peter Šemrl for valuable discussions on the organization of the book and for his enthusiastic encouragement. I also thank Victor Shulman and Primož Moravec for reading and commenting on parts of the manuscript. My son Jure was of constant help in my struggle to write decent English. The anonymous referees made some useful suggestions, and the Springer staff deserves thanks for their professional assistance.

Ljubljana/Maribor, Slovenia   Matej Brešar
January 2019

Contents

Prerequisites

A course in algebra may be a student's first encounter with abstract mathematics, but not with higher (university) mathematics. It can thus be assumed that the reader of this book has reached a certain degree of mathematical maturity. This includes familiarity with the notions that we are about to examine—at least with most of them. Our discussion will therefore be rather concise.

Sets

A **set** is a collection of objects. This is just an intuitive, informal definition, but we do not need to treat set theory at a deep, rigorous level. The objects in a set are called the **elements** of the set. If x is an element of the set X, we say that x **lies in** X (or x **belongs to** X, or X **contains** x) and write $x \in X$. Accordingly, $y \notin X$ means that y is not an element of X. For example, the number 2 lies in **the set of natural numbers**

$$\mathbb{N} = \{1, 2, 3, \ldots\},$$

which we write as $2 \in \mathbb{N}$. Some authors consider 0 as a natural number. According to our definition, however, $0 \notin \mathbb{N}$. Of course, 0 lies in the **set of integers**

$$\mathbb{Z} = \{\ldots, -2, -1, 0, 1, 2, \ldots\}.$$

If X and Y are sets and every element of X is also an element of Y, then we write $X \subseteq Y$ (or $Y \supseteq X$) and call X a **subset** of Y. The sets X and Y are **equal**, denoted $X = Y$, if $X \subseteq Y$ and $Y \subseteq X$. If $X \subseteq Y$ and $X \neq Y$, then X is said to be a **proper subset** of Y. In this case we write $X \subset Y$ or $X \subsetneq Y$ (we will use the latter notation when wishing to emphasize that $X \neq Y$). For example, $\mathbb{N}$ is a proper subset of $\mathbb{Z}$, whereas the sets $\mathbb{N}$ and $\{n \in \mathbb{Z} \mid n > 0\}$ are equal.

Let us give an example before the next definition. Denote by $\mathbb{Q}$ the **set of rational numbers**, that is,

$$\mathbb{Q} = \left\{ \frac{m}{n} \mid m, n \in \mathbb{Z}, n \neq 0 \right\},$$

and consider the set

$$\{ q \in \mathbb{Q} \mid q^2 = 2 \}.$$

Sets are usually introduced in such a way, i.e., by describing properties that are satisfied by their elements. Yet there is something unusual about this set—it has no elements! This is not entirely obvious, but is well-known (a more common formulation is that $\sqrt{2}$ is an irrational number) and not hard to prove (see Sect. 7.1 for the standard proof). We often arrive at sets that, sometimes to our surprise, have no elements. This is just one of the reasons why it is convenient to define the **empty set** as the set with no elements. This set is commonly denoted by $\emptyset$.

For any set X, we define the **power set** of X as the set of all subsets of X. It will be denoted by $\mathscr{P}(X)$. Two of its elements are X and $\emptyset$.

A set X is said to be **finite** if it is either empty or there exists a natural number n such that X has exactly n elements. We call n the **cardinality** of X and denote it by $|X|$. For example, $|\{-1, 0, 1\}| = 3$. We also define $|\emptyset| = 0$. A set is said to be **infinite** if it is not finite.

The **Cartesian product** of the sets X_1 and X_2, denoted $X_1 \times X_2$, is the set of all ordered pairs (x_1, x_2), where $x_1 \in X_1$ and $x_2 \in X_2$. More generally, given sets $X_1, \ldots, X_n$, we define their Cartesian product,

$$X_1 \times \cdots \times X_n,$$

as the set of all ordered n-tuples $(x_1, \ldots, x_n)$, where $x_i \in X_i$ for every i. If $X_i = X$ for each i, $X_1 \times \cdots \times X_n$ can be written as X^n. Probably the most familiar example is $\mathbb{R}^n$ where $\mathbb{R}$ denotes the **set of real numbers**. We can also form the Cartesian product of infinitely many sets, but we will avoid giving a formal definition.

Let X_1 and X_2 be sets. Their **intersection**, denoted $X_1 \cap X_2$, is the set of all elements that lie in both X_1 and X_2. If $X_1 \cap X_2 = \emptyset$, then we say that X_1 and X_2 are **disjoint sets**. The **union** of X_1 and X_2, denoted $X_1 \cup X_2$, is the set of all elements that lie in at least one of the sets X_1 and X_2. Finally, the **difference** of X_1 and X_2, denoted $X_1 \backslash X_2$, is the set of all elements that lie in X_1 but not in X_2. For example, if $X_1 = \{2, 5, 9\}$ and $X_2 = \{3, 5\}$, then $X_1 \cap X_2 = \{5\}$, $X_1 \cup X_2 = \{2, 3, 5, 9\}$, and $X_1 \backslash X_2 = \{2, 9\}$.

We often have to work with intersections and unions of more than just two sets. Let I be any nonempty set, finite or infinite. Suppose that to each $i \in I$ we assign a set X_i. Then we can talk about the **(indexed) family of sets** $(X_i)_{i \in I}$. Its intersection, $\bigcap_{i \in I} X_i$, is the set of all elements that lie in X_i for every $i \in I$, and its union, $\bigcup_{i \in I} X_i$, is the set of all elements that lie in X_i for at least one $i \in I$. A family of sets does not need to be indexed in order to consider its intersection and union. For example, the intersection of the family of intervals (a, b) with $a < 0 < b$ is $\{0\}$.

The elementary notions that we have just surveyed will be used throughout the book. We will occasionally mention some more advanced set-theoretic notions such as the axiom of choice and cardinal numbers, but the reader unfamiliar with them may skip the corresponding parts of the text.

Maps

A **map** (or **function**) f from a set X to a set Y is a rule that assigns to each element x of X exactly one element, denoted $f(x)$, of Y. We could give a more precise definition, but we do not need to be completely formal here.

The usual notation for a map f from X to Y is

$$f : X \to Y.$$

The sets X and Y are called the **domain** and **codomain** of f, respectively. When the domain and codomain are clear from the context, we often write

$$x \mapsto f(x)$$

instead of $f : X \to Y$. For example, $x \mapsto e^x + \sin x$ is the map given by the rule $f(x) = e^x + \sin x$, and, unless specified otherwise, we understand that f maps from $\mathbb{R}$ to $\mathbb{R}$.

Let $f : X \to Y$ and let U be a subset of X. The map $f|_U : U \to Y$ defined by

$$f|_U(u) = f(u) \quad \text{for all } u \in U$$

is called the **restriction** of f to U. We usually think of $f|_U$ as being the same map as f, only having a smaller domain.

We say that $f : X \to Y$ is **injective** (or **one-to-one**) if $f(x_1) \neq f(x_2)$ whenever $x_1 \neq x_2$, and we say that f is **surjective** (or **onto**) if for every $y \in Y$ there exists at least one $x \in X$ such that $f(x) = y$. If f is both injective and surjective, then we say that it is **bijective**. In other words, f is bijective if, for every $y \in Y$, there exists exactly one $x \in X$ such that $f(x) = y$. This makes it possible to define the map $f^{-1} : Y \to X$ with the property that for all $x \in X$ and $y \in Y$,

$$f^{-1}(y) = x \iff f(x) = y.$$

We call f^{-1} the **inverse of** f.

We will be particularly concerned with bijective maps from a set X to itself. The simplest example is the **identity map** id_X, defined by

$$\text{id}_X(x) = x \quad \text{for all } x \in X.$$

If X is a finite set, then $f : X \to X$ is injective if and only if it is surjective (and hence bijective). Proving this is an easy exercise.

Given maps $f : X \to Y$ and $g : Y \to Z$, we define a new map, denoted $g \circ f$ (or simply gf), from X to Z by

$$(g \circ f)(x) = g(f(x)) \quad \text{for all } x \in X.$$

We call $g \circ f$ the **composition** of g and f. If f is bijective, then

$$f^{-1} \circ f = \text{id}_X \quad \text{and} \quad f \circ f^{-1} = \text{id}_Y.$$

The notion of the cardinality of a finite set can be extended to infinite sets, but we will not give the definition here. We only define that arbitrary (possibly infinite) sets X and Y have the **same cardinality** if there exists a bijective map from X to Y. Note that finite sets have the same cardinality in this sense if and only if they have the same cardinality in the above defined sense (i.e., they have the same number of elements). A set X is said to be **countable** if X is either finite or has the same cardinality as the set of natural numbers $\mathbb{N}$. In the latter case, we say that X is a **countably infinite set**. It turns out that $\mathbb{Q}$ is a countably infinite set, but $\mathbb{R}$ is not.

Let $f : X \to Y$. For any subset U of X, we define

$$f(U) := \{f(u) | u \in U\}$$

(we will often, although not always, use the symbol := when giving definitions). The set $f(X)$ is called the **image** of f. Obviously, f is surjective if and only if its image is equal to its codomain. Next, for any subset V of Y, we define

$$f^{-1}(V) := \{x \in X | f(x) \in V\}.$$

Note that we always have $f(f^{-1}(V)) \subseteq V$, and that $f(f^{-1}(V)) = V$ if f is surjective.

Equivalence Relation

A **binary relation** R on a set X is a subset of the Cartesian product $X \times X$. We write xRy if $(x, y) \in R$. For example, if $X = \mathbb{R}$ and $x \leq y$ has the usual meaning, we can consider $\leq$ as the binary relation consisting of all pairs $(x, y) \in \mathbb{R} \times \mathbb{R}$ such that x is less than or equal to y. An even simpler example of a binary relation on any set X

is $=$, the equality of elements. As a subset of $X \times X$, it consists of all pairs of the form (x, x).

A binary relation R on a set X is said to be:

(a) **reflexive** if xRx for all $x \in X$;
(b) **symmetric** if xRy implies yRx;
(c) **transitive** if xRy and yRz imply xRz.

The relation $\leq$ on $\mathbb{R}$ is reflexive and transitive but is not symmetric, while the relation of equality, $=$, has all the three properties. Any such relation, i.e., a relation that is reflexive, symmetric, and transitive, is called an **equivalence relation**. These relations are usually denoted by $\sim$, so from now on we will use this symbol rather than R.

We will not give further examples of equivalence relations at this point, since we will encounter them throughout the book. Our goal now is to describe them from a different perspective. To this end, we introduce the following notion. A **partition** of a set X is a set of nonempty subsets of X with the property that every element in X lies in exactly one of these subsets. In other words, X is a **disjoint union** of these subsets, meaning that these subsets are pairwise disjoint and their union equals X. Two examples of partitions of the set $X = \{1, 2, 3, 4, 5\}$ are $\{\{1, 4\}, \{2, 5\}, \{3\}\}$ and $\{\{1, 3, 4, 5\}, \{2\}\}$.

Take now an arbitrary partition of a set X. Define the relation $\sim$ on X as follows: $x \sim y$ if and only if x and y lie in the same subset of the partition. It is immediate that $\sim$ is an equivalence relation. Thus, every partition gives rise to an equivalence relation. For example, the partition in which all the subsets consist of exactly one element gives rise to the relation of equality.

We claim that, conversely, every equivalence relation gives rise to a partition. Let $\sim$ be an equivalence relation on a set X. For every $x \in X$, we set

$$[x] := \{x' \in X \,|\, x \sim x'\}.$$

We call $[x]$ the **equivalence class** of x. The equivalence classes of different elements may be equal—as a matter of fact,

$$[x] = [y] \iff x \sim y.$$

The easy proof is left to the reader. Let us show that the equivalence classes form a partition of X. First observe that $x \sim x$ implies $x \in [x]$. This shows that all equivalence classes are nonempty and that every element of X lies in at least one of them. Suppose that an element x lies in the equivalence class $[y]$ as well as in the equivalence class $[z]$, i.e., $y \sim x$ and $z \sim x$. We must show that $[y] = [z]$. Take $u \in [y]$. Then $y \sim u$. Since we also have $y \sim x$ and $\sim$ is symmetric and transitive, it follows that $x \sim u$. Along with $z \sim x$ this implies $z \sim u$. Thus, $u \in [z]$, which proves that $[y] \subseteq [z]$. Similarly, we see that $[z] \subseteq [y]$. Hence $[y] = [z]$, completing the proof.

We can thus say that the equivalence relation and partition (or disjoint union) are equivalent concepts.

Induction and Well-Ordering

If a subset S of $\mathbb{N}$ contains 1 and has the property that $n + 1 \in S$ whenever $n \in S$, then $S = \mathbb{N}$. This is called the **principle of mathematical induction**. It is intuitively clear, and is in fact one of Peano's axioms for natural numbers.

We often use induction when we want to prove that some statement is true for every natural number n. First we prove that the statement is true for the number 1. We then assume the statement is true for the natural number n and, using this assumption, prove that it is true for the natural number $n + 1$.

We give just one standard example. Let us prove by induction that the identity

$$1 + 2 + \cdots + n = \frac{n(n+1)}{2}$$

holds for every $n \in \mathbb{N}$. The case where $n = 1$ is clear: $1 = \frac{1(1+1)}{2}$. Now assume that the identity holds for some natural number n. Then

$$1 + 2 + \cdots + n + (n+1) = \frac{n(n+1)}{2} + (n+1)$$
$$= \frac{n(n+1) + 2(n+1)}{2}$$
$$= \frac{(n+1)(n+2)}{2},$$

so the identity also holds for $n + 1$. With this our proof is complete. Indeed, we have shown that the set S of those natural numbers for which the identity holds contains 1 and has the property that $n \in S$ implies $n + 1 \in S$, so $S = \mathbb{N}$.

We have stated one form of the principle of mathematical induction. Another standard form, which yields a slightly different method of proof, reads as follows: if a subset S of $\mathbb{N}$ contains 1 and has the property that $n \in S$ whenever S contains all natural numbers smaller than n, then $S = \mathbb{N}$.

It is worth mentioning that in both forms we can replace the set $\mathbb{N}$ by any set consisting of all integers that are greater than or equal to a fixed integer n_0 (i.e., a set of the form $\{n_0, n_0 + 1, n_0 + 2, \ldots\}$ for some $n_0 \in \mathbb{Z}$). Of course, when stating the principle (in any of the two forms) in this setting, the role of 1 must be replaced by n_0.

The second property of natural numbers that will be frequently used is the **well-ordering principle**. It states that every nonempty subset of $\mathbb{N}$ has a smallest element. This means that if $S \subseteq \mathbb{N}$ and $S \neq \emptyset$, then there exists an $s_0 \in S$ such that $s_0 \leq s$ for every $s \in S$. This can be proved by induction as follows. Assume, on the contrary, that S does not have a smallest element. Then $1 \notin S$, because otherwise 1 would be the smallest element. Now take any $n \in \mathbb{N}$ and suppose that none of the numbers $1, \ldots, n - 1$ lies in S. Then $n \notin S$, because otherwise n would be the smallest element. The principle of mathematical induction (more precisely, its second form) implies that $n \notin S$ for every $n \in \mathbb{N}$. However, this means that $S = \emptyset$, contrary to our assumption.

It is easy to see that, conversely, the well-ordering principle implies the principle of mathematical induction. The two principles are therefore equivalent statements.

We say that a subset S of $\mathbb{Z}$ is **bounded below** if there exists an $n_0 \in \mathbb{Z}$ such that $n_0 \le s$ for every $s \in S$. An obvious modification of the above proof shows that the following slightly more general version of the well-ordering principle is true: every nonempty subset of $\mathbb{Z}$ that is bounded below has a smallest element.

Similarly, we say that a subset T of $\mathbb{Z}$ is **bounded above** if there exists an $n_1 \in \mathbb{Z}$ such that $n_1 \ge t$ for every $t \in T$. In this case, the set $\{-t \mid t \in T\}$ is bounded below, and so it contains a smallest element s_0, provided that it is nonempty. But then $-s_0$ lies in T and is larger than any other element in T. The following version of the well-ordering principle is thus also true: every nonempty subset of $\mathbb{Z}$ that is bounded above has a largest element.

We will often use the well-ordering principle without explicitly referring to it.

Complex Numbers

The integers and their properties will be used as an indispensable tool throughout this book. We will also quite often use the rational, real, and complex numbers, although not so much for developing, but rather for illustrating the theory. The rational and real numbers presumably need no introduction, but a brief overview of the basics on complex numbers may be welcomed by some readers. We will state several facts without providing proofs. However, they are straightforward verifications that require just a little patience.

A **complex number** is a formal expression of the form $x + yi$, where x and y are real numbers. Two complex numbers $x + yi$ and $x' + y'i$ are equal if and only if $x = x'$ and $y = y'$. The set of complex numbers is denoted by $\mathbb{C}$.

We identify the real number x by the complex number $x + 0i$, and, in this way, consider $\mathbb{R}$ to be a subset of $\mathbb{C}$. In particular, 0 and 1 are complex numbers. We further simplify the notation by writing $x - yi$ for $x + (-y)i$, $-(x + yi)$ for $-x - yi$, yi for $0 + yi$, and i for $1i$.

We will denote complex numbers by a single letter, usually z, for brevity. Thus, let $z = x + yi$ be a complex number. The real numbers x and y are called the **real part** and the **imaginary part** of z, respectively. We denote them by $\operatorname{Re}(z)$ and $\operatorname{Im}(z)$. We say that z is **purely imaginary** if $\operatorname{Re}(z) = 0$. The basic example is the number i, which we call the **imaginary unit**.

We define addition of complex numbers by the rule

$$(x + yi) + (x' + y'i) := (x + x') + (y + y')i,$$

and multiplication by

$$(x+yi)(x'+y'i) := (xx'-yy') + (xy'+x'y)i.$$

In particular, $i^2 = -1$. This is a crucial formula—note that the general formula for multiplication can be derived from it by a formal computation.

The following identities can be easily checked:

- $w + z = z + w.$
- $(u + w) + z = u + (w + z).$
- $z + 0 = z.$
- $z + (-z) = 0.$
- $zw = wz.$
- $(uw)z = u(wz).$
- $1z = z.$
- $u(w + z) = uw + uz.$

Let $z = x + yi$ be a complex number. The **complex conjugate** of z is defined to be the complex number

$$\bar{z} := x - yi.$$

We list some basic properties:

- $\bar{\bar{z}} = z.$
- $\overline{z+w} = \bar{z} + \bar{w}.$
- $\overline{zw} = \bar{z}\,\bar{w}.$
- $z + \bar{z} = 2\mathrm{Re}(z).$
- $z - \bar{z} = 2\mathrm{Im}(z)i.$
- $z\bar{z} = x^2 + y^2.$

In particular, $z\bar{z}$ is a nonzero real number, provided that $z \neq 0$. Therefore, the following is true:

- If $z = x + yi \neq 0$, then $w := \frac{1}{z\bar{z}}\bar{z} = \frac{x}{x^2+y^2} - \frac{y}{x^2+y^2}i$ satisfies $zw = 1$.

We call

$$|z| := \sqrt{z\bar{z}} = \sqrt{x^2 + y^2}$$

the **absolute value** of z. It has the same properties as the absolute value of real numbers:

- $|z| \geq 0$ and $|z| = 0$ if and only if $z = 0$.
- $|z + w| \leq |z| + |w|.$
- $|zw| = |z||w|.$

The complex number $z = x + yi$ can be represented geometrically as the point having coordinates (x, y) in the Cartesian plane. Its absolute value $|z|$ is then the distance of this point from the origin $(0, 0)$. Assuming that $z \neq 0$, denote by θ the angle from the positive x-axis to the vector representing z (see Fig. 1). We call θ the **argument** of z. Note that $x = |z| \cos \theta$ and $y = |z| \sin \theta$, and hence

$$z = |z|(\cos \theta + i \sin \theta).$$

This is called the **polar form** of z. It is especially convenient for multiplication. Indeed, using standard trigonometric identities, we readily derive the following formula:

$$(\cos \theta + i \sin \theta)(\cos \phi + i \sin \phi) = \cos(\theta + \phi) + i \sin(\theta + \phi).$$

This can be written in a more elegant way. We extend the familiar real exponential function $x \mapsto e^x$ to complex numbers as follows:

$$e^{x+iy} := e^x(\cos y + i \sin y).$$

In particular,

$$e^{iy} = \cos y + i \sin y,$$

so the above formula gets the following form:

- $e^{i\theta} e^{i\phi} = e^{i(\theta + \phi)}$.

Accordingly, the product of the complex numbers $z = |z|e^{i\theta}$ and $w = |w|e^{i\phi}$ is equal to $|z||w|e^{i(\theta + \phi)}$.

Using the polar form, it becomes easy to solve some equations involving complex numbers. For example, the following is true:

- Let $w = |w|e^{i\phi}$ be a nonzero complex number and let n be a natural number. Then $z \in \mathbb{C}$ satisfies $z^n = w$ if and only if

$$z = \sqrt[n]{|w|}\, e^{i\frac{\phi + 2k\pi}{n}}, \quad k = 0, 1, \ldots, n - 1.$$

Fig. 1 Polar form of z

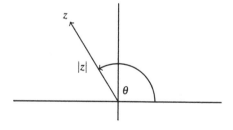

In particular, the solutions of the equation $z^n = 1$ are

$$z_k = e^{\frac{2k\pi i}{n}}, \quad k = 0, 1, \ldots, n-1.$$

These complex numbers are called the *n*th **roots of unity**. Observe that $z_k = z_1^k$ for each k.

It turns out that much more general equations are solvable. The **Fundamental Theorem of Algebra** states that for any $a_0, a_1, \ldots, a_n \in \mathbb{C}$, with $a_n \neq 0$ and $n \geq 1$, there exists at least one $z \in \mathbb{C}$ such that

$$a_n z^n + a_{n-1} z^{n-1} + \cdots + a_1 z + a_0 = 0.$$

In other words, every non-constant polynomial with complex coefficients has at least one complex root. We will prove this theorem at the very end of the book. Nevertheless, we shall assume it as a fact on a few occasions before then, since, without it, some of the abstract theory that we will develop could seem vacuous.

Frequently Used Symbols

$\mathbb{N}$	Natural numbers
$\mathbb{Z}$	Integers
$\mathbb{Q}$	Rational numbers
$\mathbb{R}$	Real numbers
$\mathbb{C}$	Complex numbers
$\mathbb{Z}_n$	Integers modulo n
$n\mathbb{Z}$	Integer multiples of n
S^*	Group of invertible elements of the monoid S
$\mathbb{C}^*$	Multiplicative group of nonzero complex numbers
$\mathbb{R}^*$	Multiplicative group of nonzero real numbers
$\mathbb{R}^+$	Multiplicative group of positive real numbers
$\mathbb{T}$	Circle group (complex numbers of absolute value 1)
$\mathrm{Sym}(X)$	Symmetric group on the set X
S_n	Symmetric group of degree n
A_n	Alternating group of degree n
D_{2n}	Dihedral group of order $2n$
$R[X]$	Ring of polynomials over R in one variable
$R[X_1, \ldots, X_n]$	Ring of polynomials over R in n variables
$R[[X]]$	Ring of formal power series over R
$C[a, b]$	Ring (algebra) of continuous functions from $[a, b]$ to $\mathbb{R}$
$M_n(R)$	Ring (algebra) of all $n \times n$ matrices with entries in R
$\mathrm{GL}_n(F)$	General linear group (invertible $n \times n$ matrices with entries in F)
$\mathrm{SL}_n(F)$	Special linear group (matrices in $\mathrm{GL}_n(F)$ with determinant 1)
$\mathbb{Z}[i]$	Ring of Gaussian integers
$\mathbb{H}$	Ring (algebra) of quaternions
Q	Quaternion group
$\mathrm{End}\ (M)$	Ring of endomorphisms of the additive group M
$\mathrm{End}_F(V)$	Algebra of endomorphisms of the vector space V over F
$F(X)$	Field of rational functions over F
$F(a_1, \ldots, a_n)$	Field obtained by adjoining $a_1, \ldots, a_n$ to F

$Z(A)$	Center of the group (ring, algebra) A
$A_1 \times \cdots \times A_n$	Direct product of the groups (rings, algebras) $A_1, \ldots, A_n$
$A_1 \oplus \cdots \oplus A_n$	Direct sum of the additive groups (vector spaces) $A_1, \ldots, A_n$
$|X|$	Cardinality of the set X
id_X	Identity map on the set X
$\mathrm{span}\, X$	Linear span of the set X
$\dim_F V$	Dimension of the vector space V over F
$\langle X \rangle$	Subgroup generated by the set X
$\langle a \rangle$	Cyclic subgroup generated by the element a
(a)	Principal ideal generated by the element a
$[x, y]$	Commutator of x and y
$\deg(f(X))$	Degree of the polynomial $f(X)$
$\mathrm{sgn}(\sigma)$	Sign of the permutation σ
$\det(A)$	Determinant of the matrix A
$b \mid a$	b divides a
$b \nmid a$	b does not divide a
$a \equiv b \pmod{n}$	n divides $a - b$
$\gcd(a, b)$	Greatest common divisor of a and b
$\varphi(n)$	Euler's phi function
$\cong$	Is isomorphic to
$\ker \varphi$	Kernel of the homomorphism φ
$\mathrm{im}\, \varphi$	Image of the homomorphism φ
aH or $a + H$	Coset of H containing a
$H \leq G$	H is a subgroup of G
$N \triangleleft G$	N is a normal subgroup of G
$I \triangleleft R$	I is an ideal of R
A/B	Quotient group (ring, vector space, algebra, module)
$[G : H]$	Index of the subgroup H in the group G
$[K : F]$	Degree of the extension K of the field F

Part I
The Language of Algebra

Chapter 1
Glossary of Basic Algebraic Structures

In this first chapter, we will introduce the algebraic structures that are the subject of this book (groups, rings, etc.), along with some accompanying fundamental notions. Elementary properties that follow easily from the definitions will be considered, and simple, basic examples will be given to help the reader understand and memorize the concepts. We will not go further than that. Our first "date" with algebraic structures will thus be rather formal and superficial, just to see what we will be dealing with.

1.1 Binary Operations

All the main concepts of algebra involve the following notion.

Definition 1.1. A **binary operation** on a nonempty set S is a map from $S \times S$ to S.

Saying that algebra is the study of binary operations is true in a way, but could be misleading. We are not interested in all binary operations we can think of, but those that arise naturally in mathematics.

Example 1.2. *Addition* on the set of integers $\mathbb{Z}$, i.e., the map

$$(m, n) \mapsto m + n,$$

is a binary operation on $\mathbb{Z}$.

Example 1.3. *Multiplication* on $\mathbb{Z}$, i.e., the map

$$(m, n) \mapsto m \cdot n,$$

is also a binary operation $\mathbb{Z}$.

Addition and multiplication are binary operations on a variety of sets, not only on $\mathbb{Z}$. We add and multiply different kinds of numbers, functions, matrices, etc. In

© Springer Nature Switzerland AG 2019
M. Brešar, *Undergraduate Algebra*, Springer Undergraduate Mathematics Series,
https://doi.org/10.1007/978-3-030-14053-3_1

algebra, however, the integers are often used as a prototype example for various situations.

Example 1.4. Let X be a nonempty set. By $\mathrm{Map}(X)$ we denote the set of all maps from X to itself. The rule

$$(g, f) \mapsto g \circ f,$$

where $\circ$ stands for *map composition*, defines a binary operation on $\mathrm{Map}(X)$.

These examples were not chosen randomly. Addition, multiplication, and composition are the most important binary operations. What about *subtraction*, i.e., the map $(m, n) \mapsto m - n$? It is also a binary operation on $\mathbb{Z}$ and on various other sets. However, it is rarely regarded as such. More precisely, we do not regard it as a "self-sustained" binary operation, but as an operation derived from addition and the notion of an additive inverse (which will be defined in Section 1.3). A similar comment can be made for *division*, which can formally be regarded as a binary operation on some sets (e.g., on the set of all nonzero real numbers). It is defined through multiplication and the notion of an inverse element.

We often drop the adjective "binary" and simply use the term "operation". Yet in algebra we sometimes consider other kinds of operations. For our purposes, it is enough to mention only the concept of an **external binary operation**. This is a map from $R \times S$ to S, where R and S are two nonempty sets.

Example 1.5. *Scalar multiplication* on $\mathbb{R}^3$ is an external binary operation. This is the map that assigns to every scalar $\lambda \in \mathbb{R}$ and every vector $\mathbf{x} = (x_1, x_2, x_3) \in \mathbb{R}^3$ the vector

$$\lambda \mathbf{x} = (\lambda x_1, \lambda x_2, \lambda x_3) \in \mathbb{R}^3.$$

Thus, $R = \mathbb{R}$ and $S = \mathbb{R}^3$ in this case.

We will not meet external binary operations until the definition of a vector space in Section 1.5. Let us return to the "ordinary" binary operation on a set S. Until further notice, we will denote it by the symbol $\star$. Thus, $\star$ is a rule that assigns to each pair of elements x and y from S an element

$$x \star y,$$

which also belongs to S. Note that, for example, the dot product $\mathbf{x} \cdot \mathbf{y}$ of vectors in $\mathbb{R}^n$ is not a binary operation since the result of the operation is a real number (a scalar) and not a vector. The cross product $\mathbf{x} \times \mathbf{y}$, on the other hand, is a binary operation on $\mathbb{R}^3$.

A binary operation $\star$ on S may induce a binary operation on a subset of S. Indeed, if T is a nonempty subset of S such that $t \star t' \in T$ whenever $t, t' \in T$, then the restriction of $\star$ to $T \times T$ is a binary operation on T. In this case, we say that T is **closed under $\star$**.

Example 1.6. The subset $\mathbb{N}$ of $\mathbb{Z}$ is closed under addition and multiplication, but not under subtraction. Therefore, addition and multiplication are binary operations on $\mathbb{N}$, whereas subtraction is not.

Not much can be said about general binary operations. Usually we assume that they satisfy certain conditions. The two most basic ones are called *associativity* and *commutativity*.

Definition 1.7. A binary operation $\star$ on S is said to be **associative** if

$$(x \star y) \star z = x \star (y \star z) \tag{1.1}$$

for all $x, y, z \in S$.

The **associative law** (1.1) is quite restrictive. Nevertheless, we will almost exclusively deal with associative operations. Not that nonassociative operations are not important. There is a whole branch of mathematics devoted to them (sometimes called "nonassociative algebra"). However, it is rarely studied at the undergraduate level—at least not systematically, special examples of nonassociative binary operations are actually often encountered. The aforementioned cross product on $\mathbb{R}^3$ is one of them.

Definition 1.8. Let $\star$ be a binary operation on S. We say that elements x and y from S **commute** if

$$x \star y = y \star x. \tag{1.2}$$

If (1.2) holds for all $x, y \in S$, then we say that $\star$ is a **commutative** binary operation.

Example 1.9. The operations of addition and multiplication on $\mathbb{Z}$ are clearly associative and commutative. On the other hand, subtraction is easily seen to be neither associative nor commutative.

Example 1.10. Take $f, g, h \in \text{Map}(X)$. We have

$$((h \circ g) \circ f)(x) = (h \circ g)(f(x)) = h(g(f(x)))$$

and

$$(h \circ (g \circ f))(x) = h((g \circ f)(x)) = h(g(f(x)))$$

for any $x \in X$, showing that the maps $(h \circ g) \circ f$ and $h \circ (g \circ f)$ are equal. The operation $\circ$ is thus associative. However, it is not commutative, unless X has only one element. Indeed, if a and b are different elements from X, then $f, g \in \text{Map}(X)$ defined by $f(x) = a$ and $g(x) = b$ for all $x \in X$ satisfy

$$f \circ g = f \neq g = g \circ f.$$

This was just one example. In fact, elements in $\text{Map}(X)$ only rarely commute.

Map composition is an operation of essential importance. Therefore, we cannot neglect noncommutativity as easily as nonassociativity. Sometimes we will assume that our operations are commutative, but sometimes not.

Another important property of a binary operation is the existence of an element that leaves every element from the set unchanged when combined with it using the operation.

Definition 1.11. Let $\star$ be a binary operation on S. An element $e \in S$ is said to be an **identity element** (or **neutral element**) for $\star$ if

$$e \star x = x \star e = x \quad \text{for all } x \in S.$$

Example 1.12. An identity element for addition in $\mathbb{Z}$ is 0.

Example 1.13. An identity element for multiplication in $\mathbb{Z}$ is 1.

Example 1.14. An identity element for composition in $\mathrm{Map}(X)$ is the identity map id_X.

An identity element does not always exist. For example, the operation of addition on the set of all natural numbers does not have one, and neither does the operation of multiplication on the set of all even integers. However, we will mostly treat operations for which an identity element exists. In such a case we can actually talk about *the* identity element.

Theorem 1.15. *If an identity element for the binary operation $\star$ exists, then it is unique.*

Proof. Suppose e and f are both identity elements for $\star$. Since e is an identity element, $e \star f = f$. On the other hand, $e \star f = e$ since f is an identity element. Consequently, $e = f$. $\qquad\qquad\qquad\qquad\qquad\qquad\qquad\qquad\qquad\qquad\qquad\qquad\qquad\quad$ $\square$

The proof shows a little bit more than what was stated. The condition that e is an identity element was not used in its full generality. It would be enough to assume that $e \star x = x$ holds for every $x \in S$. An element e satisfying this milder condition is called a **left identity element**. Similarly, it would be enough to assume that f is a **right identity element**, meaning that $x \star f = x$ holds for every $x \in S$. The proof of Theorem 1.15 thus shows that a left and a right identity element coincide. On the other hand, if a right (resp. left) identity element does not exist, there can be more left (resp. right) identity elements.

Example 1.16. If $\star$ is defined by $x \star y = y$ for all $x, y \in S$, then every element in S is a left identity element.

Exercises

1.17. Let $\mathscr{P}(X)$ be the power set of the nonempty set X. The union, intersection, and set difference are binary operations on $\mathscr{P}(X)$. Determine which among them are associative, commutative, and have a (left or right) identity element.

1.18. Determine which of the following binary operations on $\mathbb{N}$ are associative, have a (left or right) identity element, and contain two different elements that commute:

(a) $m * n = m + 2n$.

(b) $m * n = m^2 n$.

(c) $m * n = m$.

(d) $m * n = m^n$.

1.19. Let S be a set with a binary operation $\star$. If subsets T and T' of S are closed under $\star$, then so is their intersection $T \cap T'$. Find an example showing that this does not always hold for the union $T \cup T'$.

Comment. This should not be a difficult task—if $\star$ is one of the standard examples of binary operations, then $T \cup T'$ will only exceptionally be closed under $\star$. Generally speaking, algebraic properties are often preserved under intersection, but not under union.

1.20. Find all *finite* nonempty subsets of $\mathbb{Z}$ that are closed under addition.

1.21. Let T be a nonempty subset of $\mathbb{Z}$. Consider the following conditions:

(a) T is closed under subtraction.

(b) T is closed under addition.

(c) T is closed under multiplication.

Show that (a) implies both (b) and (c), whereas (b) and (c) are independent conditions (to establish the latter, find an example of a set satisfying (b) but not (c), and an example of a set satisfying (c) but not (b)). Also, find some examples of sets satisfying (a). Can you find them all?

Comment. The problem of the last question will be solved in Section 2.1, and in Section 1.6 we will see that the implication (a) $\implies$ (b) is a general algebraic fact, which does not hold only for subsets of $\mathbb{Z}$.

1.22. How many binary operations are there on a set with n elements?

1.23. Let S be a set with two elements. Find a binary operation $\star$ on S such that $(x \star y) \star z \neq x \star (y \star z)$ for all $x, y, z \in S$.

1.24. Let $f \in \mathrm{Map}(\mathbb{Z})$ be given by $f(n) = n + 1$ for every $n \in \mathbb{Z}$. Find all elements in $\mathrm{Map}(\mathbb{Z})$ that commute with f.

1.25. Denote by $\mathrm{Inj}(X)$ the set of all injective maps in $\mathrm{Map}(X)$, by $\mathrm{Sur}(X)$ the set of all surjective maps in $\mathrm{Map}(X)$, and by $\mathrm{Bij}(X)$ the set of all bijective maps in $\mathrm{Map}(X)$.

(a) Prove that the sets $\mathrm{Inj}(X)$, $\mathrm{Sur}(X)$, $\mathrm{Bij}(X)$, $\mathrm{Map}(X) \setminus \mathrm{Inj}(X)$, and $\mathrm{Map}(X) \setminus \mathrm{Sur}(X)$ are closed under $\circ$.

(b) Prove that $\mathrm{Inj}(X) = \mathrm{Sur}(X) = \mathrm{Bij}(X)$ if X is a finite set.

(c) Prove that the set $\mathrm{Map}(X) \setminus \mathrm{Bij}(X)$ is closed under $\circ$ if and only if X is a finite set.

Comment. Although dressed up in algebraic language, this is an exercise in set theory. From the algebraic perspective, the following small detail, following from (a), is important: $\circ$ is a binary operation on $\mathrm{Bij}(X)$. The set $\mathrm{Bij}(X)$ is actually a more natural framework for studying the operation of composition than the set $\mathrm{Map}(X)$. The notation $\mathrm{Bij}(X)$, however, is not standard. We will replace it by $\mathrm{Sym}(X)$ (see Example 1.54).

1.2   Semigroups and Monoids

The main objects of study in algebra are **algebraic structures**. These are sets
equipped with one or more operations that satisfy certain conditions, which we call
axioms. This is an informal description, there is no need for us to give a rigorous
definition of an algebraic structure. This notion can be understood through examples.
In this section, we will get acquainted with two elementary algebraic structures.

1.2.1   Semigroups

The simplest algebraic structures are *semigroups*, so we start with them.

Definition 1.26. A nonempty set S together with an associative binary operation $\star$
on S is called a **semigroup**.

In the case of semigroups, we thus have only one operation and only one axiom, the associative law. The study of general semigroups is essentially the study
of the associative law. A statement on semigroups can thus yield information on
such seemingly different things as addition, multiplication, and map composition.
In general, one of the main goals of algebra is to provide unified approaches to a
variety of topics that are encountered in different areas of mathematics.

A semigroup consists of a set S *and* an operation $\star$. We therefore denote it as a
pair

$$(S, \star).$$

However, it is usually clear from the context which operation $\star$ we have in mind and
then we can simply talk about the semigroup S. Moreover, it is common to attribute
properties of $\star$ to S. For example, by saying that S is commutative we really mean
that $\star$ is commutative. Similar simplifications in notation and terminology apply to
other algebraic structures.

Example 1.27. Each of the associative operations mentioned in the previous section
gives rise to a semigroup. In particular, $(\mathbb{Z}, +)$, $(\mathbb{Z}, \cdot)$, and $(\mathrm{Map}(X), \circ)$ are semigroups. Actually, one would not usually include $(\mathbb{Z}, +)$ in a list of typical examples
of semigroups. It is "too perfect" for that. Indeed, in the next section we will see that
it is even a group, not only a semigroup. A more standard example of a semigroup
is $(\mathbb{N}, +)$.

Just a brief digression before we proceed further. The way we teach algebra does
not reflect the way algebra has developed. We now start with definitions and then
look for examples. Historically, different examples were discovered first, and eventually they led to definitions. This comment applies to all areas of mathematics, but
in algebra this contrast is particularly striking.

Given three elements x_1, x_2, x_3 in a semigroup S, we have

$$(x_1 \star x_2) \star x_3 = x_1 \star (x_2 \star x_3)$$

by the very definition. We can therefore omit the parentheses and write this element as

$$x_1 \star x_2 \star x_3.$$

Can parentheses be omitted when dealing with more than three elements? That is, do identities such as

$$(x_1 \star x_2) \star (x_3 \star x_4) = (x_1 \star (x_2 \star x_3)) \star x_4$$

or

$$(x_1 \star x_2) \star ((x_3 \star x_4) \star x_5) = x_1 \star (((x_2 \star x_3) \star x_4) \star x_5)$$

hold in a semigroup? By the repeated application of the associative law the reader can readily check the validity of these two identities. Making further tests one will not find a counterexample—the parentheses are indeed redundant, the result is the same no matter where we put them. When wanting to formulate this claim in precise terms one has to face tedious notational complications. Let us avoid this and rather give a somewhat informal explanation.

Take $x_1, \ldots, x_n \in S, n \geq 2$, and set

$$s := x_1 \star (x_2 \star (x_3 \star (\cdots \star x_n))).$$

Let t be any element that can be obtained from the sequence $x_1, \ldots, x_n$ (in this order!) by making use of the operation $\star$; t is thus defined similarly as s, only the parentheses may be positioned differently. Our goal is to show that $t = s$. We proceed by induction on n. The claim is obvious for $n = 2$, so assume that $n > 2$ and the claim is true whenever less than n elements are involved. With respect to the last application of $\star$ (the last parenthesis), we can write $t = u \star v$ where u involves $x_1, \ldots, x_k$ for some $k < n$, and v involves $x_{k+1}, \ldots, x_n$. The induction assumption implies that $u = x_1 \star z$, where $z = x_2 \star (x_3 \star (\cdots \star x_k))$. Since $\star$ is associative, we can write $t = x_1 \star (z \star v)$. Now use the induction hypothesis for the expression $z \star v$, and $t = s$ follows.

We may therefore omit parentheses and write elements such as s and t simply as

$$x_1 \star x_2 \star \cdots \star x_n.$$

In particular, we have provided a justification for writing $m_1 + m_2 + \cdots + m_n$ for the sum of numbers, $m_1 m_2 \cdots m_n$ for the product of numbers, and $f_1 \circ f_2 \circ \cdots \circ f_n$ for the composition of maps.

Let $x \in S$ and let n be a natural number. We define the nth **power** of x in the same way as for numbers, i.e.,

$$x^n := \underbrace{x \star x \star \cdots \star x}_{n \text{ times}},$$

and we can easily verify the familiar formulas

$$x^{m+n} = x^m \star x^n \quad \text{and} \quad (x^m)^n = x^{mn}.$$

Incidentally, we can define $x^2 := x \star x$ even if $\star$ is not associative. To define x^3, the associativity assumption cannot be avoided since otherwise there is no guarantee that $x^2 \star x$ and $x \star x^2$ coincide. We also remark that the notation x^n is inappropriate if $\star$ is the operation of addition. But we will discuss this in the next section.

1.2.2 Monoids

Important examples of semigroups often contain an identity element.

Definition 1.28. A semigroup with an identity element is called a **monoid**.

Example 1.29. The semigroup $(\mathbb{N}, +)$ does not have an identity element and therefore is not a monoid. It becomes a monoid if we add 0 to the set $\mathbb{N}$. Thus, $(\mathbb{N} \cup \{0\}, +)$ is a monoid with identity element 0.

Example 1.30. Both $(\mathbb{N}, \cdot)$ and $(\mathbb{Z}, \cdot)$ are monoids with identity element 1.

Example 1.31. As is evident from Examples 1.10 and 1.14, $(\mathrm{Map}(X), \circ)$ is a monoid with identity element id_X.

The existence of an identity element makes is possible to define the following notions.

Definition 1.32. Let $(S, \star)$ be a monoid with identity element e.

1. An element $\ell \in S$ is called a **left inverse** of an element $x \in S$ if $\ell \star x = e$.
2. An element $r \in S$ is called a **right inverse** of an element $x \in S$ if $x \star r = e$.
3. An element $w \in S$ is called an **inverse** of an element $x \in S$ if $w \star x = x \star w = e$, i.e., w is both a left and a right inverse of x. An element for which an inverse exists is said to be **invertible**.

The following theorem indicates why left and right inverses are useful.

Theorem 1.33. *Let $(S, \star)$ be a monoid and let $x, y, z \in S$.*

(a) *If x has a left inverse, then $x \star y = x \star z$ implies $y = z$.*
(b) *If x has a right inverse, then $y \star x = z \star x$ implies $y = z$.*

Proof. (a) Let ℓ be a left inverse of x. From $x \star y = x \star z$ it follows that $\ell \star (x \star y) = \ell \star (x \star z)$. Now use the associative law for $\star$ and $y = z$ follows.
　　(b) If r is a right inverse of x, then $(y \star x) \star r = (z \star x) \star r$ yields $y = z$.　　　□

One has to be a little careful when facing the question whether an element has a (left) inverse. Is the number 2 invertible under the operation of multiplication? If we consider it as an element of, say, the monoid $\mathbb{Q}$, then the answer is clearly positive: its inverse is the number $\frac{1}{2}$. On the other hand, the answer is negative if we consider it as an element of the monoid $\mathbb{Z}$, since the inverse must belong to the monoid in question.

Every monoid has at least one invertible element, namely the identity element, which is its own inverse. It may happen that all other elements are non-invertible. The other extreme, where all elements are invertible, is certainly possible and is particularly important, but will be considered in the next section. Let us now look at a few examples where this is not the case. We start with monoids in which the operation is commutative. The notions "left inverse," "right inverse," and "inverse" then coincide. The adjectives "left" and "right" are therefore omitted.

Example 1.34. The only invertible element of the monoid $(\mathbb{N} \cup \{0\}, +)$ is 0, the only invertible element of the monoid $(\mathbb{N}, \cdot)$ is 1, and the only invertible elements of the monoid $(\mathbb{Z}, \cdot)$ are 1 and -1. In each of the monoids $(\mathbb{Q}, \cdot)$, $(\mathbb{R}, \cdot)$, and $(\mathbb{C}, \cdot)$, every element except 0 is invertible.

In the next example, it is necessary to distinguish between left and right inverses.

Example 1.35. Let $f \in \mathrm{Map}(X)$. If f has a left inverse, i.e., $g \circ f = \mathrm{id}_X$ holds for some map $g : X \to X$, then f is injective. Indeed, $f(x) = f(x')$ implies $g(f(x)) = g(f(x'))$ and hence $x = x'$. The converse is also true: if f is injective, then it has at least one left inverse g. It is defined as follows. Take $y \in X$. If y lies in the image of f, we can write it as $y = f(x)$ for some $x \in X$. In this case, we define $g(y) := x$. The element x is uniquely determined since f is injective, so this definition is unambiguous. If y does not lie in the image of f, then $g(y)$ can be defined arbitrarily—whatever we take, $g \circ f = \mathrm{id}_X$ holds. If f is not surjective, then such elements y exist and so f has several left inverses. Let us summarize:

(a) $f \in \mathrm{Map}(X)$ has a left inverse if and only if f is injective. If f is not also surjective, then it has more than one left inverse.

The latter can occur only if the set X is infinite. Indeed, if X is finite, then a map $f : X \to X$ is injective if and only if it is surjective.

An analogous statement holds for right inverses and surjective maps:

(b) $f \in \mathrm{Map}(X)$ has a right inverse if and only if f is surjective. If f is not also injective, then it has more than one right inverse.

Indeed, the existence of a right inverse clearly implies that f is surjective. Conversely, let f be surjective. Then, for every $y \in X$, there exist elements $x \in X$ such that $f(x) = y$. For each y choose such an x and define $h(y) := x$ (here we have used the axiom of choice). Then h is a right inverse of f. If f is not injective, there exists a $y \in X$ such that the equation $f(x) = y$ has more than one solution, implying that f has more than one right inverse.

The following familiar fact can be extracted either from the proof of (a) or from the proof of (b).

(c) $f \in \mathrm{Map}(X)$ has an inverse if and only if f is bijective. This inverse is unique. (We denote it by f^{-1} and call it the *inverse map* of f.)

Some elements of a monoid may thus have several left inverses, and some may have several right inverses. Corollary 1.38 below, however, states that an element cannot have more than one inverse. First we prove a theorem that tells us a bit more.

Theorem 1.36. *Let $(S, \star)$ be a monoid. If $\ell \in S$ is a left inverse of $x \in S$ and $r \in S$ is a right inverse of the same element x, then $\ell = r$.*

Proof. On the one hand, the element $\ell \star x \star r$ is equal to

$$\ell \star (x \star r) = \ell \star e = \ell,$$

and, on the other hand, it is equal to

$$(\ell \star x) \star r = e \star r = r.$$

Thus, $\ell = r$. $\square$

Elements that have only a left or only a right inverse will actually only seldom occur in what follows. Invertible elements and their inverses, however, will be considered throughout the book.

Corollary 1.37. *If x is an invertible element of a monoid $(S, \star)$, then $x \star y = e$ implies $y \star x = e$. Similarly, $y \star x = e$ implies $x \star y = e$.*

Proof. By assumption, there exists a $w \in S$ such that $x \star w = w \star x = e$. Theorem 1.36 thus tells us that each of the conditions $x \star y = e$ and $y \star x = e$ implies $y = w$. $\square$

Corollary 1.38. *An invertible element of a monoid has a unique inverse.*

Proof. If w and z are inverses of x, then w is, in particular, a left inverse of x, and z is a right inverse of x. Theorem 1.36 shows that $w = z$. $\square$

We denote the inverse of the element x by x^{-1}. Thus, x^{-1} is a uniquely determined element satisfying

$$x \star x^{-1} = x^{-1} \star x = e.$$

Note that x^{-1} is also invertible and that

$$(x^{-1})^{-1} = x.$$

Theorem 1.39. *If x and y are invertible elements of a monoid $(S, \star)$, then so is $x \star y$. Moreover, $(x \star y)^{-1} = y^{-1} \star x^{-1}$.*

Proof. We have to verify that

$$(x \star y) \star (y^{-1} \star x^{-1}) = e \quad \text{and} \quad (y^{-1} \star x^{-1}) \star (x \star y) = e.$$

Both identities become obvious by appropriately rearranging parentheses. Indeed, just write the left-hand side of the first identity as $x \star (y \star y^{-1}) \star x^{-1}$, and the desired conclusion readily follows. $\square$

It is straightforward to generalize this theorem to more than two elements. That is, if $x_1, \ldots, x_n$ are invertible, then so is $x_1 \star x_2 \star \cdots \star x_n$, and

$$(x_1 \star x_2 \star \cdots \star x_n)^{-1} = x_n^{-1} \star \cdots \star x_2^{-1} \star x_1^{-1}.$$

In the special case where all x_i are equal, we obtain that, for every invertible element x and every $n \in \mathbb{N}$, $(x^n)^{-1} = (x^{-1})^n$. This element is also denoted by x^{-n}, thus

$$x^{-n} := (x^n)^{-1} = (x^{-1})^n.$$

We have thereby introduced powers with negative exponents. What about the exponent 0? For any x, invertible or not, we define

$$x^0 := e.$$

The formulas

$$x^{m+n} = x^m \star x^n \quad \text{and} \quad (x^m)^n = x^{mn}$$

then hold for all $m, n \geq 0$, and if x is invertible they hold for all $m, n \in \mathbb{Z}$. The proof may require a little patience, but is straightforward. We leave it as an exercise for the reader.

Semigroups and monoids will, from now on, play an auxiliary role and will rarely be mentioned explicitly. However, many of the simple facts that were recorded in this section will be very frequently used.

Exercises

1.40. Show that $\mathbb{N}$ is a semigroup under the operation $m \star n = \max\{m, n\}$, as well as under the operation $m \star n = \min\{m, n\}$. Which one among them is a monoid?

1.41. Find all binary operations on the set $S = \{e, a\}$ for which S is a semigroup and e is a left identity element.

1.42. Show that $\mathbb{R}$ is a semigroup under the operation $x * y = |x|y$. Does it have a left or right identity element? For each $x \in \mathbb{R}$, find all elements commuting with x.

1.43. Show that $\mathbb{Z}$ is a monoid under the operation $m * n = m + n + mn$. Find all its invertible elements.

1.44. Prove that in each monoid with identity element e, $x \star y \star x = e$ implies that x and y are invertible and $y = x^{-2}$.

1.45. Determine which among the following conditions imply that an element of a monoid S is invertible:

(a) There exists an $y \in S$ such that $x \star y$ is invertible.
(b) There exists an $y \in S$ commuting with x such that $x \star y$ is invertible.
(c) There exists an $n \in \mathbb{N}$ such that x^n is invertible.

1.46. Prove that in a monoid with a finite number of elements, a left inverse of the element x is also its right inverse (and hence the inverse), and is of the form x^n for some $n \in \mathbb{N}$.

Hint. Since the set $\{x^m \mid m \in \mathbb{N}\}$ is finite, there exist $k \neq \ell$ such that $x^k = x^\ell$.

1.47. Let x be an invertible element of a monoid S. It is obvious that every element of the form x^n with $n \in \mathbb{Z}$ commutes with x. For some elements x it may happen that these are also the only elements commuting with it. Reconsider Exercise 1.24 and show that the map f (defined by $f(n) = n + 1$) of the monoid $\text{Map}(\mathbb{Z})$ is an example of such an element. Is the same true for $g \in \text{Map}(\mathbb{Z})$ defined by $g(n) = n + 2$?

1.48. Find an element in the monoid $\text{Map}(\mathbb{N})$ that has a left inverse and does not have a right inverse. Determine at least two of its left inverses.

1.3 Groups

The name "semigroup" carries a suggestion of incompleteness. We proceed to ("real") groups.

Definition 1.49. A monoid in which each element is invertible is called a **group**. A group is said to be **Abelian** (or **commutative**) if its binary operation is commutative.

Abelian groups are named after the Norwegian mathematician *Niels Henrik Abel* (1802–1829) who, in spite of his short life, made fundamental contributions to mathematics. Abel actually did not study groups in the modern sense of the word, yet some special groups appeared in his work. The name "group" was coined by the French mathematician *Évariste Galois* (1811–1832), who had an even shorter life. Although his definition is not exactly the same as ours, Galois is considered the founder of group theory. Abel died of disease, while Galois was killed in a duel. There is a story that, being sure that the end of his life was approaching, he wrote his mathematical testament the night before the fatal duel. It may be that this is an overly romanticized version of the unusual death of a famous person. The fact is, however, that the importance of Galois' work was recognized only many years later, and that the genius ideas of this young man had a tremendous impact on the development of mathematics.

It is interesting to note that Abel and Galois used groups only as a tool in studying a problem concerning roots of polynomials of fifth degree. We will talk about this in somewhat greater detail in Section 7.1. In general, the birth of an abstract mathematical concept is often intimately connected with some concrete problem. Unfortunately, when teaching mathematics, we often do not have enough time to explain the background and introduce new notions as if coming from nowhere.

As one would expect, the study of Abelian groups is quite different from the study of non-Abelian groups. There is also quite a difference between studying finite and infinite groups. By a **finite group** we mean a group that has a finite number of elements. The number of elements of a finite group G is called the **order of** G. So, the order of G is nothing but $|G|$, the cardinality of the set G.

Groups appear throughout mathematics. We will now only look at a few basic examples.

Example 1.50. Simple examples of Abelian groups are $(\mathbb{Z}, +)$, $(\mathbb{Q}, +)$, $(\mathbb{R}, +)$, and $(\mathbb{C}, +)$. In all of them, 0 is the identity element, and the inverse of the element a is $-a$. Since $-a \notin \mathbb{N}$ if $a \in \mathbb{N}$, the monoid $(\mathbb{N} \cup \{0\}, +)$ is not a group.

Inside every monoid S, a group is "hidden". Let S^* denote the set of all invertible elements of S.

Theorem 1.51. *If $(S, \star)$ is a monoid, then $(S^*, \star)$ is a group.*

Proof. Theorem 1.39 shows that $x \star y \in S^*$ whenever $x, y \in S^*$, proving that $\star$ is a binary operation on S^*. Since $\star$ is associative on S, it is also associative on the subset S^*. The identity element e of S is its own inverse, so it belongs to S^*. Every element in S^* is invertible by definition, and its inverse also lies in S^*.   □

Example 1.52. The monoids $(\mathbb{N} \cup \{0\}, +)$ and $(\mathbb{N}, \cdot)$ have only one invertible element, namely the identity element. The corresponding groups are thus $(\{0\}, +)$ and $(\{1\}, \cdot)$. Every group consisting of a single element, i.e., a group of order 1, is called a **trivial group**.

Example 1.53. The sets $\mathbb{Z}$, $\mathbb{Q}$, $\mathbb{R}$, and $\mathbb{C}$ are monoids under the operation of multiplication. Therefore,

$$\mathbb{Z}^* = \{1, -1\}$$

and

$$\mathbb{Q}^* = \mathbb{Q} \setminus \{0\}, \quad \mathbb{R}^* = \mathbb{R} \setminus \{0\}, \quad \text{and} \quad \mathbb{C}^* = \mathbb{C} \setminus \{0\}$$

are groups under multiplication. They all are Abelian, and only the first one is finite.

Example 1.54. The invertible elements of the monoid $\mathrm{Map}(X)$ are the bijective maps (see Example 1.35). Hence,

$$\mathrm{Sym}(X) := \mathrm{Map}(X)^* = \{f : X \to X \mid f \text{ is bijective}\}$$

is a group under the operation of composition. It is called the **symmetric group** on X. We claim that $\mathrm{Sym}(X)$ is non-Abelian, provided X contains at least three elements. Let us denote them by x_1, x_2, x_3, and define $f, g : X \to X$ by

$$f(x_1) = x_2, \quad f(x_2) = x_1, \quad \text{and } f(x) = x \text{ for all } x \in X \setminus \{x_1, x_2\},$$
$$g(x_1) = x_3, \quad g(x_3) = x_1, \quad \text{and } g(x) = x \text{ for all } x \in X \setminus \{x_1, x_3\}.$$

Note that $f, g \in \mathrm{Sym}(X)$ and

$$(f \circ g)(x_1) = x_3 \neq x_2 = (g \circ f)(x_1),$$

and so $f \circ g \neq g \circ f$.

We will be mostly concerned with the case where X is a finite set. For simplicity, we then usually denote its elements by $1, \ldots, n$, and write

$$S_n := \mathrm{Sym}(\{1, \ldots, n\}).$$

The elements of S_n, i.e., bijective maps from the set $\{1, \ldots, n\}$ to itself, are called **permutations** of $\{1, \ldots, n\}$. Actually, the term permutation is, in principle, used for elements of $\mathrm{Sym}(X)$ for any (possibly infinite) set X. It is just that this usage is more common in the special case where $X = \{1, \ldots, n\}$.

The permutation $\sigma \in S_n$ that sends k to i_k is expressed in the *two-line notation* as

$$\sigma = \begin{pmatrix} 1 & 2 & \ldots & n \\ i_1 & i_2 & \ldots & i_n \end{pmatrix}.$$

For example, $\left(\begin{smallmatrix} 1 & 2 & 3 & 4 \\ 4 & 2 & 1 & 3 \end{smallmatrix}\right)$ is the permutation in S_4 that sends 1 to 4, preserves 2, sends 3 to 1, and 4 to 3. On the other hand, $\left(\begin{smallmatrix} 1 & 2 & 3 & 4 \\ 4 & 2 & 4 & 3 \end{smallmatrix}\right)$ does not represent a permutation since a bijective map cannot send different elements 1 and 3 to the same element 4. Every element must appear exactly once in each row.

The symmetric groups S_n will be examined more closely in Section 2.5.

So far, we have denoted an abstract binary operation by $\star$. This notation was chosen for pedagogical rather than practical reasons. We wanted to have a neutral symbol that can mean $+$, $\cdot$, $\circ$, or something else. However, the repetitive writing of $\star$ is tedious and the resulting expressions are lengthy. Let us, therefore, simplify the notation. From now on, the operation in an abstract group will usually be denoted by $\cdot$, and, accordingly, we will use the terms "operation" and "multiplication" interchangeably, as synonyms for the same concept. Moreover, we will usually write $x \cdot y$ simply as xy, and call this element the **product** of the elements x and y. Similarly, we talk about the product $x_1 x_2 \cdots x_n$ of more elements. Further, an identity element, so far denoted by e, will from now on be denoted by 1. According to our convention,

$$x^0 = 1$$

holds for every element x in a group.

Let us now write the definition of a group once again, partially because of the new notation and partially in order to write all the group axioms explicitly, one after another.

Definition of a group. A nonempty set G together with a binary operation $(x, y) \mapsto xy$ is a **group** if the following properties are satisfied:

(G1) $(xy)z = x(yz)$ for all $x, y, z \in G$.
(G2) There exists an element 1 in G such that $1x = x1 = x$ for all $x \in G$.
(G3) For every $x \in G$, there exists an element x^{-1} in G such that $xx^{-1} = x^{-1}x = 1$.

If $xy = yx$ holds for all $x, y \in G$, then G is called an **Abelian group**.

Let us emphasize once again that the result of applying a binary operation on elements from the set G always lies in G. Thus, it is a part of the definition that

$$xy \in G \quad \text{for all } x, y \in G.$$

This is something one should not forget to check when proving that some set with a given operation forms a group.

Recall that an element x^{-1} from axiom (G3) is unique, and is called the **inverse** of the element x. Also, an element 1 satisfying (G2) is unique and called the **identity element** of G. Let us point out that condition (G2) is not equivalent to the condition "for each $x \in G$ there exists an element $1 \in G$ such that $1x = x1 = x$." This warning perhaps seems redundant, but it is not so uncommon for novices to confuse the two conditions.

With reference to the previous section, we now record a few elementary properties that hold in every group G.

(a) $(x^{-1})^{-1} = x$ for all $x \in G$.
(b) $(xy)^{-1} = y^{-1}x^{-1}$ for all $x, y \in G$.
(c) $x^{m+n} = x^m x^n$ and $(x^m)^n = x^{mn}$ for all $m, n \in \mathbb{Z}$ and all $x \in G$.
(d) For all $x, y, z \in G$, $xy = xz$ implies $y = z$.
(e) For all $x, y, z \in G$, $yx = zx$ implies $y = z$.
(f) For all $x, y \in G$, $xy = 1$ implies $yx = 1$, i.e., $y = x^{-1}$.

We call (d) and (e) the **cancellation laws** in a group. Let us also point out two of their important special cases.

(g) For all $x, y \in G$, each of the conditions $xy = x$ and $yx = x$ implies $y = 1$.
(h) For all $x \in G$, $x^2 = x$ implies $x = 1$.

The new notation is more transparent and as simple as can be. However, we should not forget that multiplication (= operation) can now mean many different things in concrete situations, including addition—which is slightly annoying. Imagine that we have developed a theory for abstract groups, and then we ask ourselves what it tells us for, let us say, the group $(\mathbb{Z}, +)$. We are then forced to make the "translation" from one notation to another, like substituting $-x$ for x^{-1}, $x + x$ for x^2, etc. It is better if we prepare systematically for this, since groups in which the operation is addition are of extreme importance in mathematics. We will, therefore, sometimes denote the operation in a group by the addition symbol $+$. In this case, we will call it an **additive group**. Analogously, we talk about a **multiplicative group**, but we will not use this term so often since our groups will be multiplicative by default.

Additive groups are usually Abelian. One can find exceptions in the literature, but not very often. To simplify the exposition, let us make the following terminological convention:

In this book, all additive groups are assumed to be Abelian.

Thus, whenever the operation will be denoted by $+$, we will automatically assume that

$$x + y = y + x$$

holds for all x and y. The identity element in an additive group is denoted by 0 and is called the **zero element**. The element $x + y$ is, of course, called the **sum** of x and y. The inverse of x is denoted by $-x$, and called the **additive inverse** of x. Thus, given any x, $-x$ is the unique element satisfying

$$x + (-x) = 0.$$

In an additive group, we define **subtraction** in a familiar way:

$$x - y := x + (-y).$$

The element $x - y$ is called the **difference** of x and y. Further, in an additive group we write nx instead of x^n. For any natural number n we thus have

$$nx = \underbrace{x + x + \cdots + x}_{n \text{ times}} \quad \text{and} \quad (-n)x = \underbrace{-x - x - \cdots - x}_{n \text{ times}},$$

and the identity $x^0 = 1$ gets the form

$$0x = 0.$$

Note that 0 on the left-hand side is the integer zero, while 0 on the right-hand side is the zero element of an additive group. Finally,

$$(-n)x = n(-x) = -(nx)$$

holds for every integer n and every element x.

In an additive group G, the statements (a)–(d) read as follows:

(a') $-(-x) = x$ for all $x \in G$.
(b') $-(x + y) = -x - y$ for all $x, y \in G$.
(c') $(m + n)x = mx + nx$ and $n(mx) = (nm)x$ for all $m, n \in \mathbb{Z}$ and all $x \in G$.
(d') For all $x, y, z \in G$, $x + y = x + z$ implies $y = z$.

Exercises

1.55. Determine which of the following sets of numbers are groups under the usual addition:

(a) $\{\frac{n}{2} \mid n \in \mathbb{Z}\}$.
(b) $\{x \in \mathbb{R} \mid x \geq 0\}$.
(c) $\{x \in \mathbb{R} \mid x \notin \mathbb{Q}\} \cup \{0\}$.
(d) $\{z \in \mathbb{C} \mid \operatorname{Im}(z) \in \mathbb{Z}\}$.
(e) $\{z \in \mathbb{C} \mid \operatorname{Re}(z) \cdot \operatorname{Im}(z) = 0\}$.
(f) $\{p + q\sqrt{2} \mid p, q \in \mathbb{Q}\}$.

1.56. Determine which of the following sets of numbers are groups under the usual multiplication:

(a) $\{2^n \mid n \in \mathbb{Z}\}$.
(b) $\{x \in \mathbb{R} \mid x > 0\}$.

(c) $\{x \in \mathbb{R} \mid x < 0\}$.
(d) $\{x \in \mathbb{R} \mid 0 < x < 1\}$.
(e) $\{z \in \mathbb{C} \mid |z| \geq 1\}$.
(f) $\{p + q\sqrt{2} \mid p, q \in \mathbb{Q}\} \setminus \{0\}$.

1.57. Find three different complex numbers u, z, w such that the set $\{u, z, w\}$ is a group under the usual multiplication.

1.58. Let G be the set of all non-constant linear functions, i.e., functions $f : \mathbb{R} \to \mathbb{R}$ of the form $f(x) = ax + b$ with $a, b \in \mathbb{R}$ and $a \neq 0$. Show that G is a group under the operation of composition. Is it Abelian?

1.59. Show that the open (real) interval $(-1, 1)$ is a group under the operation $x * y = \frac{x+y}{1+xy}$.

1.60. Use Theorem 1.51 to find examples of finite and infinite sets of numbers that are groups under the operation $x \star y = x + y + xy$.

1.61. Give the proofs of the statements (a)–(f) (without immediately looking at Section 1.2!).

1.62. Prove that, for every element x of a *finite* group G, there exists an $n \in \mathbb{N}$ such that $x^n = 1$.

Comment. The smallest integer n with this property is called the **order of the element** x. Later we will see that this is a very important notion in group theory.

1.63. In every group, $xy = 1$ implies $yx = 1$. Prove that $xyz = 1$ implies $zyx = 1$ only in an Abelian group.

1.64. Suppose a finite Abelian group G does not contain elements $x \neq 1$ such that $x^2 = 1$. Prove that the product of all elements in G equals 1.

1.65. Prove that a group G is Abelian if $x^2 = 1$ holds for every $x \in G$.

1.66. Prove that every group with four elements is Abelian.

Comment. Later we will show, as a byproduct of the general theory, that every group with less than six elements is Abelian. You can try to prove this now!

1.67. Let $(S, \star)$ be a semigroup. Prove that S is a group if and only if for all $a, b \in S$, the equations $a \star x = b$ and $y \star a = b$ have solutions in S.

1.4 Rings and Fields

This book is mostly about groups, rings, and fields. This section introduces the latter two algebraic structures.

1.4.1 Rings

The additive groups $\mathbb{Z}, \mathbb{Q}, \mathbb{R}$, and $\mathbb{C}$ were listed as basic examples of Abelian groups. They all are also equipped with multiplication under which they are not groups, but merely monoids. Moreover, addition and multiplication are connected through the distributive law. Therefore, $\mathbb{Z}, \mathbb{Q}, \mathbb{R}$, and $\mathbb{C}$ satisfy all the conditions of the following definition.

Definition 1.68. A nonempty set R together with two binary operations $(x, y) \mapsto x + y$ (addition) and $(x, y) \mapsto xy$ (multiplication) is a **ring** if the following properties are satisfied:

(R1) R is an Abelian group under addition.
        [This means that $(x + y) + z = x + (y + z)$ for all $x, y, z \in R, x + y = y + x$ for all $x, y \in R$, there exists an element $0 \in R$ such that $x + 0 = x$ for all $x \in R$, and for every $x \in R$ there exists an element $-x \in R$ such that $x + (-x) = 0.$]
(R2) R is a monoid under multiplication.
        [This means that $(xy)z = x(yz)$ for all $x, y, z \in R$ and that there exists an element $1 \in R$ such that $1x = x1 = x$ for all $x \in R$. We call 1 the **unity** of R.]
(R3) The **distributive laws** hold:

$$(x + y)z = xz + yz, \quad z(x + y) = zx + zy$$

        for all $x, y, z \in R$.

A ring is thus a triple

$$(R, +, \cdot).$$

It consists of the set R and both operations $+$ and $\cdot$. Usually, however, we simply talk about the ring R. For example, when speaking about the ring $\mathbb{Z}$, we have in mind the usual addition and multiplication of the integers. The notation $(R, +, \cdot)$ is used when it is not entirely clear what the operations on R are, or when we wish to emphasize that we consider R as a ring.

Another standard name for the element 1 is an **identity element**, as in groups. Our choice—a unity—is more or less equally common. It should be mentioned that in some texts the existence of 1 is not included among the ring axioms (that is, "monoid" in (R2) is replaced by "semigroup"). This has some advantages. The study of **rings without unity** is not much more demanding, but covers more examples. To give just one, the set of all *even* integers is a ring without unity under the usual addition and multiplication. However, one always arrives at small but tedious technical difficulties when a ring has no unity. Our definition therefore makes things easier for beginners.

Similarly as for groups, we can derive some elementary properties of rings immediately from the axioms. Note, first of all, that everything that was noticed for groups holds for rings with respect to addition, and everything that was noticed for monoids holds for rings with respect to multiplication. The next theorem describes properties that connect addition and multiplication. Besides the distributive laws, in the proofs we will also use the cancellation law in additive groups.

Theorem 1.69. *In any ring R, the following statements hold for all $x, y, z \in R$:*

(a) $0x = x0 = 0$.
(b) $(-x)y = x(-y) = -(xy)$.
(c) $(x - y)z = xz - yz$ and $z(x - y) = zx - zy$.
(d) $(-x)(-y) = xy$.
(e) $(-1)x = x(-1) = -x$.

Proof. From

$$0x = (0 + 0)x = 0x + 0x$$

we obtain $0x = 0$. Similarly, we see that $x0 = 0$. This yields

$$0 = x0 = x(y + (-y)) = xy + x(-y),$$

showing that $x(-y) = -(xy)$. Similarly, we establish the other part of (b). Using the distributive law we derive (c) from (b). Also, (d) follows from (b):

$$(-x)(-y) = -(x(-y)) = -(-(xy)) = xy.$$

Finally, (b) also implies (e).   □

Example 1.70. The simplest example of a ring is the **zero ring** (or **trivial ring**). This is a ring containing only one element, which we denote by 0. The elements 1 and 0 coincide in the zero ring. Conversely, if $1 = 0$ holds in a ring R, then $x = 1x = 0x = 0$ for every $x \in R$, meaning that R is the zero ring. Any ring containing more than one element is called a **nonzero ring**. In such a ring, 1 and 0 are thus different elements.

Axiom (R3) involves two distributive laws. If multiplication is commutative, then one is enough.

Definition 1.71. A **commutative ring** is a ring in which multiplication is commutative.

Example 1.72. The rings $\mathbb{Z}, \mathbb{Q}, \mathbb{R}$, and $\mathbb{C}$ are commutative.

The theory of commutative rings is quite different from the theory of general, not necessarily commutative rings. Each of the two theories has its own goals and methods. However, they have a common basis which will be discussed in Part I. In Part II, we will focus on commutative rings. We can also divide rings into finite and infinite ones, although one usually does not emphasize this division so strongly as in the case of groups. Indeed, finite rings do not play such a significant role in mathematics as finite groups. Still, there are some important finite rings which will be frequently considered in later chapters.

Students usually first encounter noncommutative multiplication when learning about the product of matrices. If we wish to make a ring out of a set of matrices, we have to restrict ourselves to square matrices in order to guarantee that this set will be closed under multiplication. For simplicity, we will now consider only 2×2 matrices. General matrix rings will be discussed in Section 2.7.

Example 1.73. Let $M_2(\mathbb{R})$ denote the set of all real matrices of size 2×2. Its elements are thus matrices $\begin{bmatrix} a_{11} & a_{12} \\ a_{21} & a_{22} \end{bmatrix}$, where a_{ij} are real numbers. We define addition and multiplication in $M_2(\mathbb{R})$ in the standard way:

$$\begin{bmatrix} a_{11} & a_{12} \\ a_{21} & a_{22} \end{bmatrix} + \begin{bmatrix} b_{11} & b_{12} \\ b_{21} & b_{22} \end{bmatrix} := \begin{bmatrix} a_{11} + b_{11} & a_{12} + b_{12} \\ a_{21} + b_{21} & a_{22} + b_{22} \end{bmatrix}$$

and

$$\begin{bmatrix} a_{11} & a_{12} \\ a_{21} & a_{22} \end{bmatrix} \cdot \begin{bmatrix} b_{11} & b_{12} \\ b_{21} & b_{22} \end{bmatrix} := \begin{bmatrix} a_{11}b_{11} + a_{12}b_{21} & a_{11}b_{12} + a_{12}b_{22} \\ a_{21}b_{11} + a_{22}b_{21} & a_{21}b_{12} + a_{22}b_{22} \end{bmatrix}.$$

As is well-known and easy to check, both operations are associative and the distributive laws are fulfilled. The identity element for addition is the zero matrix $0 = \begin{bmatrix} 0 & 0 \\ 0 & 0 \end{bmatrix}$, and the identity element for multiplication is the identity matrix $I = \begin{bmatrix} 1 & 0 \\ 0 & 1 \end{bmatrix}$. The additive inverse of the matrix $\begin{bmatrix} a_{11} & a_{12} \\ a_{21} & a_{22} \end{bmatrix}$ is the matrix $\begin{bmatrix} -a_{11} & -a_{12} \\ -a_{21} & -a_{22} \end{bmatrix}$. Thus, $M_2(\mathbb{R})$ is a ring. It is easy to see that it is not commutative. For example, the matrices

$$A = \begin{bmatrix} 1 & 0 \\ 0 & 0 \end{bmatrix} \quad \text{and} \quad B = \begin{bmatrix} 0 & 1 \\ 0 & 0 \end{bmatrix}$$

satisfy $AB = B$ and $BA = 0$.

This example also shows that the product of two nonzero elements in a ring can be 0. This makes computations in general rings quite different from computations with numbers.

Definition 1.74. An element x of a ring R is a **zero-divisor** if $x \neq 0$ and there exists a $y \neq 0$ in R such that $xy = 0$ or $yx = 0$.

Strictly speaking, we say that x is a **left zero-divisor** if $xy = 0$, and a **right zero-divisor** if $yx = 0$. A zero-divisor is thus an element that is either a left or a right zero-divisor. The adjectives "left" and "right" are obviously unnecessary in commutative rings.

Many important rings, such as for example $\mathbb{Z}$, $\mathbb{Q}$, $\mathbb{R}$, and $\mathbb{C}$, have no zero-divisors. In such rings, the following **cancellation laws** hold: if $x \neq 0$ and $xy = xz$ (or $yx = zx$), then $y = z$. This is because $xy = xz$ can be written as $x(y - z) = 0$.

1.4.2 Division Rings and Fields

A ring is a monoid under multiplication, so we can talk about its invertible elements. We remark that an invertible element of a ring is also commonly called a **unit**, but we will stick with the term invertible. By Theorem 1.51, the set R^* of all invertible elements of the ring R is a group under multiplication. It can be very small (e.g., $\mathbb{Z}^* = \{1, -1\}$), or very big, containing all elements except 0 (e.g., $\mathbb{R}^* = \mathbb{R} \setminus \{0\}$). Rings with this latter property are of special interest.

Definition 1.75. A nonzero ring in which every nonzero element is invertible is called a **division ring**. A commutative division ring is called a **field**.

An example of a division ring which is not commutative will be given later, in Section 2.8. Finding fields is much easier.

Example 1.76. The prototype examples of fields are $\mathbb{Q}$, $\mathbb{R}$, and $\mathbb{C}$.

In some areas of mathematics, these three fields are also the only ones that matter. Not in algebra, however. We will encounter various further examples in the later chapters.

The following theorem follows from Theorem 1.33, but we give the proof since it is very short.

Theorem 1.77. *An invertible element of a ring is not a zero-divisor. Therefore, division rings, and in particular fields, have no zero-divisors.*

Proof. Let x be an invertible element of a ring R. Suppose that $y \in R$ is such that $xy = 0$. Multiplying from the left by x^{-1} we obtain $y = 0$. Similarly, $yx = 0$ implies $y = 0$. $\qquad\square$

Example 1.78. The ring $\mathbb{Z}$ also does not have zero-divisors, but is not a field.

Although fields are special examples of rings, field theory is quite different from ring theory. Even deep results about rings are often obvious for fields. Therefore, if one has to choose the most illustrative example of a ring, $\mathbb{Z}$ is a much better choice than $\mathbb{Q}$, $\mathbb{R}$, or $\mathbb{C}$.

Exercises

1.79. Determine which of the following sets of numbers are rings under the usual addition and multiplication:

(a) $\{\frac{m}{2n+1} \mid m, n \in \mathbb{Z}\}$.
(b) $\{\frac{2n+1}{m} \mid m, n \in \mathbb{Z}, m \neq 0\}$.
(c) $\{m + 2ni \mid m, n \in \mathbb{Z}\}$.
(d) $\{2m + ni \mid m, n \in \mathbb{Z}\}$.
(e) $\{m + n\sqrt{2} \mid m, n \in \mathbb{Z}\}$.
(f) $\{p + q\sqrt{2} \mid p, q \in \mathbb{Q}\}$.

Which among these rings are fields?

1.80. Let R be a ring without unity. Show that the set $\mathbb{Z} \times R$ becomes a ring (with unity!) if we equip it with the following addition and multiplication:

$$(m, x) + (n, y) := (m + n, x + y),$$
$$(m, x) \cdot (n, y) := (mn, nx + my + xy)$$

(here, mn is the product of the integers m and n, xy is the product of the elements x and y from R, and nx has the usual meaning explained at the end of Section 1.3).

Comment. This construction makes it possible to reduce some problems on rings without unity to rings in our sense of the word. Usually we identify elements x in R with elements $(0, x)$ in $\mathbb{Z} \times R$, and accordingly consider R as a subset of $\mathbb{Z} \times R$. This makes sense since the operations in $\mathbb{Z} \times R$ are then extensions of the operations in R.

1.81. Let R be a ring. Which among the following conditions is equivalent to the condition that R is commutative?

(a) $(x + y)^2 = x^2 + 2xy + y^2$ for all $x, y \in R$.
(b) $x^2 - y^2 = (x - y)(x + y)$ for all $x, y \in R$.
(c) For all $x, y \in R$, $xy = 1$ implies $yx = 1$.
(d) For all $x, y, z, w \in R$, $xy = zw$ implies $yx = wz$.

1.82. Let $\mathscr{P}(X)$ be the power set of the nonempty set X. Prove that $\mathscr{P}(X)$ becomes a commutative ring if we define the sum of elements as the *symmetric difference*, i.e., $A + B := (A \setminus B) \cup (B \setminus A)$, and the product as the intersection, i.e., $A \cdot B := A \cap B$.

1.83. An element x of a ring R is called an **idempotent** if $x^2 = x$. A **Boolean ring** is a ring in which every element is an idempotent. Note that the previous exercise provides an example. Prove that every Boolean ring is commutative and that every element x from a Boolean ring coincides with its additive inverse (i.e., $x + x = 0$).

Comment. It turns out that the commutativity of a ring R follows already from the existence of any integer $n \geq 2$ such that $x^n = x$ for every $x \in R$. Proving this, however, is not easy and requires some prior knowledge. For some small n, like $n = 3$ or $n = 4$, finding an elementary proof can be a challenging yet entertaining exercise.

1.84. Prove that the commutativity of addition, $x + y = y + x$, follows from the other ring axioms.

1.85. Prove that every ring with zero-divisors contains nonzero elements x and y such that $xy = yx = 0$.

1.86. Prove that in every ring without zero-divisors, $x^2 = 1$ implies $x = 1$ or $x = -1$. Provide an example showing that this is not true in all rings.

1.87. Prove that a ring is a division ring if all its nonzero elements have a left inverse.

1.88. Let x and y be ring elements. Prove that $1 - xy$ is invertible if and only if $1 - yx$ is invertible.

Hint. The proof is simple—but one may only realize this after seeing it. Try to find a connection between the inverse of $1 - xy$ and the inverse of $1 - yx$. The formula you are searching for can be guessed by making use of the well-known formula for the sum of the geometric series, i.e., $(1 - q)^{-1} = 1 + q + q^2 + \ldots$ (just ignore the question of convergence throughout). Keep in mind that this can only be used to guess the right approach—it cannot be used in a proof. Nevertheless it provides a nice illustration of how knowledge of one part of mathematics can be unexpectedly helpful in solving a problem from another part.

1.5 Vector Spaces and Algebras

A special feature of the two algebraic structures that will be introduced in this section is the involvement of an external binary operation, called "scalar multiplication".

1.5.1 Vector Spaces

The reader is presumably familiar with the next definition from linear algebra (perhaps not at the same level of abstraction, however).

Definition 1.89. Let F be a field. A nonempty set V together with a binary operation $(u, v) \mapsto u + v$ (addition) and an external binary operation $(\lambda, v) \mapsto \lambda v$ from $F \times V$ to V, called **scalar multiplication**, is a **vector space over** F if the following properties are satisfied:

(V1) V is an Abelian group under addition.
(V2) $\lambda(u + v) = \lambda u + \lambda v$ for all $\lambda \in F$ and $u, v \in V$.
(V3) $(\lambda + \mu)v = \lambda v + \mu v$ for all $\lambda, \mu \in F$ and $v \in V$.
(V4) $\lambda(\mu v) = (\lambda \mu)v$ for all $\lambda, \mu \in F$ and $v \in V$.
(V5) $1v = v$ for all $v \in V$.

Elements of V are called **vectors** and elements of F are called **scalars**. A vector of the form λv is called a **scalar multiple** of the vector v.

Vector spaces occur in many areas of mathematics, but mostly only over the fields $\mathbb{R}$ and $\mathbb{C}$. These are called **real vector spaces** and **complex vector spaces**, respectively. We will see later that vector spaces over other fields also arise naturally in some contexts.

Let us add some explanation to the definition. We have denoted the additions in V and in F by the same symbol $+$. In (V3), both of them actually appear.

Similarly, two different multiplications, scalar multiplication and multiplication of scalars, appear in (V4). The symbol 1 in (V5) denotes the unity of F, i.e., the element satisfying $1\lambda = \lambda$ for all $\lambda \in F$. We thus require that this element also satisfies $1v = v$ for every vector v.

The following statements hold in any vector space V over F:

(a) $\lambda 0 = 0$ for every $\lambda \in F$; here, 0 denotes the zero element of V.
(b) $0v = 0$ for every $v \in V$; here, 0 on the left-hand side denotes the zero element of F, and 0 on the right-hand side denotes the zero element of V.
(c) For all $\lambda \in F$ and $v \in V$, $\lambda v = 0$ implies $\lambda = 0$ or $v = 0$.
(d) For all $\lambda \in F$ and $v \in V$, $(-\lambda)v = -(\lambda v) = \lambda(-v)$.

The proofs are simple and similar to those of elementary properties of rings. We leave them to the reader.

Let us record a few examples of vector spaces. First, the most standard one.

Example 1.90. Let F be a field and let $n \in \mathbb{N}$. The set F^n, i.e., the Cartesian product of n copies of F, becomes a vector space by defining operations componentwise:

$$(u_1, u_2, \ldots, u_n) + (v_1, v_2, \ldots, v_n) := (u_1 + v_1, u_2 + v_2, \ldots, u_n + v_n),$$
$$\lambda(v_1, v_2, \ldots, v_n) := (\lambda v_1, \lambda v_2, \ldots, \lambda v_n).$$

This can be easily checked. As a special case, $F = F^1$ is a vector space over itself. In the most familiar cases where $F = \mathbb{R}$ and $n = 2$ or $n = 3$, the operations can be represented geometrically. This geometric background may be useful for building intuition even when we deal with abstract vector spaces over arbitrary fields.

Example 1.91. A vector space whose only element is 0 is called the **zero** (or **trivial**) **vector space**.

Example 1.92. We may consider $\mathbb{C}$, just as any other field, as a vector space over itself. On the other hand, $\mathbb{C}$ is also a vector space over the field $\mathbb{R}$ if we define the scalar multiplication by $t(x + yi) := tx + tyi$ for all $t \in \mathbb{R}$ and $x + yi \in \mathbb{C}$. Similarly, we can view $\mathbb{C}$ or $\mathbb{R}$ as a vector space over $\mathbb{Q}$.

Example 1.93. A function $f : \mathbb{R} \to \mathbb{R}$ that can be written in the form

$$f(x) = a_0 + a_1 x + \cdots + a_n x^n, \ x \in \mathbb{R},$$

for some $a_i \in \mathbb{R}$, is called a **real polynomial function**. The set $\mathscr{P}$ of all real polynomial functions is a vector space under the usual addition of functions and multiplication of a function by a real number.

Remark 1.94. Let G be an arbitrary additive group. For any $n \in \mathbb{Z}$ and $x \in G$, we have defined $nx \in G$. The map $(n, x) \mapsto nx$ can be considered as an external binary operation from $\mathbb{Z} \times G$ to G. Observe that it has all the properties that are required for a vector space, i.e.,

$$n(x + y) = nx + ny, \ (m + n)x = mx + nx, \ n(mx) = (nm)x, \ \text{and} \ 1x = x.$$

Nevertheless, we do not say that G is a vector space over $\mathbb{Z}$ since $\mathbb{Z}$ is not a field. It is good to keep in mind, however, that every additive group is "close" to a vector space.

1.5.2 Algebras

Many important rings are simultaneously vector spaces. If their ring multiplication is appropriately synchronized with scalar multiplication, we talk about *algebras*.

Definition 1.95. Let F be a field. A nonempty set A together with two binary operations $(x, y) \mapsto x + y$ (addition) and $(x, y) \mapsto xy$ (multiplication), and an external binary operation $(\lambda, x) \mapsto \lambda x$ from $F \times A$ to A (scalar multiplication), is an **algebra over** F (or an F**-algebra**) if the following properties are satisfied:

(A1) A is a vector space under addition and scalar multiplication.
(A2) A is a ring under addition and multiplication.
(A3) $\lambda(xy) = (\lambda x)y = x(\lambda y)$ for all $\lambda \in F$ and $x, y \in A$.

The word "algebra" thus has even more meanings than mentioned in the Preface. Algebras are also a class of algebraic structures, so the study of algebras is a part of (abstract) algebra.

The vector spaces from Examples 1.90–1.93 can be endowed with multiplications making them algebras.

Example 1.96. The vector space F^n becomes an algebra over F by defining the componentwise multiplication:

$$(u_1, u_2, \ldots, u_n)(v_1, v_2, \ldots, v_n) := (u_1 v_1, u_2 v_2, \ldots, u_n v_n).$$

Checking the axioms is easy. For instance, the unity is $(1, 1, \ldots, 1)$. The simplest case where $n = 1$ shows that F is an algebra over itself.

Example 1.97. The **zero** (or **trivial**) **algebra** is the algebra containing only 0.

An algebra over $\mathbb{R}$ is called a **real algebra**, and an algebra over $\mathbb{C}$ is called a **complex algebra**. Thus, $\mathbb{R}^n$ is a real and $\mathbb{C}^n$ is a complex algebra.

Example 1.98. We already know that $\mathbb{C}$ is a ring and a real vector space. Clearly, it is also a real algebra.

Example 1.99. Equipping the vector space $\mathscr{P}$ of real polynomial functions with the usual multiplication of functions, $\mathscr{P}$ becomes a real algebra.

Example 1.100. The ring $M_2(\mathbb{R})$ from Example 1.73 becomes a real algebra if we equip it with the usual scalar multiplication:

$$\lambda \begin{bmatrix} a_{11} & a_{12} \\ a_{21} & a_{22} \end{bmatrix} := \begin{bmatrix} \lambda a_{11} & \lambda a_{12} \\ \lambda a_{21} & \lambda a_{22} \end{bmatrix}$$

for all $\lambda, a_{ij} \in \mathbb{R}$. Similarly, the set $M_2(\mathbb{C})$ of all 2×2 complex matrices is a complex algebra.

The study of algebras is ring-theoretic in nature. The difference between the terms "theory of rings" and "theory of rings and algebras" is only in emphasis.

Exercises

1.101. Determine which of the following sets are real vector spaces under the usual addition of vectors and the usual scalar multiplication:

(a) $\{(x, y) \in \mathbb{R}^2 \mid x \geq 0, \ y \geq 0\}$.
(b) $\{(x, y) \in \mathbb{R}^2 \mid x \in \mathbb{Q}\}$.
(c) $\{(x, y) \in \mathbb{R}^2 \mid x = y\}$.
(d) $\{(x, y, z) \in \mathbb{R}^3 \mid x = y = -z\}$.
(e) $\{(x, y, z) \in \mathbb{R}^3 \mid x = y = z^2\}$.
(f) $\{(x, y, z, w) \in \mathbb{R}^4 \mid x = w, \ y = z\}$.

Which of these vector spaces are real algebras under the multiplication defined in Example 1.96? In (c)–(f), we can replace $\mathbb{R}$ by any field F. Do the above questions then have the same answers?

1.102. Let $\mathscr{P}$ be the set of all real polynomial functions. By f' we denote the derivative of $f \in \mathscr{P}$. Determine which of the following sets are real vector spaces under the usual addition of functions and multiplication of a function by a real number:

(a) $\{f \in \mathscr{P} \mid f(0) = 0\}$.
(b) $\{f \in \mathscr{P} \mid f'(0) = 0\}$.
(c) $\{f \in \mathscr{P} \mid |f(0)| = |f'(0)|\}$.
(d) $\{f \in \mathscr{P} \mid f(-x) = f(x) \text{ for all } x \in \mathbb{R}\}$.
(e) $\{f \in \mathscr{P} \mid f(-x) = -f(x) \text{ for all } x \in \mathbb{R}\}$.
(f) $\{f \in \mathscr{P} \mid |f(-x)| = |f(x)| \text{ for all } x \in \mathbb{R}\}$.

Which of these vector spaces are real algebras under the usual multiplication of functions?

1.103. Prove that the set of all invertible elements of an algebra is closed under multiplication by nonzero scalars.

1.104. Which elements of the algebra of all real polynomial functions $\mathscr{P}$ are invertible? Does $\mathscr{P}$ have zero-divisors?

1.105. Prove that every nonzero element of the algebra F^n from Example 1.96 is either invertible or a zero-divisor.

1.106. Take any matrix $A \in M_2(\mathbb{C})$ that is not a scalar multiple of the identity matrix. Prove that $B \in M_2(\mathbb{C})$ commutes with A if and only if B is of the form $B = \alpha A + \beta I$ for some scalars $\alpha, \beta \in \mathbb{C}$.

Hint. Use the theorem on the Jordan normal form (see Section 5.6). For complex matrices of size 2×2, this theorem states that for every matrix A there exists an invertible matrix P such that $P A P^{-1}$ is either of the form $\begin{bmatrix} \lambda & 0 \\ 0 & \mu \end{bmatrix}$ for some $\lambda, \mu \in \mathbb{C}$ or of the form $\begin{bmatrix} \lambda & 1 \\ 0 & \lambda \end{bmatrix}$ for some $\lambda \in \mathbb{C}$. (Incidentally, we can replace $\mathbb{C}$ by $\mathbb{R}$ in the formulation of the exercise, but this makes it a bit harder.)

1.107. Let $A \in M_2(\mathbb{R})$ be such that A^2 is not a scalar multiple of the identity matrix. Prove that $B \in M_2(\mathbb{R})$ commutes with A if and only if it commutes with A^2.

Hint. Calculations can be avoided by using a well-known theorem from linear algebra.

1.108. Can every ring $(R, +, \cdot)$ be made into an algebra over some field? That is, does there exist a field F and an external operation $F \times R \rightarrow R$ such that $(R, +, \cdot)$, along with this additional operation, is an algebra over F?

1.6   Substructures

The groups $(\mathbb{R}, +)$ and $(\mathbb{C}, +)$ are obviously related: they share the same operation and the set $\mathbb{R}$ is a subset of the set $\mathbb{C}$. We use the term "subgroup" for describing such a relation. Thus, the group $(\mathbb{R}, +)$ is a subgroup of the group $(\mathbb{C}, +)$. Analogously, we say that the ring $(\mathbb{R}, +, \cdot)$ is a subring of the ring $(\mathbb{C}, +, \cdot)$. As these two rings happen to be fields, we also say that $(\mathbb{R}, +, \cdot)$ is a subfield of $(\mathbb{C}, +, \cdot)$. These simple examples hopefully make the concept of "substructure" intuitively clear. We proceed rigorously and systematically.

1.6.1   Subgroups

We start with groups and a prototype definition of a substructure.

Definition 1.109. A subset H of a group G is called a **subgroup** of G if H is itself a group under the operation of G.

In more precise terms, H is a subgroup of G if it is a group under the *restriction* of the operation of G to $H \times H$. This is formally more correct, but the way we stated the definition is simpler and can hardly cause any misunderstanding. Definitions of other substructures will therefore be stated in a similar simple manner; just keep in mind that the word "restriction" would be appropriate at some points.

Since a subgroup H is itself a group, we can speak either about the group H or the subgroup H. The choice of term depends on the context, yet there are no strict rules.

We use the notation $H \leq G$ to denote that H is a subgroup of G. For example,

$$\{1\} \leq G \quad \text{and} \quad G \leq G$$

holds for any group G. If $H \leq G$ and $H \neq G$, then H is called a **proper subgroup** of G. The subgroup $\{1\}$ is called the **trivial subgroup** (if G is an additive group, it is of course written as $\{0\}$).

Every subgroup H of G contains the identity element 1. Indeed, being a group, H must have its own identity element $e \in H$. However, since e is an element of G and is equal to e^2, the only possibility is that $e = 1$. Thus, a group and a subgroup share the same identity element, and the trivial subgroup is contained in any other subgroup.

How to verify whether a subset H of a group G is a subgroup? Apparently the answer should be clear from the definition, but one does not need to blindly follow the group axioms to show that "H is itself a group". For example, the associativity of multiplication (=operation) on H automatically follows from the associativity on G. On the other hand, it cannot be taken for granted that multiplication is a binary operation on H. One thus has to make sure that H is closed under multiplication, i.e., that $x, y \in H$ implies $xy \in H$. Further, we know that the inverse of any $x \in H$ exists since x is also an element of G. However, does it lie in H?

The next theorem gives two subgroup criteria.

Theorem 1.110. *Let H be a nonempty subset of a group G. The following conditions are equivalent:*

(i)   *H is a subgroup of G.*
(ii)   *$xy^{-1} \in H$ whenever $x, y \in H$.*
(iii)   *H is closed under multiplication and $x^{-1} \in H$ whenever $x \in H$.*

Proof. (i) $\Longrightarrow$ (ii). This is clear since H is itself a group.

(ii) $\Longrightarrow$ (iii). Take $x \in H$. By (ii), $1 = xx^{-1} \in H$, and hence $x^{-1} = 1x^{-1} \in H$. For any $x, y \in H$, we thus have $xy = x(y^{-1})^{-1} \in H$ since $y^{-1} \in H$.

(iii) $\Longrightarrow$ (i). The condition that H is closed under multiplication means that multiplication is a binary operation on H. It is certainly associative since G is a group. For every $x \in H$, we have $x^{-1} \in H$, and so $1 = xx^{-1}$ belongs to H as H is closed under multiplication. The inverse of an element of H lies in H by assumption. $\square$

An obvious advantage of criterion (ii) is that it deals with only one condition. In practice, however, it does not really matter whether we use (ii) or (iii). Verifying that a subset is a subgroup is rarely time consuming.

Let us also mention an alternative version of (ii):

(ii') $x^{-1}y \in H$ whenever $x, y \in H$.

This criterion is also equivalent to the others.

Example 1.111. The set of all nonzero complex numbers, $\mathbb{C}^* = \mathbb{C} \setminus \{0\}$, is a group under multiplication (see Example 1.53). Applying Theorem 1.110, one easily verifies that the following sets are subgroups of $\mathbb{C}^*$:

$$\mathbb{R}^* = \mathbb{R} \setminus \{0\},$$
$$\mathbb{R}^+ = \{x \in \mathbb{R} \mid x > 0\},$$
$$\mathbb{T} = \{z \in \mathbb{C} \mid |z| = 1\},$$
$$U_4 = \{1, -1, i, -i\}.$$

These groups will frequently appear throughout the book. We call $\mathbb{T}$ the **circle group**. Its elements can be written as e^{iy}, $y \in \mathbb{R}$, and can be geometrically viewed as points of the unit circle (i.e., the circle of radius one centered at the origin).

We remark that $\mathbb{R}^+$ is also a subgroup of $\mathbb{R}^*$, and U_4 is a subgroup of $\mathbb{T}$. In general, if $H \le G$ and K is a subset of H, then $K \le G$ and $K \le H$ are equivalent conditions.

Exercise 1.136 asks to prove that a finite nonempty subset of a group is a subgroup if (and only if) it is closed under multiplication. This is not true for infinite subsets. For example, the sets

$$\{x \in \mathbb{R} \mid x > 1\} \quad \text{and} \quad \{z \in \mathbb{C}^* \mid |z| \le 1\}$$

are closed under multiplication, but are not subgroups of $\mathbb{C}^*$. The first one does not even contain 1, and the second one does not contain inverses of all of its elements.

In an additive group, criterion (iii) reads as

$$x + y, -x \in H \text{ whenever } x, y \in H,$$

and criterion (ii) reads as

$$x - y \in H \text{ whenever } x, y \in H.$$

Thus, a nonempty subset of an additive group is a subgroup if and only if it is closed under subtraction.

Example 1.112. The set of all even integers, denoted $2\mathbb{Z}$, is a subgroup of the group $(\mathbb{Z}, +)$, whereas the set of all odd integers is not. In Section 2.1, we will describe all subgroups of $(\mathbb{Z}, +)$.

Example 1.113. Let G be a group. The set

$$Z(G) := \{c \in G \mid cx = xc \text{ for all } x \in G\},$$

i.e., the set of elements in G that commute with every element in G, is called the **center of the group** G. It is clear that $1 \in Z(G)$ and that G is Abelian if and only if $Z(G) = G$. Note that $Z(G)$ is closed under multiplication. Further, multiplying $cx = xc$ from the left and right by c^{-1} we see that $c^{-1} \in Z(G)$ whenever $c \in Z(G)$.

Hence, $Z(G)$ is a subgroup of G. Concrete examples will be met throughout the text, especially in the exercises of Chapter 2. As we will see, the center of some non-Abelian groups is just the trivial subgroup.

We say that elements x and y of a group G are **conjugate elements** (in G) if there exists an $a \in G$ such that $y = axa^{-1}$; this can also be written as $x = a^{-1}ya$, so x and y occur symmetrically in this definition. We also say that y is a **conjugate of** x (and that x is a conjugate of y). Later we will see that conjugate elements are similar to each other in many ways. But now we return to subgroups.

Example 1.114. If H is a subgroup of a group G and a is an element in G, then

$$aHa^{-1} := \{aha^{-1} \mid h \in H\}$$

is also a subgroup of H. The proof is simple and is left to the reader. Every subgroup of the form aHa^{-1} is called a **conjugate** of H. If a subgroup H' is a conjugate of H, then we say that H and H' are **conjugate subgroups**. Since $H' = aHa^{-1}$ implies $H = a^{-1}H'a$, H is also a conjugate of H' in this case.

1.6.2 Subrings

Definitions of various substructures are very similar to each other. In the case of rings, however, the unity element needs special attention.

Definition 1.115. A subset S of a ring R is called a **subring** of R if S is itself a ring under the operations of R and contains the unity 1 of R.

A subring S also contains the zero element 0 of R, not only the unity 1. This is because S is, in particular, a subgroup under addition and its identity element for addition can only be 0 (see the previous subsection). On the other hand, the requirement that S is a ring under the operations of R does not imply that $1 \in S$.

Example 1.116. The set $\mathscr{R}$ of all matrices of the form $\left[\begin{smallmatrix} x & 0 \\ 0 & 0 \end{smallmatrix}\right]$, where $x \in \mathbb{R}$, is a ring under the same operations as the ring $M_2(\mathbb{R})$, but its unity, namely the matrix $\left[\begin{smallmatrix} 1 & 0 \\ 0 & 0 \end{smallmatrix}\right]$, does not coincide with the unity of $M_2(\mathbb{R})$, which is $\left[\begin{smallmatrix} 1 & 0 \\ 0 & 1 \end{smallmatrix}\right]$. Therefore, $\mathscr{R}$ is not a subring of $M_2(\mathbb{R})$.

Theorem 1.117. *A subset S of a ring R is a subring of R if and only if the following conditions hold:*

(a) *S is a subgroup under addition,*
(b) *S is closed under multiplication, and*
(c) *$1 \in S$.*

Proof. The "only if" part is clear since a subring is itself a ring. Assume that (a), (b), and (c) hold. Since multiplication is associative in R, it is also associative in S. Similarly, the distributive laws hold in R, and hence also in S. Therefore, S is a ring under the operations of R.   □

Recall that S is a subgroup under addition if and only if $x - y \in S$ for all $x, y \in S$. To prove that S is a subring, one thus has to check that $x - y$ and xy belong to S whenever $x, y \in S$, and that S contains 1.

Example 1.118. The chain of subsets

$$\mathbb{Z} \subset \mathbb{Q} \subset \mathbb{R} \subset \mathbb{C}$$

is also a chain of subrings. Thus, $\mathbb{Z}$ is a subring of the rings $\mathbb{Q}$, $\mathbb{R}$, and $\mathbb{C}$, $\mathbb{Q}$ is a subring of $\mathbb{R}$ and $\mathbb{C}$, and $\mathbb{R}$ is a subring of $\mathbb{C}$.

Example 1.119. The **center of a ring** R is defined in the same way as the center of a group, i.e.,

$$Z(R) := \{c \in R \mid cx = xc \text{ for all } x \in R\}.$$

One easily checks that $Z(R)$ is a subring of R. We will encounter concrete examples later.

1.6.3 Subfields

Subfields are defined similarly as other substructures. However, as will be explained below, they are studied in a somewhat different way.

Definition 1.120. A subset F of a field K is called a **subfield** of K if F is itself a field under the operations of K.

A subfield F automatically contains the unity 1 of K, so there is no need to require this in the definition. Indeed, F contains its own unity e, which satisfies $e^2 = e$ and hence $e(1 - e) = 0$. Since $e \neq 0$ and fields have no zero-divisors (see Theorem 1.77), this yields $e = 1$.

Theorem 1.121. *A subset F of a field K is a subfield of K if and only if the following conditions hold:*

(a) *F is a subgroup under addition,*
(b) *F is closed under multiplication,*
(c) *$1 \in F$, and*
(d) *if $x \in F$ and $x \neq 0$, then $x^{-1} \in F$.*

Proof. The "only if" part is clear. To prove the "if" part, we first note that (a)–(c) imply that F is a subring of K (see Theorem 1.117), and after that, that a subring satisfying (d) is a a subfield. □

The condition that F contains 1 can be replaced by the condition that F contains at least one nonzero element x, since along with (b) and (d) this yields $1 = xx^{-1} \in F$.

Example 1.122. The field $\mathbb{R}$ is a subfield of the field $\mathbb{C}$. Similarly, $\mathbb{Q}$ is a subfield of both $\mathbb{R}$ and $\mathbb{C}$.

The next definition indicates a different perspective for viewing subfields.

Definition 1.123. A field K is an **extension field** (or simply an **extension**) of a field F if F is a subfield of K.

Thus, $\mathbb{C}$ is an extension of $\mathbb{R}$. Since the complex numbers are constructed from the reals, this sounds more natural than saying that $\mathbb{R}$ is a subfield of $\mathbb{C}$. Similarly, saying that $\mathbb{Q}$ is a subfield of $\mathbb{R}$ is technically correct, but does not correspond to the way we learn about these fields—we first get familiar with $\mathbb{Q}$ and only later extend it to $\mathbb{R}$. In field theory, we usually start with a field which we know and understand, but has some deficiencies. Therefore we try to construct its extension which has properties that the original field lacks. The usual approach in the study of other algebraic structures is different. For instance, in group theory we are typically given a group and our goal is to understand its structure. It is often too difficult to handle the whole group at once, so we study its smaller parts—its subgroups. By collecting information on various subgroups we can obtain some insight into the original group.

1.6.4 Subspaces

We assume the reader is already familiar with the following notion.

Definition 1.124. A subset U of a vector space V is called a **subspace** of V if U is itself a vector space under the operations of V.

More formal terms are "vector subspace" and "linear subspace," but we will use only "subspace".

Theorem 1.125. *Let U be a nonempty subset of a vector space V over a field F. The following conditions are equivalent:*

(i)   *U is a subspace of V.*
(ii)   *For all $\lambda, \mu \in F$, $\lambda u + \mu w \in U$ whenever $u, w \in U$.*
(iii)   *U is closed under addition and for all $\lambda \in F$, $\lambda u \in U$ whenever $u \in U$.*

Proof. (i) $\Longrightarrow$ (ii). This is clear since a subspace is itself a vector space.

(ii) $\Longrightarrow$ (iii). Taking first $\lambda = \mu = 1$ and then $\mu = 0$ in (ii), we get (iii).

(iii) $\Longrightarrow$ (i). Taking $\lambda = -1$ in (iii), we see that U contains additive inverses of all of its elements. Therefore, U is a subgroup under addition. By assumption, scalar multiplication is an operation from $F \times U$ to U. Since V is a vector space, its subset U also satisfies all the vector space axioms. $\square$

Example 1.126. Besides $\{0\}$ and $\mathbb{R}^2$, the only subspaces of $\mathbb{R}^2$ are the lines through the origin. Similarly, the subspaces of $\mathbb{R}^3$ are $\{0\}$, $\mathbb{R}^3$, all lines through the origin, and all planes through the origin. We assume the reader knows this well. At any rate, it can easily be verified via geometric consideration.

1.6.5 Subalgebras

The definition of a subalgebra should by now be self-evident.

Definition 1.127. A subset B of an algebra A is called a **subalgebra** of A if B is itself an algebra under the operations of A and contains the unity 1 of A.

In other words, a subalgebra is a subspace which is also a subring. The next theorem therefore needs no proof.

Theorem 1.128. *A subset B of an algebra A is a subalgebra of A if and only if the following conditions hold:*

(a) *B is closed under addition,*
(b) *B is closed under multiplication,*
(c) *for all $\lambda \in F$, $\lambda x \in B$ whenever $x \in B$, and*
(d) *$1 \in B$.*

Example 1.129. Consider $\mathbb{C}$ as a real algebra (Example 1.98). Then $\mathbb{R}$ is a subalgebra of $\mathbb{C}$.

Example 1.130. The **center of an algebra** is defined in the same way as the center of a ring (or of a group). It is clearly a subalgebra (not only a subring). The obvious examples of elements in the center are scalar multiples of 1; sometimes these are also the only examples.

Let us add a slightly more colorful example.

Example 1.131. The set of all upper triangular real matrices, i.e., matrices of the form $\begin{bmatrix} a_{11} & a_{12} \\ 0 & a_{22} \end{bmatrix}$, where $a_{ij} \in \mathbb{R}$, is a subalgebra of the algebra $M_2(\mathbb{R})$.

1.6.6 The Intersection (and the Union) of Substructures

Intersection is a "friendly" operation with respect to substructures, whereas taking unions is not. This is what we would like to point out in this short concluding subsection. We begin with the following statement:

• The intersection of subgroups is a subgroup.

This holds not only for the intersection of two subgroups, but for the intersection of an arbitrary, possibly infinite family of subgroups. The proof is a simple application of Theorem 1.110. Indeed, if x and y are elements from the intersection of a family of subgroups, then each of them is contained in any of these subgroups, hence the same holds for xy^{-1}, which is therefore contained in the intersection of all subgroups.

Similarly, the following statements hold:

- The intersection of subrings is a subring.
- The intersection of subfields is a subfield.
- The intersection of subspaces is a subspace.
- The intersection of subalgebras is a subalgebra.

Nothing like this holds for the union operation. The union of two subgroups is in fact only exceptionally a subgroup, see Exercise 1.137. The same can be said for the union of subrings, subfields, subspaces, and subalgebras.

Exercises

1.132. Find a subgroup of the symmetric group S_3 with two elements.

1.133. Find a subgroup of the symmetric group S_3 with three elements.

1.134. The first exercises of the previous three sections can now be interpreted differently. For instance, Exercise 1.55 asks which of the given sets are subgroups of $(\mathbb{C}, +)$. Find by yourself a further example of a subgroup of this group, as well as an example of a set which is not a subgroup but satisfies some of the subgroup conditions. Ask yourself similar questions with regard to Exercises 1.56, 1.79, 1.101, and 1.102.

1.135. Find an example of a group G and a subset X of G which is not a subgroup, but has the following property: for all $x \in X$ and all $n \in \mathbb{Z}$, $x^n \in X$.

1.136. Let H be a *finite* nonempty subset of a group G. Show that the assumption that H is closed under multiplication already implies that H is a subgroup.

Hint. Exercise 1.46.

1.137. Let H and K be subgroups of a group G such that their union $H \cup K$ is also a subgroup. Prove that then $H \subseteq K$ or $K \subseteq H$.

1.138. Find a group G with four elements that contains subgroups H_1, H_2, H_3 such that $G = H_1 \cup H_2 \cup H_3$ and $H_i \cap H_j = \{1\}$ for all $i \neq j$.

1.139. Let A be a group (resp. ring). For every $a \in A$, let $C_A(a)$ denote the set of all elements in A that commute with a. We call $C_A(a)$ the **centralizer of** a (special cases were already considered in Exercises 1.106 in 1.107).

(a) Show that $C_A(a)$ is a subgroup (resp. subring).
(b) Show that $C_A(a) \subseteq C_A(a^2)$.
(c) Show that $C_A(a) = C_A(a^2)$ if $a^{2k+1} = 1$ for some $k \in \mathbb{N}$.
(d) Find a group (resp. ring) A and an element $a \in A$ such that $C_A(a) \subsetneq C_A(a^2)$.

Comment. A somewhat more general concept is the **centralizer of a subset** $S \subseteq A$. It is defined as $C_A(S) := \bigcap_{a \in S} C_A(a)$, i.e., as the set of all elements in A that commute with every element in S. Note that $C_A(S)$ is also a subgroup (resp. subring). Obviously, $C_A(A)$ is the center of A (which, however, is traditionally denoted by $Z(A)$, originating from Zentrum, the German word for center).

1.140. Let R be a nonzero ring. Show that the set R^* of all invertible elements in R is not a subring.

1.141. Let S be a subring of the ring $\mathbb{R}$. Show that $S[i] := \{s+ti \mid s, t \in S\}$ is a subring of the ring $\mathbb{C}$. Moreover, $S[i]$ is a subfield of $\mathbb{C}$ if S is a subfield of $\mathbb{R}$. Can you now provide an example of a proper subfield of $\mathbb{C}$ that contains $\sqrt{2}$ and i?

1.142. Find a subset of the vector space $\mathbb{R}^2$ which is a subgroup under addition, but is a not a subspace.

1.143. Find a subset of the algebra $M_2(\mathbb{R})$ which is a subring, but is not a subalgebra.

1.144. For every $a \in \mathbb{R}$, let A_a denote the set of all matrices of the form $\begin{bmatrix} x & y \\ ay & x \end{bmatrix}$, where $x, y \in \mathbb{R}$. Show that A_a is a commutative subalgebra of the algebra $M_2(\mathbb{R})$. Find all a such that A_a has no zero-divisors, and show that A_a is a field for every such a.

1.145. Find a commutative subalgebra of the algebra $M_2(\mathbb{R})$ that has zero-divisors and is not an algebra of the form A_a from the previous exercise.

1.7 Generators

To gain some intuition, we begin by mentioning a few simple linear algebra examples that are presumably already known to the reader. Take the vector space $\mathbb{R}^3$. The only subspace that contains the vectors $(1, 0, 0)$, $(0, 1, 0)$, and $(0, 0, 1)$ is $\mathbb{R}^3$ itself. We therefore say that these three vectors generate $\mathbb{R}^3$. Needless to say, these are not the only such vectors. For example, the vectors $(1, 0, 0)$, $(1, 1, 0)$, and $(1, 1, 1)$ also generate $\mathbb{R}^3$. The vectors $(1, 0, 0)$ and $(0, 1, 0)$ generate the plane $z = 0$, i.e., the subspace $\{(x, y, 0) \mid x, y \in \mathbb{R}\}$. Indeed, this plane is the smallest subspace of $\mathbb{R}^3$ that contains both vectors.

We can similarly talk about generators in other algebraic structures.

1.7.1 Basic Definitions

Let X be a subset of a group G. What is the smallest subgroup of G that contains X? The meaning of "smallest" here is intuitively clear, but must be made precise. We are interested in a subgroup K of G that satisfies the following two conditions:

(a) K contains X, and
(b) if H is a subgroup of G that contains X, then $K \subseteq H$.

Such a subgroup is obviously unique, provided it exists. And it does exist! Namely, K is simply the intersection of all subgroups of G that contain X. Indeed, this intersection is a subgroup (as remarked at the end of the previous section), and it clearly satisfies both conditions (a) and (b). It may happen that G is the only subgroup of G that contains X. In this case, $K = G$.

The subgroup K satisfying (a) and (b) is called the **subgroup generated by** X. It will be denoted by $\langle X \rangle$. If X consists of elements x_i, we also say that $\langle X \rangle$ is the **subgroup generated by the elements** x_i. In the case where $\langle X \rangle = G$, we say that G **is generated by** X, or that X **generates** G. The elements in X are then called **generators of** G, and X is called a **set of generators of** G.

Every group G satisfies $\langle G \rangle = G$, which is hardly useful information. One is usually interested in sets of generators that are as small as possible. If a group G is generated by some finite set X, then we say that G is **finitely generated**. Finite groups are obviously finitely generated. A simple example of a finitely generated infinite group is $(\mathbb{Z}, +)$ (see Example 1.147 below). Studying a group-theoretic problem in finitely generated groups may be more demanding than studying it in finite groups, but easier than in general groups.

Everything we have said so far for groups can be said for other algebraic structures. For example, for a subset X of a ring R, we define the **subring generated by** X as a subring K satisfying (a) K contains X, and (b) if S is a subring of R that contains X, then $K \subseteq S$. In fact, in the above paragraphs on generators in groups, we can simply substitute "ring" for "group" and "subring" for "subgroup", and make appropriate notational adjustments. Similar obvious changes must be made when introducing generators in fields, vector spaces, and algebras.

However, all these are just plain definitions. What does the subgroup (subring etc.) generated by X look like? This question is the theme of the next subsections. Let us consider here only the case where X is the empty set $\emptyset$. The subgroup generated by $\emptyset$ is the trivial subgroup $\{1\}$. Indeed, it satisfies both conditions (a) and (b). Similarly, the trivial subspace $\{0\}$ is the subspace generated by $\emptyset$. Since, by definition, every subring contains the unity 1, the subring generated by $\emptyset$ is equal to the subring generated by 1. In this statement, we can replace "subring" with "subfield" or "subalgebra".

1.7.2 Group Generators

Let X be a nonempty subset of a group G. We claim that $\langle X \rangle$, the subgroup of G generated by X, consists of all elements in G of the form

$$y_1 y_2 \cdots y_n, \quad \text{where } y_i \in X \text{ or } y_i^{-1} \in X. \tag{1.3}$$

It is clear that the set of all such elements is closed under multiplication, and from

$$(y_1 y_2 \cdots y_n)^{-1} = y_n^{-1} \cdots y_2^{-1} y_1^{-1}$$

we see that it also contains inverses of its elements. Theorem 1.110 therefore tells us that this set is a subgroup. It certainly contains X; on the other hand, every subgroup

H of G that contains X also contains every element of the form (1.3). Hence, $\langle X \rangle$ is indeed the set of all elements of this form.

If $X = \{x_1, \ldots, x_n\}$, we also write $\langle x_1, \ldots, x_n \rangle$ for $\langle X \rangle$.

The elements y_i in the expression $y_1 y_2 \cdots y_n$ are not necessarily different. Thus, for example, the subgroup $\langle x, y \rangle$ consists of elements such as

$$1, \ x, \ y, \ yx^{-1}, \ x^2 y^{-1} x, \ y^{-3} x y^5 x^{-4}, \text{etc.}$$

If G is Abelian, every such element can be written as $x^i y^j$ for some $i, j \in \mathbb{Z}$.

Example 1.146. The set of all positive rational numbers $\mathbb{Q}^+$ is a group under multiplication. It is generated by the set of all natural numbers, i.e., $\langle \mathbb{N} \rangle = \mathbb{Q}^+$. The subgroup $\langle 2, 3, 5 \rangle$, for example, consists of all numbers of the form $2^i 3^j 5^k$, where $i, j, k \in \mathbb{Z}$.

In an additive group, elements in $\langle X \rangle$ are written as $x_1 + \cdots + x_n$, where either $x_i \in X$ or $-x_i \in X$. Equivalently, and more clearly, we can write them as

$$k_1 x_1 + k_2 x_2 + \cdots + k_n x_n,$$

where $x_i \in X$ and $k_i \in \mathbb{Z}$.

Example 1.147. The group $(\mathbb{Z}, +)$ is generated by a single element, namely $\mathbb{Z} = \langle 1 \rangle$. Groups with this special property that have a single generator are called *cyclic*. They will be studied in Section 3.1. Returning to our concrete example, we also have $\mathbb{Z} = \langle -1 \rangle$. Besides 1 and -1, no other element generates the whole group $\mathbb{Z}$. For example, the subgroup $\langle 2 \rangle$ is equal to $2\mathbb{Z}$, the group of even integers. We also have $\langle -2 \rangle = 2\mathbb{Z}$, $\langle 4, 6 \rangle = 2\mathbb{Z}$, etc. A subgroup can be generated by many different sets.

1.7.3 Ring Generators

Let now X be a nonempty subset of a ring R. Denote by $\overline{X}$ the subgroup under addition generated by all products of elements from $X \cup \{1\}$. Thus, $\overline{X}$ consists of elements of the form

$$k_1 x_{11} \cdots x_{1m_1} + k_2 x_{21} \cdots x_{2m_2} + \cdots + k_n x_{n1} \cdots x_{nm_n},$$

where $x_{ij} \in X \cup \{1\}$ and $k_i \in \mathbb{Z}$. From Theorem 1.117, we immediately infer that $\overline{X}$ is a subring. Clearly, $\overline{X}$ contains X and is contained in any subring that contains X. Thus, $\overline{X}$ is the subring generated by X.

Example 1.148. Let R be the ring of complex numbers $\mathbb{C}$.

(a) The subring generated by the element 1 is the ring of integers $\mathbb{Z}$. Since every subring of R contains 1, it also contains all integers. Therefore, $\mathbb{Z}$ is the smallest subring of $\mathbb{C}$ in the sense that it is contained in any other subring.

(b) The subring generated by the element i consists of all complex numbers of the form $m + ni$, where $m, n \in \mathbb{Z}$. We call it the ring of **Gaussian integers** and denote it by $\mathbb{Z}[i]$.

1.7.4 Field Generators

Let X be a nonempty subset of a field F. As above, we denote by $\overline{X}$ the subring of F generated by X. There is no reason why $\overline{X}$ should be a subfield. We claim that the subfield generated by X is the set consisting of all elements of the form

$$uv^{-1}, \quad \text{where } u, v \in \overline{X} \text{ and } v \neq 0.$$

Clearly, every subfield that contains X must also contain all these elements. Therefore, we only have to verify that this set is a subfield. Using the identity

$$uv^{-1} - wz^{-1} = (uz - vw)(vz)^{-1}, \tag{1.4}$$

we see that it is a subgroup under addition. Checking the other conditions from Theorem 1.121 is straightforward.

Let us add a little comment. Some readers may have the impression that they could not have found the formula (1.4) by themselves. However, if one writes $\frac{u}{v}$ instead of uv^{-1} and $\frac{w}{z}$ instead of wz^{-1}, and then uses the standard rule for subtracting fractions, the formula presents itself.

In Example 1.148, we have treated the complex numbers as a ring. Let us now treat them as a field.

Example 1.149. Let F be the field of complex numbers $\mathbb{C}$.

(a) The subfield generated by the element 1 is the field of rational numbers $\mathbb{Q}$. Consequently, every subfield of $\mathbb{C}$ contains the rational numbers.

(b) The subfield generated by the element i is denoted $\mathbb{Q}(i)$ and called the field obtained by adjoining i to $\mathbb{Q}$. It consists of all complex numbers of the form $p + qi$, where $p, q \in \mathbb{Q}$. One can prove this directly by checking that the set of these numbers forms a subfield which is contained in every subfield containing i. On the other hand, we can follow the above argument and arrive at the same conclusion with the help of the ring of Gaussian integers $\mathbb{Z}[i]$.

1.7.5 Vector Space Generators

We assume the reader is familiar with basic vector space theory. Nevertheless, recalling some notions and facts may be useful.

Let V be a vector space over a field F. A vector of the form

$$\lambda_1 v_1 + \lambda_2 v_2 + \cdots + \lambda_n v_n, \quad \text{where } \lambda_i \in F,$$

is called a **linear combination** of the vectors $v_1, v_2, \ldots, v_n \in V$. Let X be a nonempty subset of V. We denote the set of all linear combinations of vectors in X by span X and call it the **linear span of** X. Note that span X is a subspace containing X and is contained in every subspace that contains X. Therefore, span X is the subspace generated by X (for $X = \emptyset$ we define span $X = \{0\}$ and so this case is no exception). A finitely generated vector space, i.e., a vector space that is generated by some finite set, is called a **finite-dimensional vector space**. Thus, V is finite-dimensional if there exist finitely many vectors $u_1, \ldots, u_m \in V$ such that $V = \text{span}\{u_1, \ldots, u_m\}$. A vector space is **infinite-dimensional** if it is not finite-dimensional.

For vector spaces, unlike for other algebraic structures, we can quickly derive decisive and far-reaching results concerning generators. To this end, we need some further definitions.

A subset S of a vector space V is said to be **linearly dependent** if there exist different vectors $v_1, v_2, \ldots, v_n \in S$ such that

$$\lambda_1 v_1 + \lambda_2 v_2 + \cdots + \lambda_n v_n = 0$$

for some scalars $\lambda_1, \lambda_2, \ldots, \lambda_n \in F$, not all of which are 0. If S is not linearly dependent, then we say that it is **linearly independent**. An infinite set is thus linearly independent if all of its finite subsets are linearly independent. If $S = \{v_1, \ldots, v_n\}$ and all v_i are different, then instead of saying that S is linearly dependent (resp. independent), we also say that the vectors $v_1, \ldots, v_n$ are linearly dependent (resp. independent).

The next theorem is the key to understanding finite-dimensional vector spaces.

Theorem 1.150. *Let V be a finite-dimensional vector space. If $V = \text{span}\{u_1, \ldots, u_m\}$ and $v_1, \ldots, v_n \in V$ are linearly independent, then $m \geq n$.*

Proof. By assumption, we can write $v_1 = \lambda_1 u_1 + \cdots + \lambda_m u_m$ for some scalars λ_i. Since $v_1 \neq 0$, we may assume, without loss of generality, that $\lambda_1 \neq 0$. Multiplying the above identity by λ_1^{-1} we see that u_1 is a linear combination of the vectors $v_1, u_2, \ldots, u_m$. This implies that $V = \text{span}\{v_1, u_2, \ldots, u_m\}$. Therefore,

$$v_2 = \gamma v_1 + \mu_2 u_2 + \cdots + \mu_m v_m$$

for some scalars γ and μ_i. Since v_1 and v_2 are linearly independent, not all μ_i are 0. We may assume that $\mu_2 \neq 0$. Hence, u_2 is a linear combination of $v_1, v_2, u_3, \ldots, u_m$, from which we infer that $V = \text{span}\{v_1, v_2, u_3, \ldots, u_m\}$. We continue this process and at each step substitute v_i for u_i without changing the linear span. If m was smaller than n, then after m steps we would arrive at $V = \text{span}\{v_1, v_2, \ldots, v_m\}$, which is a contradiction to $v_{m+1} \notin \text{span}\{v_1, v_2, \ldots, v_m\}$. $\square$

A subset B of a vector space V is called a **basis** of V if $V = \text{span}\,B$ and B is linearly independent. This is equivalent to the condition that each vector in V can be written as a linear combination of vectors in B in a unique way.

Every vector space has a basis. The method of proof for the general case is explained in Exercise 1.174. The finite-dimensional case is much easier. Indeed, just take any finite set B such that span $B = V$ but span $B_0 \neq V$ for every proper subset B_0 of B, and note that it is linearly independent and hence a basis (if $V = \{0\}$, then $B = \emptyset$). A finite-dimensional space V thus has a finite basis, and Theorem 1.150 implies that the number of vectors in any basis is the same. This number is called the **dimension** of V and is denoted $\dim_F V$, or simply $\dim V$. If $\dim_F V = n$, then we say that V is n-**dimensional.** In such a space, every linearly independent set B with n elements is a basis. Indeed, taking any vector $v \in V \setminus B$ we see from Theorem 1.150 that the set $B \cup \{v\}$ is linearly dependent, from which it follows that v lies in span B. Therefore, $V = $ span B and so B is a basis.

Example 1.151. The vector space F^n is n-dimensional. The vectors

$$e_1 := (1, 0, \ldots, 0), \; e_2 := (0, 1, 0, \ldots, 0), \; \ldots, e_n := (0, \ldots, 0, 1)$$

form its simplest basis. It is called the *standard basis* of F^n.

Example 1.152. The set of complex numbers $\mathbb{C}$ is a 2-dimensional real vector space. Its *standard basis* is $\{1, i\}$. When considered as a complex vector space, $\mathbb{C}$ is of course 1-dimensional.

Example 1.153. The real vector space of all 2×2 matrices $M_2(\mathbb{R})$ is 4-dimensional. Its *standard basis* consists of the following matrices:

$$E_{11} := \begin{bmatrix} 1 & 0 \\ 0 & 0 \end{bmatrix}, \; E_{12} := \begin{bmatrix} 0 & 1 \\ 0 & 0 \end{bmatrix}, \; E_{21} := \begin{bmatrix} 0 & 0 \\ 1 & 0 \end{bmatrix}, \; E_{22} := \begin{bmatrix} 0 & 0 \\ 0 & 1 \end{bmatrix}.$$

They are called **matrix units**.

Example 1.154. The vector space of all real polynomial functions $\mathscr{P}$ is infinite-dimensional. The polynomial functions

$$1, \, x, \, x^2, \, x^3, \ldots$$

form its *standard basis*. Indeed, every real polynomial function can be written as a linear combination of these polynomial functions in a unique way.

1.7.6 Algebra Generators

Let A be an algebra over a field F and let X be a nonempty subset of A. Similarly as in the case of ring generators, we see that the subalgebra generated by X consists of all elements of the form

$$\lambda_1 x_{11} \cdots x_{1m_1} + \lambda_2 x_{21} \cdots x_{2m_2} + \cdots + \lambda_n x_{n1} \cdots x_{nm_n},$$

where $x_{ij} \in X \cup \{1\}$ and $\lambda_i \in F$. Observe that this is the linear span of the subring generated by X.

Example 1.155. The algebra of real polynomial functions $\mathscr{P}$ is generated by a single element, namely x.

By saying that an *algebra is finite-dimensional* we of course mean that it is finite-dimensional as a vector space.

Example 1.156. The matrix algebra $M_2(\mathbb{R})$ is finite-dimensional, of dimension 4 (see Example 1.153). Thus, as a vector space, it is generated by four elements. As an algebra, however, it is generated by only two elements, for example by the matrix units E_{12} and E_{21}. Indeed, since $E_{11} = E_{12}E_{21}$ and $E_{22} = E_{21}E_{12}$, every matrix can be written as a linear combination of the matrices E_{12}, E_{21}, $E_{12}E_{21}$, and $E_{21}E_{12}$. For comparison, the matrix units E_{11} and E_{22} generate "only" the subalgebra of all diagonal matrices, i.e., matrices of the form $\left[\begin{smallmatrix} \lambda & 0 \\ 0 & \mu \end{smallmatrix}\right]$, where $\lambda, \mu \in \mathbb{R}$.

One can also consider $M_2(\mathbb{R})$ as a ring. The subring generated by the matrices E_{12} and E_{21} is then the ring of all matrices with integer entries.

Exercises

1.157. Let X be a nonempty subset of a group G. Denote by X^{-1} the set of inverses of elements in X. Is $\langle X \rangle = \langle X^{-1} \rangle$?

1.158. Let H and K be subgroups of an additive group G. Show that the subgroup generated by their union $H \cup K$ consists of all elements of the form $h + k$, where $h \in H$ and $k \in K$.

Comment. We call this subgroup the sum of H and K, and denote it by $H + K$. We will discuss this in more detail later.

1.159. Find two elements in the symmetric group S_3 that generate the whole group. Is S_3 generated by a single element?

1.160. Show that the group $(\mathbb{R}^*, \cdot)$ is generated by the interval $[-2, -1]$.

1.161. Prove that the group $(\mathbb{Q}, +)$ is not finitely generated. What about $(\mathbb{Q}^*, \cdot)$?

1.162. Determine the subgroups of $(\mathbb{Z}, +)$ generated by the sets $\{6, 9\}$, $\{63, 90\}$, and $\{28, 99\}$. Can you determine the subgroup generated by an arbitrary pair of integers m and n?

Comment. Perhaps it is easier to guess than to prove the answer to the latter question. After reading Section 2.1, providing a proof should also be easy.

1.163. Determine the subgroup of $(\mathbb{C}^*, \cdot)$ generated by the element $z_n := e^{\frac{2\pi i}{n}}$, where $n \in \mathbb{N}$. Determine also the subgroup generated by all z_n.

1.164. Determine the subring of the ring $\mathbb{Q}$ generated by the element $\frac{1}{2}$.

1.165. Determine the subring of the ring $\mathbb{Q}$ generated by the set $\left\{ \frac{1}{2n+1} \mid n \in \mathbb{N} \right\}$.

1.166. The set of complex numbers $\mathbb{C}$ can be considered either as an additive group, a ring, a real vector space, a real algebra, or a field. Determine the subgroup, the subring, the subspace, the subalgebra, and the subfield generated by the element $1 + 2i$.

1.167. The set of all real polynomial functions $\mathscr{P}$ can be considered either as an additive group, a ring, a real vector space, or a real algebra. Determine the subgroup, the subring, the subspace, and the subalgebra generated by the elements x^2 and $2x^3$.

1.168. The set of all 2×2 real matrices $M_2(\mathbb{R})$ can be considered either as an additive group, a ring, a real vector space, or a real algebra. Determine the subgroup, the subring, the subspace, and the subalgebra generated by the elements $\left[\begin{smallmatrix} 1 & 0 \\ 0 & 0 \end{smallmatrix} \right]$ in $\left[\begin{smallmatrix} 0 & 0 \\ 3 & 0 \end{smallmatrix} \right]$.

1.169. Determine the subfield of $\mathbb{R}$ generated by the elements $\sqrt{2}$ and $\sqrt{3}$.

1.170. Determine the subfield of $\mathbb{R}$ generated by the element $\sqrt[3]{2}$.

Comment. In Chapter 7, we will learn how to solve problems like this without computation. At this stage, however, you must rely on definitions and common sense.

1.171. Show that the set $\mathscr{R}$ of all real polynomial functions f satisfying $f(0) = f(1)$ is a subalgebra of the algebra of all real polynomial functions $\mathscr{P}$. Find a (finite) set with the least number of elements that generates $\mathscr{R}$.

1.172. Prove that a nonzero element x of a finite-dimensional algebra A is either a zero-divisor or invertible. Prove also that the invertibility of x follows already from the existence of a left (or right) inverse of x.

Hint. If A is n-dimensional, then the elements $1, x, \dots, x^n$ are linearly dependent.

1.173. Let V be a (possibly infinite-dimensional) vector space. Prove that a set $B \subseteq V$ is a basis of V if and only if B is a *maximal linearly independent subset* of V; by this we mean that B is linearly independent and there *does not* exist a linearly independent subset B' of V such that $B \subsetneqq B'$.

1.174. Prove that an arbitrary (possibly infinite-dimensional) vector space V has a basis.

Sketch of proof. The proof requires familiarity with the following important result from set theory.

Zorn's lemma. *Let $\mathscr{S}$ be a partially ordered set. If every chain in $\mathscr{S}$ has an upper bound in $\mathscr{S}$, then $\mathscr{S}$ contains a maximal element.*

We now give the necessary definitions. A **partially ordered set** is a set $\mathscr{S}$ together with a binary relation $\leq$ which is reflexive $(x \leq x)$, anti-symmetric $(x \leq y$ and $y \leq x$

imply $x = y$), and transitive ($x \leq y$ and $y \leq z$ imply $x \leq z$). A subset $\mathscr{C}$ of a partially ordered set $\mathscr{S}$ is called a **chain** if for all $y, y' \in \mathscr{C}$, either $y \leq y'$ or $y' \leq y$. An **upper bound** for $\mathscr{C}$ is an element $u \in \mathscr{S}$ such that $y \leq u$ for all $y \in \mathscr{C}$. We say that $b \in \mathscr{S}$ is a **maximal element** of $\mathscr{S}$ if there does not exist a $b' \in \mathscr{S}$ such that $b \neq b'$ and $b \leq b'$.

The proof that V has a basis is a rather straightforward application of Zorn's lemma. Partially order the set $\mathscr{S}$ of all linearly independent subsets of V by set inclusion (i.e., define $X \leq X' \iff X \subseteq X'$). If $\mathscr{C} = \{Y_i \mid i \in I\}$ is a chain in $\mathscr{S}$, then $U := \bigcup_{i \in I} Y_i$ is a linearly independent set (why?), and hence an upper bound for $\mathscr{C}$ in $\mathscr{S}$. Zorn's lemma implies the existence of a maximal element B in $\mathscr{S}$, which is a basis of V by the result of the preceding exercise.

Comment. Zorn's lemma is an extremely useful tool in mathematics. However, we will not use it in the main text since we will not provide its proof. Let us only mention that the proof depends on the axiom of choice; as a matter of fact, Zorn's lemma and the axiom of choice are equivalent statements.

1.8 Direct Products and Sums

Constructions of new objects from old ones occur throughout algebra. The simplest of them are based on the concept of componentwise operation. This short section is devoted to their introduction. In later parts of the book it will become evident that these constructions are of fundamental importance in algebra.

1.8.1 Direct Products of Groups

Let $G_1, \ldots, G_s$ be groups. Their Cartesian product $G_1 \times \cdots \times G_s$ becomes a group by defining the operation in the most natural way:

$$(x_1, \ldots, x_s)(y_1, \ldots, y_s) := (x_1 y_1, \ldots, x_s y_s).$$

Indeed, the associative law for this operation follows from the associative laws for the operations in the groups G_i, the identity element 1 is $(1, \ldots, 1)$, and the inverse of $(x_1, \ldots, x_s)$ is $(x_1^{-1}, \ldots, x_s^{-1})$.

A comment about notation may be in order. We denoted identity elements of possibly different groups by the same symbol 1. Writing the identity element of the group G more unambiguously by 1_G, the equality $1 = (1, \ldots, 1)$ becomes

$$1_{G_1 \times \cdots \times G_s} = (1_{G_1}, \ldots, 1_{G_s}).$$

This may be very precise, but looks complicated. It will be much easier if we use symbols like 1 and 0 for different things. We just have to be aware that their meaning depends on the context.

We call $G_1 \times \cdots \times G_s$, together with the above defined (componentwise) operation, the **direct product of the groups** $G_1, \ldots, G_s$. This is the easiest way to "glue" different groups G_i together. Some of the properties of these groups are thereby preserved. For example, the direct product of Abelian groups is an Abelian group, the direct product of finite groups is a finite group, etc.

If G_i are additive groups, then we change both notation and terminology. Instead of $G_1 \times \cdots \times G_s$ we write $G_1 \oplus \cdots \oplus G_s$, instead of "direct product of groups" we use the term **direct sum of groups**, and instead of $\cdot$ we denote the operation by $+$. The above definition then reads

$$(x_1, \ldots, x_s) + (y_1, \ldots, y_s) := (x_1 + y_1, \ldots, x_s + y_s).$$

In the special case where each G_i is equal to $\mathbb{R}$, we thus obtain the standard formula for the sum of vectors in $\mathbb{R}^s$.

The reader should be warned that different texts use different notation. Some write $G_1 \oplus \cdots \oplus G_s$ for the direct product of arbitrary groups, and some write $G_1 \times \cdots \times G_s$ when the groups G_i are additive.

Finally, we mention that a more precise term for what we have introduced is the **external direct product** (or **sum**) of groups. The internal product (sum) will be introduced later, in Section 4.5.

1.8.2 Direct Products of Rings

After introducing direct products of groups, it should be easy to guess what the definitions of direct products (or sums) of other algebraic structures are. The **direct product of rings** $R_1, \ldots, R_s$ is, of course, the set $R_1 \times \cdots \times R_s$ together with the operations defined by

$$(x_1, \ldots, x_s) + (y_1, \ldots, y_s) := (x_1 + y_1, \ldots, x_s + y_s)$$

and

$$(x_1, \ldots, x_s)(y_1, \ldots, y_s) := (x_1 y_1, \ldots, x_s y_s).$$

It is easy to check that this is a ring with zero element $0 = (0, \ldots, 0)$ and unity $1 = (1, \ldots, 1)$. (Some denote this ring by $R_1 \oplus \cdots \oplus R_s$ and call it the direct sum of $R_1, \ldots, R_s$.)

If R_i are nonzero rings and $s \geq 2$, then $R_1 \times \cdots \times R_s$ has zero-divisors. For example, the product of $(x_1, 0, \ldots, 0)$ and $(0, x_2, 0, \ldots, 0)$ is 0. The direct product of two or more fields is thus not a field. The notion of the direct product therefore does not make much sense in field theory, which is, as already mentioned, in many ways quite different from the theories of other algebraic structures.

1.8.3 Direct Sums of Vector Spaces

Let $V_1, \ldots, V_s$ be vector spaces over a field F. Their Cartesian product $V_1 \times \cdots \times V_s$, together with the operations given by

$$(u_1, \ldots, u_s) + (v_1, \ldots, v_s) := (u_1 + v_1, \ldots, u_s + v_s),$$
$$\lambda(v_1, \ldots, v_s) := (\lambda v_1, \ldots, \lambda v_s),$$

is a vector space over F. We call it the **(external) direct sum of the vector spaces** $V_1, \ldots, V_s$ and denote it by $V_1 \oplus \cdots \oplus V_s$. The vector space F^s is thus the direct sum of s copies of the 1-dimensional vector space F.

1.8.4 Direct Products of Algebras

Algebras are simultaneously rings and vector spaces. The **direct product of algebras** $A_1, \ldots, A_s$ over a field F is thus defined to be the set $A_1 \times \cdots \times A_s$ together with the operations defined by

$$(x_1, \ldots, x_s) + (y_1, \ldots, y_s) := (x_1 + y_1, \ldots, x_s + y_s),$$
$$(x_1, \ldots, x_s)(y_1, \ldots, y_s) := (x_1 y_1, \ldots, x_s y_s),$$
$$\lambda(x_1, \ldots, x_s) := (\lambda x_1, \ldots, \lambda x_s).$$

Endowed with these operations, $A_1 \times \cdots \times A_s$ is an algebra over F.

We have restricted ourselves to direct products and sums of finitely many objects primarily for simplicity. We can start with an infinite family of groups, rings, etc., and introduce operations in their Cartesian product in the same way as above. Here is a sample.

Example 1.175. Regard the field $\mathbb{R}$ as an algebra over itself. The direct product of countably infinite copies of the algebra $\mathbb{R}$ is the Cartesian product $\mathbb{R} \times \mathbb{R} \times \ldots$ together with the operations defined by

$$(x_1, x_2, \ldots) + (y_1, y_2, \ldots) := (x_1 + y_1, x_2 + y_2, \ldots),$$
$$(x_1, x_2, \ldots)(y_1, y_2, \ldots) := (x_1 y_1, x_2 y_2, \ldots),$$
$$\lambda(x_1, x_2, \ldots) := (\lambda x_1, \lambda x_2, \ldots).$$

The set $\mathbb{R} \times \mathbb{R} \times \ldots$ can be identified with the set of all real sequences, and these operations are the usual operations with sequences.

Exercises

1.176. Prove that if groups $G_1, \ldots, G_s$ are finitely generated, then so is their direct product $G_1 \times \cdots \times G_s$.

1.177. Prove that

$$Z(A_1 \times \cdots \times A_s) = Z(A_1) \times \cdots \times Z(A_s),$$

i.e., the center of the direct product is equal to the direct product of the centers. Here, A_i can be either groups, rings, or algebras.

1.178. If H_i is a subgroup of a group G_i, $i = 1, \ldots, s$, then $H_1 \times \cdots \times H_s$ is a subgroup of $G_1 \times \cdots \times G_s$. However, not every subgroup of $G_1 \times \cdots \times G_s$ is necessarily of this form. Find a different example (for some concrete groups G_i).

1.179. Let G be an additive group. A subset B of G is called a **basis of** G if the following two conditions are satisfied:

- for every $x \in G$, there exist $b_1, \ldots, b_k \in B$ and $n_1, \ldots, n_k \in \mathbb{Z}$ such that $x = n_1 b_1 + \cdots + n_k b_k$,
- for all distinct $b_1, \ldots, b_k \in B$ and all $n_1, \ldots, n_k \in \mathbb{Z}$, $n_1 b_1 + \cdots + n_k b_k = 0$ implies $n_i = 0$ for every i.

This definition is the same as that of a basis of a vector space, only the role of scalars is replaced by the integers (compare Remark 1.94). However, unlike vector spaces, additive groups only rarely have a basis. Those that do are called **free Abelian groups**.

(a) Prove that $\mathbb{Z}^s := \mathbb{Z} \oplus \cdots \oplus \mathbb{Z}$, the direct sum of s copies of the group $(\mathbb{Z}, +)$, is free.
(b) Prove that a finite Abelian group cannot be free.
(c) Is the group $(\mathbb{Q}, +)$ free?

1.180. Determine the subring of $\mathbb{Z} \times \mathbb{Z}$ generated by the element $(1, -1)$.

1.181. Determine the subring of $\mathbb{R} \times \mathbb{R}$ generated by the element $(\sqrt{2}, 0)$.

1.182. Let F be a field. Determine all subspaces of the algebra $F \times F$. Which among them are subalgebras? Perhaps it is even easier to answer a more general question: which subsets of a 2-dimensional algebra are subalgebras?

1.183. Let e denote the constant sequence of 1's, and let e_n denote the sequence whose nth term is 1 and all other terms are 0. Note that $B := \text{span}\{e, e_1, e_2, \ldots\}$ consists of all sequences that are constant from some term onward. Prove that B is a subalgebra of the algebra from Example 1.175. Is B finitely generated as an algebra?

Chapter 2
Examples of Groups and Rings

The importance of examples in algebra cannot be emphasized enough. Not only because they help us to understand the theory—the theory exists because of examples! Abstract mathematical concepts are introduced after realizing that various, often very different, objects have certain common properties, and that studying these properties alone is more efficient than studying concrete examples where particularities tend to overshadow the essence. An abstract theory should therefore yield a better understanding of motivating examples. But this it not supposed to be its only goal. A successful theory usually becomes more and more independent of its roots, and eventually starts to be regarded as interesting and valuable in its own right. Its success, however, heavily depends on its applicability to other topics. Thus, again, there are examples, often different from the initial ones, that justify the theory.

This chapter is a survey of the most prominent examples of groups and rings (as well as algebras). We will also arrive at some examples of fields, although not that many. Constructions of fields depend on certain theoretical tools that will not be considered earlier than in Chapter 7. Vector spaces are studied in linear algebra, and therefore their examples will not be examined carefully.

Roughly speaking, the first half of the chapter is devoted to commutative structures and the second half to noncommutative structures. A ring is, in particular, an additive group, and its invertible elements form a multiplicative group. Examples of rings and groups will therefore be intertwined. Examples of groups that are not intimately connected with rings will be considered in Sections 2.5 and 2.6.

One of the goals of the chapter is to show that groups and rings occur throughout mathematics. Exposure to algebraic ideas is therefore useful for mathematicians working in all areas.

2.1 The Integers

The set of natural numbers $\mathbb{N}$ is only a semigroup under addition. In algebra, we therefore prefer the set of integers $\mathbb{Z}$, which is a group under addition, and, moreover,

© Springer Nature Switzerland AG 2019
M. Brešar, *Undergraduate Algebra*, Springer Undergraduate Mathematics Series,
https://doi.org/10.1007/978-3-030-14053-3_2

a ring when endowed with multiplication. Even when we are interested in a problem concerning the natural numbers, it is often more convenient to consider it in the setting of the integers.

We remark that, throughout the book, the symbol $\mathbb{Z}$ will be sometimes used for the ring $(\mathbb{Z}, +, \cdot)$, sometimes for the group $(\mathbb{Z}, +)$, and sometimes just for the set $\mathbb{Z}$. However, it should be easy to figure out from the context which role of $\mathbb{Z}$ we have in mind.

We start with a basic result, known as the **division algorithm for integers**, from which all others will be, either directly or indirectly, derived.

Theorem 2.1. *Let $m \in \mathbb{Z}$ and $n \in \mathbb{N}$. Then there exist unique $q, r \in \mathbb{Z}$ such that*

$$m = qn + r \quad and \quad 0 \leq r < n.$$

Proof. Note, first of all, that for a large enough integer ℓ we have $\ell n > m$. Any such ℓ is larger than any integer from the set

$$S := \{k \in \mathbb{Z} \mid kn \leq m\}.$$

This means that S is bounded above. It is nonempty, since otherwise $kn > m$ would hold for all (including negative) integers k. Applying (a version of) the well-ordering principle, we conclude that S has a largest element, which we denote by q. Then $q + 1 \notin S$, and so

$$qn \leq m < (q + 1)n = qn + n.$$

Thus, $r := m - qn$ satisfies $0 \leq r < n$, which is the desired conclusion.

To prove the uniqueness, assume that there are $q, q, r, r' \in \mathbb{Z}$ such that

$$m = qn + r, \quad 0 \leq r < n \quad and \quad m = q'n + r', \quad 0 \leq r' < n.$$

Without loss of generality, we may assume that $r \geq r'$. Comparing the two expressions for m, we obtain $(q' - q)n = r - r'$. However, since $0 \leq r - r' \leq r < n$ and $q' - q$ is an integer, this is possible only if $r = r'$ and $q' = q$. $\qquad\square$

We call q the **quotient** and r the **remainder** of the division of m by n.

As the first application of Theorem 2.1, we will describe all subgroups of $(\mathbb{Z}, +)$. Let us write

$$n\mathbb{Z} := \{nx \mid x \in \mathbb{Z}\}.$$

Here, n can be any integer. However, since $n\mathbb{Z} = (-n)\mathbb{Z}$, we normally take $n \in \mathbb{N} \cup \{0\}$. If $n \in \mathbb{N}$, then n is the smallest natural number in $n\mathbb{Z}$. This observation is the key to the proof of the less obvious "only if" part of the next theorem.

Theorem 2.2. *A subset H of $\mathbb{Z}$ is a subgroup under addition if and only if $H = n\mathbb{Z}$ for some $n \in \mathbb{N} \cup \{0\}$.*

Proof. The set $n\mathbb{Z}$ is closed under addition and contains additive inverses of its elements. Therefore, it is a subgroup.

Now take any subgroup H of $\mathbb{Z}$. If $H = \{0\}$, then $H = n\mathbb{Z}$ for $n = 0$. Suppose $H \neq \{0\}$. Since H contains additive inverses of its elements, we have

$$H \cap \mathbb{N} \neq \emptyset.$$

By the well-ordering principle, $H \cap \mathbb{N}$ contains a smallest element. Let us call it n. Since H is a subgroup and $n \in H$, we have $n\mathbb{Z} \subseteq H$. It remains to prove the converse inclusion. Take $m \in H$. By Theorem 2.1, there exist $q, r \in \mathbb{Z}$ such that

$$r = m - qn \quad \text{and} \quad 0 \leq r < n.$$

As H is a subgroup containing m and qn, H also contains r. From $r < n$ and the choice of n it follows that r cannot be a natural number. Hence, $r = 0$ and so $m = qn \in n\mathbb{Z}$. $\qquad\square$

The $n = 0$ case corresponds to the trivial subgroup $\{0\}$, and the $n = 1$ case corresponds to the whole group $\mathbb{Z}$. Every proper nontrivial subgroup of $(\mathbb{Z}, +)$ thus consists of integer multiples of some natural number $n \geq 2$. For example, $2\mathbb{Z} = \{0, \pm 2, \pm 4, \dots\}$ contains all multiples of 2, i.e., all even integers. The subgroup $n\mathbb{Z}$ is generated by the element n. We can therefore formulate Theorem 2.2 as follows: all subgroups of the group $(\mathbb{Z}, +)$ are generated by a single element. In what follows, we will illustrate the applicability of this result to a topic in elementary mathematics.

We say that an integer $k \neq 0$ **divides** an integer m if $m = qk$ for some integer q. In this case, we write

$$k \mid m.$$

We also say that k is a **divisor** of m or that m is **divisible** by k. If k does not divide m, we write $k \nmid m$. Obviously, $1 \mid m$ and $-1 \mid m$ for every $m \in \mathbb{Z}$, $k \mid 0$ for every $k \neq 0$, and $k \nmid 1$ unless $k = 1$ or $k = -1$. Note also that $k \mid m$ if and only if $m \in k\mathbb{Z}$.

Let m and n be integers, not both 0. An integer $k \neq 0$ that divides both m and n is called a **common divisor** of m and n. A natural number d which is a common divisor of m and n and is divisible by any other common divisor of m and n is called the **greatest common divisor** of m and n. We then write $d = \gcd(m, n)$. It is clear that there can be only one greatest common divisor, but does it exist at all? To answer this question, we first make a general remark on subgroups. If H and K are subgroups of an additive group G, then so is the set

$$H + K := \{h + k \mid h \in H, k \in K\}.$$

Indeed, from

$$(h + k) - (h' + k') = (h - h') + (k - k')$$

we see that the difference of two elements from $H + K$ again lies therein. Thus, as a special case,

$$m\mathbb{Z} + n\mathbb{Z} = \{mx + ny \mid x, y \in \mathbb{Z}\}$$

is a subgroup of $\mathbb{Z}$.

Theorem 2.3. *For each pair of integers m and n, not both 0, $\gcd(m, n)$ exists. Moreover, there exist $x, y \in \mathbb{Z}$ such that $\gcd(m, n) = mx + ny$.*

Proof. Since $m\mathbb{Z} + n\mathbb{Z}$ is a subgroup of $(\mathbb{Z}, +)$, Theorem 2.2 shows that there exists a $d \in \mathbb{N} \cup \{0\}$ such that

$$d\mathbb{Z} = m\mathbb{Z} + n\mathbb{Z}.$$

Clearly, $d \neq 0$ for $m \neq 0$ or $n \neq 0$. As $d \in d\mathbb{Z}$, it can be written as $d = mx + ny$ for some $x, y \in \mathbb{Z}$. This implies that d is divisible by any common divisor c of m and n. Indeed, from $m = cz$ and $n = cw$ it follows that $d = c(zx + wy)$. But is d a common divisor? From

$$m \in m\mathbb{Z} \subseteq m\mathbb{Z} + n\mathbb{Z} = d\mathbb{Z}$$

we see that $d \mid m$. Similarly, $d \mid n$, and so $d = \gcd(m, n)$. $\square$

Now, after proving the existence, we see that $\gcd(m, n)$ can be equivalently defined as the largest among all common divisors of m and n. The proof we gave can be seen as a nice little illustration of the efficiency of abstract algebra. Later, in Chapter 5, we will use the same method of proof in a considerably more general situation. It should be remarked, however, that there are other proofs that use only the language of elementary mathematics. After all, Theorem 2.3 can be extracted from *Euclid's* famous book *Elements* written circa 300 B.C., more than two millennia before the introduction of the notion of a group. More precisely, Euclid described the following procedure, known as **Euclidean algorithm**, for computing the greatest common divisor of two natural numbers m and n. All we have to do is to repeatedly apply Theorem 2.1, as follows:

$$
\begin{aligned}
m &= q_1 n + r_1, & 0 &< r_1 < n, \\
n &= q_2 r_1 + r_2, & 0 &< r_2 < r_1, \\
r_1 &= q_3 r_2 + r_3, & 0 &< r_3 < r_2, \\
&\ \ \vdots \\
r_{k-2} &= q_k r_{k-1} + r_k, & 0 &< r_k < r_{k-1}, \\
r_{k-1} &= q_{k+1} r_k + 0.
\end{aligned}
$$

Indeed, the remainder must eventually be zero since each r_i is smaller than r_{i-1}. The last equality shows that $r_k \mid r_{k-1}$, the second to last shows that $r_k \mid r_{k-2}$, etc. At the end we get that $r_k \mid n$ and $r_k \mid m$, so r_k is a common divisor of m and n. Now let c be any common divisor of m and n. From the first equality we see that $c \mid r_1$, from the second that $c \mid r_2$, etc. Finally, we get $c \mid r_k$. This means that

$$r_k = \gcd(m, n).$$

We leave it as an exercise to the reader to show that r_k can be written as $mx + ny$ for some $x, y \in \mathbb{Z}$.

We say that two integers m and n, not both 0, are **relatively prime** if $\gcd(m, n) = 1$. That is, the only common divisors of m and n are 1 and -1.

Corollary 2.4. *Two integers m and n are relatively prime if and only if there exist $x, y \in \mathbb{Z}$ such that $mx + ny = 1$.*

Proof. From $mx + ny = 1$ it follows that a natural number different from 1 cannot be a common divisor of m and n, meaning that $\gcd(m, n) = 1$. The converse follows from Theorem 2.3. $\qquad\square$

Remark 2.5. The greatest common divisor of more than two integers is defined in the same way. Thus, $d \in \mathbb{N}$ is the **greatest common divisor of the integers** $n_1, \ldots, n_k$ (not all 0) if d is their common divisor (i.e., $d \mid n_i$ for each i) and d is divisible by any other common divisor of $n_1, \ldots, n_k$. By adapting the proof of Theorem 2.3, we see that d exists and is of the form

$$d = n_1 x_1 + \cdots + n_k x_k$$

for some $x_i \in \mathbb{Z}$. We say that $n_1, \ldots, n_k$ are relatively prime if $d = 1$. This is equivalent to the condition that $n_1 x_1 + \cdots + n_k x_k = 1$ for some $x_i \in \mathbb{Z}$. However, it is not equivalent to the condition that $n_1, \ldots, n_k$ are pairwise relatively prime. For example, the integers 2, 3, 6 are relatively prime, but not pairwise relatively prime.

We say that $p \in \mathbb{N}$ is a **prime number** (or a **prime**) if $p \neq 1$ and the only natural numbers that divide p are 1 and p. A natural number $n \neq 1$ that is not prime is called a **composite number**.

The next result also appears in *Elements* and is named after Euclid.

Corollary 2.6. (Euclid's Lemma) *Let p be a prime number. If $p \mid mn$, then $p \mid m$ or $p \mid n$.*

Proof. Suppose $p \nmid m$. Since p is a prime, this implies that p and m are relatively prime. Corollary 2.4 thus tells us that $px + my = 1$ for some $x, y \in \mathbb{Z}$. Multiplying this by n we obtain $pxn + mny = n$. By assumption, $mn = pc$ for some $c \in \mathbb{N}$, and hence $p(xn + cy) = n$. Thus, $p \mid n$. $\qquad\square$

The next theorem was also known already to Euclid, and surely is not unfamiliar to the reader who may try to prove it before reading the proof below. In the statement, we use the convention that a product of primes can consist of a single factor.

Theorem 2.7. (Fundamental Theorem of Arithmetic) *Every natural number $n \neq 1$ is a product of primes. This factorization is unique up to the order of the factors.*

Proof. We prove the first statement by induction on n. The $n = 2$ case is trivial, so assume that $n > 2$ and that the statement is true for all natural numbers larger than 1 and smaller than n. There is nothing to prove if n is a prime, so let $n = k\ell$ with $1 < k, \ell < n$. By our assumption, each of k and ℓ is a product of primes, and therefore the same is true for n.

To prove the uniqueness, assume that n can be expressed as a product of primes in two ways,

$$n = p_1 p_2 \cdots p_s \quad \text{and} \quad n = q_1 q_2 \cdots q_t. \tag{2.1}$$

Since p_1 is a prime and divides $q_1 q_2 \cdots q_t$, Corollary 2.6 shows that p_1 divides either q_1 or $q_2 \cdots q_t$. In the latter case it must divide either q_2 or $q_3 \cdots q_t$. Continuing in this manner, we see that p_1 divides at least one of the q_i. Since the order of the factors is irrelevant to us, we may assume that p_1 divides q_1. However, since q_1 is a prime and $p_1 \neq 1$, this is possible only if $p_1 = q_1$. From (2.1) it thus follows that

$$p_2 \cdots p_s = q_2 \cdots q_t.$$

We can now repeat the argument and conclude that, without loss of generality, $p_2 = q_2$. Continuing, we get $p_3 = q_3$, $p_4 = q_4$, etc. If s was different from t, say $s > t$, then we would arrive at $p_{t+1} \cdots p_s = 1$, which is a contradiction. Therefore, $s = t$ and both factorizations in (2.1) are the same.   □

The primes in the factorization $n = p_1 p_2 \cdots p_s$ are not necessarily distinct (e.g., $4 = 2 \cdot 2$). Collecting together those that are equal, we obtain the factorization

$$n = p_1^{k_1} p_2^{k_2} \cdots p_r^{k_r},$$

where the primes p_i are distinct and the natural numbers k_i are uniquely determined.

In algebra, we often try to decompose objects into indecomposable parts. Theorem 2.7 shows that this can be accomplished for elements of the ring of integers. In Chapter 5, we will extend this theorem to considerably more general rings.

We close this section with another classical theorem and Euclid's timeless proof.

Theorem 2.8. *There are infinitely many primes.*

Proof. Suppose that there is only a finite number, n, of primes. Let $p_1, p_2, \ldots, p_n$ be the only primes. Consider the number

$$p_1 p_2 \cdots p_n + 1.$$

Theorem 2.7 shows that it is divisible by some prime p_j. Therefore,

$$p_j k = p_1 p_2 \cdots p_n + 1$$

for some natural number k. Rewriting this equality in the form

$$p_j (k - p_1 \cdots p_{j-1} p_{j+1} \cdots p_n) = 1,$$

we arrive at a contradiction since p_j cannot divide 1.   □

There is something fascinating about primes. Mathematicians have studied them for centuries and discovered many deep and beautiful theorems. Nevertheless, many old and seemingly simple problems on primes, understandable to high school students, are still unsolved. Let us mention two of them. The **Goldbach conjecture** states that every even natural number different from 2 is equal to a sum of two primes. The **twin prime conjecture** states that there are infinitely many primes p such that $p + 2$ is also a prime. Mathematics is developing fast and who knows, maybe one of these problems will be solved in our lifetime.

Exercises

2.9. Revisit Exercise 1.162. Is it easier now?

2.10. Compute the greatest common divisor of the day and the year of your birthday, and express it in the form described in Theorem 2.3.

2.11. Let m and n be natural numbers. By the Fundamental Theorem of Arithmetic, there exist distinct primes $p_1, \ldots, p_r$ and nonnegative (not necessarily positive!) integers $k_1, \ldots, k_r$ and $\ell_1, \ldots, \ell_r$ such that $m = p_1^{k_1} \cdots p_r^{k_r}$ and $n = p_1^{\ell_1} \cdots p_r^{\ell_r}$. Express the greatest common divisor of m and n in terms of the p_i, k_i, and ℓ_i. Consider the same problem for the **least common multiple** of m and n, i.e., the smallest natural number that is divisible by both m and n.

2.12. Let a be an element of a group G. Suppose there exist relatively prime integers m and n such that $a^m = a^n = 1$. Prove that $a = 1$.

2.13. Let G be a group, $H \leq G$, and $a \in G$. Suppose that $a^m = 1$ for some $m \in \mathbb{Z}$ and that n is the smallest natural number such that $a^n \in H$. Prove that $n \mid m$.

2.14. Find all primes p such that $7p + 4$ is the square of some integer.

2.15. A **Mersenne prime** is a prime that can be written as $2^p - 1$ for some $p \in \mathbb{N}$. Prove that p must itself be a prime and that the number $n := 2^{p-1}(2^p - 1)$ has the property that the sum of its positive divisors is $2n$.

2.16. For each $m \geq 2$, let P_m denote the set of all primes less than or equal to m. Prove that

$$\prod_{p \in P_m} \frac{1}{1 - \frac{1}{p}} \geq \sum_{n=1}^{m} \frac{1}{n}$$

and hence derive that there are infinitely many primes (here, $\prod$ denotes the product).
Hint. Geometric series.

2.17. Prove that there are infinitely many primes of the form $4n + 3$.
Hint. Assume that $p_1, \ldots, p_n$ are the only such primes and consider the number $4p_1 \cdots p_n - 1$.

2.18. Prove that, for every $n \in \mathbb{N}$, there exist n consecutive natural numbers none of which is a prime.
Hint. Find a "big" number N such that each of the numbers

$$N + 2, N + 3, \ldots, N + (n + 1)$$

has an obvious nontrivial divisor.

2.2 The Integers Modulo n

Imagine an evening, the clock has just chimed seven. What will the time be in four hours? The answer is eleven, since $7 + 4 = 11$. Now, what will the time be in six hours? The answer is one and again we arrived at it by using addition, although the sum of the integers 7 and 6 is not 1. However, we did not perform the addition in the group $\mathbb{Z}$, but in the group $\mathbb{Z}_{12}$. Let us explain!

Fix a natural number n. We say that the integers a and b are **congruent modulo** n if $n \mid a - b$. In this case, we write

$$a \equiv b \pmod{n}.$$

For example, $13 \equiv 1 \pmod{12}$ and $19 \equiv -5 \pmod{12}$. The reader can easily check that a and b are congruent modulo n if and only if they have the same remainder when divided by n.

We point out three simple properties of congruence:

(r) $a \equiv a \pmod{n}$,
(s) $a \equiv b \pmod{n} \implies b \equiv a \pmod{n}$,
(t) $a \equiv b \pmod{n}$ and $b \equiv c \pmod{n} \implies a \equiv c \pmod{n}$.

The first two are entirely obvious, and the last one follows from $a - c = (a - b) + (b - c)$. Of course, (r) stands for reflexivity, (s) for symmetry, and (t) for transitivity. Hence, the congruence modulo n is an *equivalence relation* on $\mathbb{Z}$. Let $[a]$ denote the equivalence class containing a. We thus have

$$[a] = [a'] \iff a \equiv a' \pmod{n}.$$

The equivalence class $[0]$ consists of all integers that are divisible by n, $[1]$ consists of all integers that have the remainder 1 when divided by n, etc.

Denote the set of all equivalence classes by $\mathbb{Z}_n$. It has n elements, namely

$$\mathbb{Z}_n = \big\{[0], [1], \ldots, [n-1]\big\}.$$

We will endow $\mathbb{Z}_n$ with the operations of addition and multiplication. In doing this, we will face, for the first of many times, the problem of showing that something is *well-defined*. This problem occurs whenever we want to define a map or an operation on a set in which elements can be written in different ways. Let us illustrate by an example.

Example 2.19. Define $f : \mathbb{Q} \to \mathbb{Q}$ by the rule

$$f\left(\frac{r}{s}\right) = \frac{r^2}{s^2}.$$

This is the function that sends every element into its square, that is, $f(x) = x^2$ for every $x \in \mathbb{Q}$. Let us slightly modify the rule and define

$$\mathscr{F}\left(\frac{r}{s}\right) = \frac{r^2}{s}.$$

Is everything all right with this definition? For example, what is $\mathscr{F}(0.5)$? Since $0.5 = \frac{1}{2}$, our immediate answer may be that $\mathscr{F}(0.5) = \frac{1^2}{2} = 0.5$. However, we can also write 0.5 as, for instance, $\frac{-3}{-6}$, which leads to a different answer: $\mathscr{F}(0.5) = \frac{(-3)^2}{-6} = -1.5$. The rule $\mathscr{F}(\frac{r}{s}) = \frac{r^2}{s}$ thus does not make any sense, or, using the mathematical language, $\mathscr{F}$ is not well-defined (and simply is not a function). On the other hand, there is nothing wrong with the first rule $f(\frac{r}{s}) = \frac{r^2}{s^2}$. Indeed, if we write $x \in \mathbb{Q}$ in two ways, firstly as $x = \frac{r}{s}$ and secondly as $x = \frac{r'}{s'}$, then $rs' = r's$, which gives $(rs')^2 = (r's)^2$ and hence $\frac{r^2}{s^2} = \frac{r'^2}{s'^2} = x^2$. Thus, for every $x \in \mathbb{Q}$ we have $f(x) = x^2$, independently of the way x is represented as a fraction.

Just as the same rational number can be represented by different fractions, the same element in $\mathbb{Z}_n$ (i.e., the same equivalence class) can be represented by different integers, namely,

$$\ldots = [a - n] = [a] = [a + n] = [a + 2n] = \ldots$$

Therefore, when given a rule that presumably defines a function f on $\mathbb{Z}_n$, we must check that $f([a]) = f([a'])$ whenever $[a] = [a']$. Then we can say that f is well-defined. Similarly, a binary operation $\star$ on $\mathbb{Z}_n$ must satisfy $[a] \star [b] = [a'] \star [b']$ whenever $[a] = [a']$ and $[b] = [b']$.

Theorem 2.20. *If we endow $\mathbb{Z}_n$ with addition given by*

$$[a] + [b] := [a + b],$$

$\mathbb{Z}_n$ becomes an Abelian group.

Proof. To check that the operation is well-defined, we must prove that the conditions $[a] = [a']$ and $[b] = [b']$ imply that $[a + b] = [a' + b']$, or, in other words, that n divides $(a + b) - (a' + b')$ whenever n divides both $a - a'$ and $b - b'$. Since

$$(a + b) - (a' + b') = (a - a') + (b - b'),$$

we see that this is indeed true.

The associative law for addition in $\mathbb{Z}_n$ follows from the associative law for addition in $\mathbb{Z}$. Indeed,

$$([a] + [b]) + [c] = [a + b] + [c] = [(a + b) + c]$$
$$= [a + (b + c)] = [a] + [b + c] = [a] + ([b] + [c]).$$

Similarly, we check the commutativity. The zero element in $\mathbb{Z}_n$ is $[0]$, and the additive inverse of $[a]$ is $[-a]$. $\qquad\square$

The group $(\mathbb{Z}_n, +)$ is generated by a single element, namely by the element $[1]$. Thus, it is a cyclic group, just like $(\mathbb{Z}, +)$ (see Example 1.147). In Section 3.1, we will see that there are essentially no other cyclic groups.

Theorem 2.21. *If we endow the additive group $\mathbb{Z}_n$ with multiplication given by*

$$[a] \cdot [b] := [ab],$$

$\mathbb{Z}_n$ *becomes a commutative ring.*

Proof. Again we must check that the new operation is well-defined. That is, we have to prove that $[ab] = [a'b']$ (i.e., $n \mid ab - a'b'$) whenever $[a] = [a']$ and $[b] = [b']$ (i.e., $n \mid a - a'$ and $n \mid b - b'$). This follows from

$$ab - a'b' = (a - a')b + a'(b - b').$$

The associativity, commutativity, and distributive laws can be verified in a similar way as the associativity of addition in the proof of Theorem 2.20. The unity is the element $[1]$. $\qquad\square$

We call $(\mathbb{Z}_n, +)$ the **group of integers modulo** n and $(\mathbb{Z}_n, +, \cdot)$ the **ring of integers modulo** n. The symbol $\mathbb{Z}_n$ will thus have a double meaning (similarly as $\mathbb{Z}$), and we must take care not to confuse the two, i.e., the group $\mathbb{Z}_n$ and the ring $\mathbb{Z}_n$. They will both appear frequently throughout the book.

Writing elements in $\mathbb{Z}_n$ as $[a]$ would eventually become tiresome. Let us simplify the notation: instead of $[a]$, where $0 \le a < n$, we will write just a. Thus,

$$\mathbb{Z}_n = \{0, 1, \ldots, n - 1\}.$$

The obvious danger with the new notation is that one could mix up elements in $\mathbb{Z}_n$ with elements in $\mathbb{Z}$. On those occasions where $\mathbb{Z}_n$ and $\mathbb{Z}$ will be treated simultaneously, we will therefore return to to the original notation. Of course, calculations in $\mathbb{Z}_n$ are essentially different from those in $\mathbb{Z}$. The sum $a + b$ in $\mathbb{Z}_n$ is equal to the *remainder* of the usual sum of the integers a and b when divided by n. For example, in $\mathbb{Z}_{12}$ we have $7 + 6 = 1$. Similarly, the product ab in $\mathbb{Z}_n$ is the remainder of the usual product of the integers a and b when divided by n. Thus, in $\mathbb{Z}_{12}$ we have $5 \cdot 7 = 11$ and $4 \cdot 3 = 0$. The latter shows that $\mathbb{Z}_n$ may have zero-divisors. In fact, if n is a composite number, then it certainly does have them (why?). If $n = 1$, $\mathbb{Z}_n$ is the zero ring. What about if n is a prime?

Let us first consider the simplest case, where $n = 2$. The ring $\mathbb{Z}_2$ has only two elements, 0 and 1, so the list of all calculations is very short:

$$0 + 0 = 0, \quad 0 + 1 = 1, \quad 1 + 0 = 1, \quad 1 + 1 = 0,$$

$$0 \cdot 0 = 0, \quad 0 \cdot 1 = 0, \quad 1 \cdot 0 = 0, \quad 1 \cdot 1 = 1.$$

The formula $1 + 1 = 0$ stands out, all others are the same as in $\mathbb{Z}$. The only nonzero element in $\mathbb{Z}_2$ is 1, and is invertible. This means that $\mathbb{Z}_2$ is a field.

We will show that $\mathbb{Z}_p$ is a field for every prime p. To this end, we need a lemma which is interesting in its own right. But first a definition.

Definition 2.22. An **integral domain** is a nonzero commutative ring without zero-divisors.

Recall that fields have no zero-divisors, so they are the obvious examples of integral domains—and also the only examples among finite rings.

Lemma 2.23. *A finite integral domain R is a field.*

Proof. Take any $a \neq 0$ in R. Our goal is to show that the equation $ax = 1$ has a solution in R. Since R has no zero-divisors, it satisfies the cancellation law, so $ax = ay$ implies $x = y$. This means that the map $x \mapsto ax$ from R to R is injective. But then it is also surjective since R is a finite set. In particular, its image contains the element 1, which is the desired conclusion. $\square$

The ring $\mathbb{Z}$ is the simplest example of an infinite integral domain that is not a field. There are many more, maybe the reader can think of some? Anyway, we will study integral domains in greater depth later.

Theorem 2.24. *A natural number p is a prime if and only if $\mathbb{Z}_p$ is a field.*

Proof. If p is 1 or a composite number, then $\mathbb{Z}_p$ is the zero ring or has zero-divisors, so it is not a field. Let p be a prime. By Lemma 2.23, we only have to prove that $\mathbb{Z}_p$ has no zero-divisors. To this end, it is more convenient to return to the original notation $[a]$ for elements in $\mathbb{Z}_p$. Thus, suppose that $[a] \cdot [b] = [0]$ for some $a, b \in \mathbb{Z}$. Then $[ab] = [0]$ and so $ab = kp$ for some $k \in \mathbb{Z}$. Euclid's Lemma (Corollary 2.6) implies that $p \mid a$ or $p \mid b$, meaning that $[a] = 0$ or $[b] = 0$. $\square$

Our short list of examples of fields has thus been enriched by essentially different items, the finite fields $\mathbb{Z}_p$. Incidentally, there are further examples of finite fields. They will be described in Section 7.6.

Exercises

2.25. Determine all n for which $f : \mathbb{Z}_n \to \mathbb{Z}_n$, $f([a]) = [|a|]$, is a well-defined function.

2.26. Determine all n for which $[a] \star [b] = [|ab|]$ is a well-defined binary operation on $\mathbb{Z}_n$.

2.27. Find all subgroups of the group $\mathbb{Z}_6$.

2.28. The order of an element $a \in \mathbb{Z}_n$ is the smallest natural number r satisfying $ra = 0$ (see Exercise 1.62). Determine the orders of all elements of the groups $\mathbb{Z}_5$ and $\mathbb{Z}_6$.

2.29. For every $a \in \mathbb{Z}_{12}$, describe $\langle a \rangle$, the subgroup generated by a.

2.30. Let $d \mid n$. Prove that $\mathbb{Z}_n$ has a subgroup with d elements.

Comment. This subgroup is unique. Proving this is one of the exercises in Section 3.1. But you can try to work it out now!

2.31. Determine all $n \geq 2$ for which the sum of all elements in $\mathbb{Z}_n$ is 0.

2.32. By adapting the proof of Lemma 2.23, show that a nonzero finite ring without zero-divisors is a division ring.

Comment. A much deeper result is **Wedderburn's theorem** from 1905 stating that a finite division ring is necessarily commutative. It can be proved in many ways, but all known proofs use tools that are not available to us at this stage.

2.33. Prove the following generalization of the statement of the preceding exercise: if R is any ring and $a \in R$ is such that the set $\{a^n \mid n \in \mathbb{N}\}$ is finite, then a is either a zero-divisor or invertible, with inverse equal to a power of a.

Hint. Exercise 1.46.

2.34. Prove that the ring $\mathbb{Z}_n$ contains a nonzero element a satisfying $a^2 = 0$ if and only if there exists a prime p such that $p^2 \mid n$.

2.35. Prove that an element $[a]$ of the ring $\mathbb{Z}_n$ is invertible if and only if a and n are relatively prime integers.

Comment. Again we have denoted an element in $\mathbb{Z}_n$ by $[a]$, to keep the notation a for an integer. At first glance the statement may seem ambiguous since $[a] = [a + kn]$ for every $k \in \mathbb{Z}$. However, the condition that a and n are relatively prime is readily equivalent to the condition that $a + kn$ and n are relatively prime.

Let us also mention the connection with **Euler's phi function** (named after the great mathematician *Leonhard Euler* (1707– 1783)), known from number theory. This is the function from $\mathbb{N}$ to $\mathbb{N}$ defined by

$$\varphi(n) := \text{ the number of natural numbers } k \leq n \text{ such that } \gcd(k, n) = 1.$$

For example, $\varphi(6) = 2$ since 1 and 5 are the only numbers among 1, 2, 3, 4, 5, 6 that are relatively prime to 6. Clearly, $\varphi(1) = 1$ and $\varphi(p) = p - 1$ for every prime p. The statement of this exercise implies that $\varphi(n)$ is equal to the order of the group of invertible elements of $\mathbb{Z}_n$:

$$\varphi(n) = |\mathbb{Z}_n^*|.$$

One of the basic properties of φ is that $\varphi(k\ell) = \varphi(k)\varphi(\ell)$ whenever k and ℓ are relatively prime, from which one can derive **Euler's product formula**

$$\varphi(n) = n \prod_{p \mid n} \left(1 - \frac{1}{p}\right),$$

where the product runs over all primes p dividing n. Proving this is more a number-theoretic than an algebraic exercise. Nevertheless, the reader is encouraged to tackle it.

2.36. Determine all n such that $\mathbb{Z}_n$ has exactly one zero-divisor, and all n such that $\mathbb{Z}_n$ has exactly two zero-divisors.

2.37. Find an example of an infinite ring that has exactly 24 invertible elements, and an example of an infinite ring that has exactly 40 invertible elements.

Hint. Direct product of rings.

2.3 Rings of Functions

Let X be a nonempty set. We define the *sum* and *product of functions* $f : X \to \mathbb{R}$ and $g : X \to \mathbb{R}$ as the functions $f + g : X \to \mathbb{R}$ and $fg : X \to \mathbb{R}$ given by

$$(f + g)(x) := f(x) + g(x)$$

and

$$(fg)(x) := f(x)g(x)$$

for all $x \in X$. These are the usual addition and multiplication of functions that we normally use without thinking. Endowed with these two operations, the set of all functions from X to $\mathbb{R}$ becomes a *commutative ring*. Indeed, the associativity, commutativity, and distributive laws follow from the corresponding properties of real numbers. The zero element is the constant function 0, i.e., the function sending every $x \in X$ to 0. The additive inverse of a function f is the function $x \mapsto -f(x)$. The unity is the constant function 1.

In the preceding paragraph, we can replace $\mathbb{R}$ by any commutative ring R. However, let us stick with $\mathbb{R}$ to avoid excessive generality. As a matter of fact, even the ring of all functions from X to $\mathbb{R}$ is in a way "too big" to be very interesting. We usually consider its subrings consisting of some special functions. In different parts of mathematics, we often encounter theorems stating that a certain set of functions is closed under sums, differences, and products, and contains constant functions. Such a set is then a ring (i.e., a subring of the ring of all functions). A nice example is $C(X)$, the ring of all *continuous* functions from X to $\mathbb{R}$. Of course, the set X here is no longer arbitrary; we can take X to be, for example, $\mathbb{R}$ or a closed interval $[a, b]$ (we can write $C[a, b]$ instead of $C([a, b])$). Similarly, we can talk about the ring of all bounded functions $B(X)$, the ring of all continuously differentiable functions $C^1(X)$, the ring of all smooth functions $C^\infty(X)$, etc. Another important example is the ring of (real) polynomial functions $\mathscr{P}$, which has already been mentioned a few times. Polynomials are extremely important in algebra, but we will start their systematic study in the next section.

We can also multiply functions by scalars. For $\lambda \in \mathbb{R}$ and $f : X \to \mathbb{R}$, we define $\lambda f : X \to \mathbb{R}$ by

$$(\lambda f)(x) := \lambda f(x) \quad \text{for all } x \in X.$$

The ring of all functions from X to $\mathbb{R}$ thereby becomes an algebra. All rings from the previous paragraph are also algebras, and actually it is more common to consider them as algebras than just rings. There are good reasons for this in some mathematical areas, but it may also simply be more convenient. For example, the algebra of polynomial functions $\mathscr{P}$ can be described as the subalgebra of the algebra of all functions from $\mathbb{R}$ to $\mathbb{R}$ generated by the identity function $\mathrm{id}_\mathbb{R}$. This statement is no longer true if we replace the terms "algebra" and "subalgebra" by "ring" and "subring".

The case where $X = \mathbb{N}$ is of special interest. Functions from X to $\mathbb{R}$ are then real sequences. That is, a function $f : \mathbb{N} \to \mathbb{R}$ is identified with the sequence $(x_n) = (x_1, x_2, \dots)$, where $x_i = f(i)$. The algebra of all real sequences has already been considered in Example 1.175. It is easy to see that the operations from this example coincide with the operations defined in this section. We thus have two different views of the same algebra. Let us record two interesting examples of its subalgebras. The first one is the algebra of all convergent sequences, which we denote by c. It is indeed a subalgebra since the linear combination and product of two convergent sequences is convergent. The second example is the algebra of all bounded sequences, commonly denoted by ℓ^∞ (rather than $B(\mathbb{N})$).

The above notation, namely $C(X)$, $B(X)$, $C^1(X)$, $C^\infty(X)$, c, and ℓ^∞, is used in mathematical analysis. To be precise, in analysis these symbols denote algebras (or sometimes just vector spaces) that are also endowed with certain topologies. We have indicated only the algebraic aspect.

Finally, a word on the case where X is a finite set. If $|X| = n$, say $X = \{1, \dots, n\}$, then a function $f : X \to \mathbb{R}$ can be represented as an n-tuple $(x_1, \dots, x_n)$, where $x_i = f(i)$. The algebra of all functions from $\{1, \dots, n\}$ to $\mathbb{R}$ can thus be identified with the algebra $\mathbb{R}^n$ from Example 1.96.

Exercises

2.38. Let $\mathrm{Map}(\mathbb{R})$ be the set of all functions from $\mathbb{R}$ to $\mathbb{R}$. As we know, $\mathrm{Map}(\mathbb{R})$ is an Abelian group under the usual addition of functions. Instead of the usual multiplication, take function composition as the second binary operation. Prove that $\mathrm{Map}(\mathbb{R})$ then satisfies all the ring axioms except one of the distributive laws (which one?).

2.39. Show that the ring $C[a, b]$ has zero-divisors. Is the same true for the ring of polynomial functions $\mathscr{P}$?

2.40. True or False:

(a) $f \in C[a, b]$ is an invertible element of the ring $C[a, b]$ if and only if f has no zeros on $[a, b]$.
(b) $f \in C[a, b]$ is a zero-divisor of the ring $C[a, b]$ if and only if f has a zero on $[a, b]$.

2.41. It is well-known that every convergent sequence of real numbers is bounded. Thus, $c \subseteq \ell^\infty$. Find a sequence in $\ell^\infty \setminus c$ whose square lies in c.

2.42. Prove that the subalgebra of $C[a, b]$ generated by a non-constant function f is infinite-dimensional.

2.43. Recall that an element that is equal to its square is called an idempotent. Does the ring $C[a, b]$ contain idempotents different from 0 and 1?

2.44. Is the algebra of all real sequences generated by its idempotents?

2.4 Polynomial Rings

In the previous section and earlier, the ring of real polynomial functions appeared in our discussion. One of the reasons for choosing this particular ring was that we were assuming that, due to previous experience, the readers feel comfortable with real polynomial functions. However, in abstract algebra we usually do not treat polynomials as functions, and we are not interested only in polynomials whose coefficients are real (or some other) numbers.

2.4.1 Polynomials in One Variable

To define the notion of a polynomial, we must first choose a symbol (a letter), which we call a **variable** or an **indeterminate**; let this be X. The term variable is actually slightly misleading because we associate it with functions, but it is commonly used. Let R be an arbitrary ring. The formal expression

$$f(X) = a_0 + a_1 X + \cdots + a_n X^n, \quad \text{where } n \geq 0 \text{ and } a_i \in R,$$

is called a **polynomial** in X with **coefficients** a_i in R. We usually omit terms with zero coefficients and use some other notational simplifications that can hardly cause confusion. For example, instead of

$$0 + 1X + 0X^2 + (-a)X^3 + 0X^4$$

we write $X - aX^3$, and we write $f(X) = 0$ if every $a_k = 0$. We call the coefficient a_0 the **constant term** of $f(X)$, and a_n the **leading coefficient** of $f(X)$, provided it is not zero. We say that $f(X)$ is a **monic polynomial** if its leading coefficient is 1. A polynomial with only one nonzero term is called a **monomial**.

The symbol X^k basically plays the role of an index, indicating the kth position. We could introduce $f(X)$ as a sequence

$$(a_0, a_1, \ldots, a_n, 0, 0, \ldots),$$

and, accordingly, view polynomials as infinite sequences of elements in R that have only finitely many nonzero terms. However, this notation is much less convenient for multiplying polynomials. Let us therefore keep the original notation, which we will sometimes replace with

$$f(X) = \sum_{k \geq 0} a_k X^k.$$

Here it should be understood that only finitely many a_k can be different from zero. This notation makes it easier for us to state the next definitions.

We say that the polynomials $\sum_{k \geq 0} a_k X^k$ and $\sum_{k \geq 0} \bar{a}_k X^k$ are **equal** if $a_k = \bar{a}_k$ for all $k \geq 0$. The **sum** of the polynomials

$$f(X) = \sum_{k \geq 0} a_k X^k \quad \text{and} \quad g(X) = \sum_{k \geq 0} b_k X^k$$

is defined as the polynomial

$$f(X) + g(X) := \sum_{k \geq 0} (a_k + b_k) X^k,$$

and their **product** as the polynomial

$$f(X) \cdot g(X) := \sum_{k \geq 0} c_k X^k,$$

where

$$c_k = a_0 b_k + a_1 b_{k-1} + \cdots + a_{k-1} b_1 + a_k b_0.$$

This may look complicated at first glance, but it is what we get by using the formula

$$a_i X^i \cdot b_j X^j = (a_i b_j) X^{i+j}$$

together with the distributive laws. If $a_k = 0$ for $k > m$ and $b_k = 0$ for $k > n$, then $c_k = 0$ for $k > n + m$, and the formula for the product reads

$$(a_0 + a_1 X + \cdots + a_m X^m) \cdot (b_0 + b_1 X + \cdots + b_n X^n)$$
$$= a_0 b_0 + (a_0 b_1 + a_1 b_0) X + \cdots + (a_{m-1} b_n + a_m b_{n-1}) X^{m+n-1} + a_m b_n X^{m+n}.$$

Computational rules for these abstract polynomials are thus the same as those for "high school polynomials".

Endowed with these operations, the set of all polynomials with coefficients in R becomes a ring. Showing this is an easy exercise which we omit. We only advise the readers to ask themselves how to prove the associative law for multiplication and what are the zero element and the unity. We denote this ring by $R[X]$ and call it the **polynomial ring over** R, or, more precisely, the polynomial ring *in one variable* over R. We will usually restrict ourselves to the case where R is commutative, in which case $R[X]$ is commutative too. If R is an algebra over a field F, then so is $R[X]$ if we define scalar multiplication by

$$\lambda \cdot \sum_{k \geq 0} a_k X^k := \sum_{k \geq 0} (\lambda a_k) X^k$$

for all $\lambda \in F$ and $a_i \in R$. However, this aspect will not be of particular importance for us. More often than not, we will talk about the ring $R[X]$ even when it could be considered an algebra.

Remark 2.45. The above definitions of the sum and product also make sense for expressions of the form $f(X) = \sum_{k \geq 0} a_k X^k$ and $g(X) = \sum_{k \geq 0} b_k X^k$ for which we do not assume that the coefficients a_k and b_k are 0 from some point onward. Such expressions are called **formal power series**. They also form a ring which contains the polynomial ring $R[X]$ as a subring. We denote it by $R[[X]]$. For example, $\sum_{k \geq 0} X^k = 1 + X + X^2 + \ldots$ lies in $R[[X]]$, but not in $R[X]$. Its inverse in $R[[X]]$, however, is $1 - X \in R[X]$.

Let us return to polynomials over a ring R. We say that the polynomial $f(X) = a_0 + a_1 X + \cdots + a_n X^n \in R[X]$ has **degree** n if $a_n \neq 0$. We write this as

$$\deg(f(X)) = n.$$

The degree of the polynomial 0 is not defined. A **constant polynomial** is a polynomial that is either 0 or has degree 0. It is common to identify constant polynomials with elements in R and in this way consider R as a subring of $R[X]$. A **non-constant polynomial** is, of course, a polynomial of degree at least 1. A polynomial of degree 1, 2, and 3 is called **linear**, **quadratic**, and **cubic**, respectively.

Theorem 2.46. *If R is a ring without zero-divisors, then so is $R[X]$. Moreover, if $f(X), g(X) \in R[X]$ are both nonzero, then*

$$\deg(f(X)g(X)) = \deg(f(X)) + \deg(g(X)).$$

Proof. Let $f(X)$ be a polynomial of degree m with leading coefficient a_m, and $g(X)$ a polynomial of degree n with leading coefficient b_n. Since R has no zero-divisors, $a_m b_n \neq 0$. From the formula for the product of polynomials we thus see that $f(X)g(X)$ is a polynomial of degree $m + n$ with leading coefficient $a_m b_n$. In particular, $f(X)g(X) \neq 0$. □

In the next definitions, it is natural to assume that the ring R is commutative (see Exercise 2.52 for an explanation). Given

$$f(X) = a_0 + a_1 X + \cdots + a_n X^n \in R[X]$$

and an element $x \in R$, we define $f(x)$ as the element in R obtained by substituting x for the variable X in $f(X)$. That is,

$$f(x) := a_0 + a_1 x + \cdots + a_n x^n.$$

We call $f(x)$ the **evaluation** of $f(X)$ at x. This definition still makes sense in a more general situation where x is an element of a (commutative) ring R' containing R as a subring (then, of course, $f(x) \in R'$). If $f(x) = 0$, then x is said to be a **root** or **zero** of $f(X)$.

Example 2.47. The polynomial $X^2 + 1 \in \mathbb{R}[X]$ has no root in $\mathbb{R}$. However, it has two roots in $\mathbb{C}$, namely i and $-i$.

For each polynomial $f(X) \in R[X]$, we can consider the associated **polynomial function**

$$x \mapsto f(x).$$

This is a function from R to R which is, by its very definition, determined by the polynomial $f(X)$. Is the converse also true, that is, is the polynomial uniquely determined by its associated polynomial function? The answer is negative in general.

Example 2.48. The polynomials 0 and $X + X^2$, considered as elements of $\mathbb{Z}_2[X]$, have the same associated polynomial function, namely the zero function $x \mapsto 0$ for every $x \in \mathbb{Z}_2$.

This simple example explains why it is important to distinguish between polynomials and polynomial functions. On the other hand, an example like this cannot be found in polynomial rings such as $\mathbb{R}[X]$. From the algebraic viewpoint, the difference between the polynomial ring $\mathbb{R}[X]$ and the ring of real polynomial functions $\mathscr{P}$ is merely formal.

Let this be enough for now about polynomials in *one* variable. They will be studied in greater depth in Part II. We will be especially concerned with polynomials over fields and their connection with field extensions. Polynomials in *several* variables is a more advanced topic which is usually not treated in detail in a first course in abstract algebra. In view of its importance, however, we should at least touch upon it.

2.4.2 Polynomials in Several Variables

What is the ring of polynomials in several variables? Let us first answer this informally, through examples. The following is an example of a polynomial in two variables X and Y with real coefficients:

$$f(X, Y) = 5X^2Y^4 - \sqrt{2}X^8Y + \frac{2}{3}X^6 + 7.$$

Its **coefficients** are 5, $-\sqrt{2}$, $\frac{2}{3}$, and 7. The first one among its four **monomials** has degree $2 + 4 = 6$, the second one 9, the third one 6 again, and the last one 0. Hence, $f(X, Y)$ has **degree** 9, since 9 is the highest degree of its monomials. The addition of polynomials in two (or more) variables is defined in a self-evident manner. For example,

$$(X^2Y + X^2 - 1) + (-4X^2Y + XY - X^2 + 6) = -3X^2Y + XY + 5.$$

The definition of multiplication is based on the rule

$$(aX^pY^q) \cdot (bX^rY^s) = (ab)X^{p+r}Y^{q+s}$$

and distributive laws. Thus, for example,

$$(XY + 2)(X^2Y^3 - XY^2) = X^3Y^4 + X^2Y^3 - 2XY^2,$$

as the reader can easily check.

Let us now proceed to the formal definition. Even the definition of the polynomial ring in one variable took some space, so one might expect that a tedious task is ahead of us. Fortunately, there is an easy and elegant way: the **polynomial ring in two variables over** R is defined as

$$R[X, Y] := \big(R[X]\big)[Y].$$

Stated in words, $R[X, Y]$ is the polynomial ring in the variable Y over the polynomial ring in the variable X. A polynomial in X and Y can thus be written as

$$\sum_{j \geq 0}\Big(\sum_{i \geq 0} a_{ij}X^i\Big)Y^j,$$

where only finitely many $a_{ij} \in R$ are different from 0. Using the distributive laws and omitting parentheses we obtain the usual form

$$\sum_{j \geq 0}\sum_{i \geq 0} a_{ij}X^iY^j.$$

Note that the addition and multiplication from the formal definition coincide with the addition and multiplication indicated in the above examples.

We remark that $R[X, Y]$ could also be defined as $\big(R[Y]\big)[X]$. Indeed, strictly speaking, $\big(R[X]\big)[Y]$ does not coincide with $\big(R[Y]\big)[X]$. However, the differences are inessential, so we neglect them and consider these two rings to be identical.

The **polynomial ring in** n **variables** is defined iteratively:

$$R[X_1, \ldots, X_{n-1}, X_n] := \big(R[X_1, \ldots, X_{n-1}]\big)[X_n].$$

The meaning of the notions of a coefficient, monomial, and degree should be clear from the above examples, so we omit the formal definitions. If R is an algebra, then clearly $R[X_1, \ldots, X_n]$ can also be made into an algebra.

Let R be a commutative ring. Given $x_1, \ldots, x_n \in R$ and a polynomial $f(X_1, \ldots, X_n)$ with coefficients in R, we define $f(x_1, \ldots, x_n)$ in the obvious way, by substituting each X_i by x_i. If $f(x_1, \ldots, x_n) = 0$, then we say that the n-tuple $(x_1, \ldots, x_n)$ is a **root** or **zero** of $f(X_1, \ldots, X_n)$.

Polynomials in several variables are important not only in abstract algebra. They appear in several areas of mathematics, but especially in *algebraic geometry*. What could be geometric about, say, the polynomial

$$f(X_1, X_2, X_3) = X_1^2 + X_2^2 + X_3^2 - 1 \in \mathbb{R}[X_1, X_2, X_3]?$$

Its roots! The set of its roots is the unit sphere in $\mathbb{R}^3$. Classical algebraic geometry studies roots of polynomials in several variables.

Let us consider another example. Take a natural number $n \geq 2$. The set of all roots of the polynomial

$$\Phi(X_1, X_2, X_3) = X_1^n + X_2^n - X_3^n \in \mathbb{R}[X_1, X_2, X_3]$$

is a surface in $\mathbb{R}^3$. Does it contain a point (a, b, c) such that all its components are natural numbers? That is, do there exist $a, b, c \in \mathbb{N}$ such that

$$a^n + b^n = c^n?$$

This equation is certainly solvable for $n = 2$. Its solutions are called *Pythagorean triples*, the simplest example being $(3, 4, 5)$. In 1637, *Pierre de Fermat* (1607–1665), a French lawyer who did mathematics as a hobby, wrote in the margin of a book that for *any* $n \geq 3$ there are *no* such natural numbers a, b, c, and added "...I have discovered a truly marvelous proof of this, which this margin is too narrow to contain". This margin note was discovered many years after his death, and became known as *Fermat's Last Theorem*. We will never know the truth, but most likely Fermat was mistaken about his proof. The question of whether or not the theorem is true became one of the most famous problems in the history of mathematics, unsuccessfully tackled by many great mathematicians as well as by mathematical amateurs who were hoping for a flash of genius. Finally, in 1995, more than 350 years after Fermat's note, the English mathematician *Andrew Wiles* published a proof of Fermat's Last Theorem. He used various mathematical tools, including geometric and algebraic ones. Not too many mathematicians possess the technical knowledge needed to understand all the details of this proof (which is over 100 pages long). Its discovery was not just a triumph of one man, it was a triumph of contemporary mathematics.

The statement of Fermat's Last Theorem is, of course, entirely elementary, and does not need the introduction of the polynomial $\Phi(X_1, X_2, X_3)$. Wiles' proof was actually based on the elliptic curve $y^2 = x(x - a^n)(x + b^n)$. Thus, it was the set of roots of another polynomial that played a crucial role. By mentioning the polynomial $\Phi(X_1, X_2, X_3)$ we just wanted to give a small indication of how different mathematical areas are intertwined, and how problems can be approached from different angles.

Exercises

2.49. Let $n \in \mathbb{N}$. Find a polynomial $f(X) \in \mathbb{Z}_2[X]$ such that $f(X)(1 + X) = 1 + X^n$.

2.50. Find all $n \in \mathbb{N}$ for which there exists a polynomial $f(X) \in \mathbb{Z}_3[X]$ satisfying $f(X)(1 + X) = 1 + X^n$.

2.51. Let R be a ring without zero-divisors. Note that only constant polynomials can be invertible in the polynomial ring $R[X]$ (which ones?). Find an invertible non-constant polynomial in $\mathbb{Z}_4[X]$.

2.52. Let R be a commutative ring and let $f(X), g(X), h(X) \in R[X]$ be such that $f(X) = g(X)h(X)$. Prove that then $f(x) = g(x)h(x)$ holds for every $x \in R$.

Comment. This is why it is natural to define the evaluation $f(x)$ for $f(X) \in R[X]$ only when R is commutative.

2.53. Explain why different polynomials in $\mathbb{R}[X]$ have different associated polynomial functions. Generalize to polynomials in several variables.

2.54. Prove that for every finite commutative ring R with n elements, there exists a polynomial $f(X) \in R[X]$ of degree n whose associated polynomial function is the zero function.

2.55. Let R be a commutative ring and let $a \in R$ be such that $2a$ is invertible. Prove that $aX^2 + bX + c \in R[X]$ has a root in R if and only if there exists a $d \in R$ such that $d^2 = b^2 - 4ac$. Show also that, in this case, both $x_1 := (2a)^{-1}(-b + d)$ and $x_2 := (2a)^{-1}(-b - d)$ are its roots, and that, moreover, they are its only roots in R if R has no zero-divisors.

2.56. What are the roots of $f(X) = X^2 + X + 1$ in $\mathbb{Z}_{19}$? Find all prime p such that $f(X)$ has exactly one root in $\mathbb{Z}_p$, and at least one prime $p \neq 2$ such that $f(X)$ has no root in $\mathbb{Z}_p$.

2.57. Show that the last assertion of Exercise 2.55 does not always hold in rings with zero-divisors. Find, for example, a commutative ring R in which $2 := 1 + 1$ is invertible and the polynomial X^2 has more than two roots in R.

2.58. Let R be an arbitrary (possibly noncommutative) ring. Describe $Z(R[X])$, the center of the ring $R[X]$, in terms of the center of R.

2.59. The **derivative** of the formal power series $f(X) = \sum_{k \geq 0} a_k X^k \in \mathbb{R}[[X]]$ is defined as
$$f'(X) = \sum_{k \geq 1} k a_k X^{k-1} \in \mathbb{R}[[X]].$$
When is $f'(X) = f(X)$?

2.60. Prove that $\sum_{k \geq 0} a_k X^k$ is an invertible element of the ring $R[[X]]$ if and only if a_0 is an invertible element of the ring R.

2.61. A polynomial in n variables is said to be **homogeneous** if all its monomials have the same degree. Prove that $0 \neq f(X_1, \ldots, X_n) \in \mathbb{R}[X_1, \ldots, X_n]$ is a homogeneous polynomial of degree d if and only if $f(\lambda x_1, \ldots, \lambda x_n) = \lambda^d f(x_1, \ldots, x_n)$ for all $\lambda, x_1, \ldots, x_n \in \mathbb{R}$.

2.62. Let F be a field. The algebra $F[X, Y]$ is generated by two elements, namely X and Y. Is it generated by a single element?

2.63. Let F be a field. For each $f(X, Y) \in F[X, Y]$, let $f(Y, X)$ denote the polynomial obtained by switching X and Y in $f(X, Y)$ (for example, if $f(X, Y) = \lambda X Y^3 + \mu Y$, then $f(Y, X) = \lambda X^3 Y + \mu X$). If $f(Y, X) = f(X, Y)$, then we say that $f(X, Y)$ is a **symmetric polynomial**. Prove that the set of all symmetric polynomials is the subalgebra of the algebra $F[X, Y]$ generated by the polynomials $s_1(X, Y) := X + Y$ and $s_2(X, Y) := XY$.

Comment. We have restricted ourselves here to polynomials in two variables only for simplicity. A polynomial $f(X_1, \ldots, X_n)$ in n variables is symmetric if

$$f(X_{\sigma(1)}, \ldots, X_{\sigma(n)}) = f(X_1, \ldots, X_n)$$

for every permutation σ of $\{1, \ldots, n\}$. Symmetric polynomials form a subalgebra of the algebra $F[X_1, \ldots, X_n]$. The **Fundamental Theorem of Symmetric Polynomials** states that this subalgebra is generated by the polynomials

$$s_1(X_1, \ldots, X_n) := X_1 + X_2 + \cdots + X_n,$$
$$s_2(X_1, \ldots, X_n) := X_1 X_2 + X_1 X_3 + \cdots + X_{n-1} X_n,$$
$$\vdots$$
$$s_n(X_1, \ldots, X_n) := X_1 X_2 \cdots X_n.$$

They are called the **elementary symmetric polynomials**.

2.5 Symmetric Groups

Recall that S_n denotes the group of all permutations of the set

$$\mathbb{N}_n := \{1, \ldots, n\}$$

(see Example 1.54). We call it the **symmetric group of degree** n. The operation in S_n is composition. However, we will write $\sigma \pi$ rather than $\sigma \circ \pi$, and call this element the product of the permutations σ and π from S_n. The identity element of S_n is the identity map $\mathrm{id}_{\mathbb{N}_n}$, which we will write simply as 1.

How many permutations of $\mathbb{N}_n$ are there? A map from a finite set to itself is bijective if and only if it is injective. We thus have to count the number of injective maps from $\mathbb{N}_n$ to itself. Any such map, let us call it σ, can send 1 to any element of $\mathbb{N}_n$. Thus, there are n possible choices for $\sigma(1)$. Next, $\sigma(2)$ can be any element of $\mathbb{N}_n$ except $\sigma(1)$, so the number of possible choices for $\sigma(2)$ is $n - 1$. For $\sigma(3)$ we have $n - 2$ possibilities, etc. Consequently, the number of injective maps from $\mathbb{N}_n$ to itself is $n \cdot (n - 1) \cdots 2 \cdot 1$. That is to say,

$$|S_n| = n!$$

which is something to keep in mind.

A permutation that interchanges two elements i and j from $\mathbb{N}_n$ and leaves the others unchanged is called a **transposition**. We denote it by $(i\,j)$. It is clear that $(i\,j)^2 = 1$, so a transposition is its own inverse. It can be easily checked that the transpositions $(1\,2)$ and $(1\,3)$ do not commute (see also Example 1.54). The group S_n is thus non-Abelian whenever $n \geq 3$.

The next theorem indicates why transpositions are important: the group S_n is generated by them. We remark that, by convention, the product of no factors is 1 (we need this to cover the $n = 1$ case).

Theorem 2.64. *Every permutation $\sigma \in S_n$ is a product of transpositions.*

Proof. We proceed by induction on n. There is nothing to prove for $n = 1$. Thus, let $n > 1$ and assume that the theorem is true for S_{n-1}. If $\sigma(n) = n$, then we may view σ as an element of S_{n-1} and the desired conclusion follows. If $\sigma(n) = i \neq n$, then the product $(i\,n)\sigma$ maps n to n, and so $(i\,n)\sigma = \tau_1 \cdots \tau_r$ for some transpositions τ_i. Multiplying this from the left by $(i\,n)$ and using $(i\,n)^2 = 1$ we obtain the expression of σ as a product of transpositions. $\qquad\qquad\square$

Unfortunately, this factorization is far from being unique. For example, $(1\,2)(1\,3)$ can also be written as $(1\,3)(2\,3)$, $(2\,3)$ as $(1\,2)(1\,3)(1\,2)$, 1 as $(1\,2)(1\,2)$, etc. A permutation can thus be expressed as a product of transpositions in many different ways. The next theorem shows, however, that all these expressions have something in common. But first we need some preliminaries.

Let $f(X_1, \ldots, X_n)$ be a polynomial in n variables over a ring. For each $\sigma \in S_n$, we introduce the polynomial $(\sigma \cdot f)(X_1, \ldots, X_n)$ as follows:

$$(\sigma \cdot f)(X_1, \ldots, X_n) := f(X_{\sigma(1)}, \ldots, X_{\sigma(n)})$$

(compare Exercise 2.63). Using this definition for a permutation $\pi \in S_n$ and the polynomial $(\sigma \cdot f)(X_1, \ldots, X_n)$, we obtain

$$(\pi \cdot (\sigma \cdot f))(X_1, \ldots, X_n) = (\sigma \cdot f)(X_{\pi(1)}, \ldots, X_{\pi(n)}). \qquad (2.2)$$

To make the next step easier to understand, we write, temporarily, Y_i for $X_{\pi(i)}$. By the very definition,

$$(\sigma \cdot f)(Y_1, \ldots, Y_n) = f(Y_{\sigma(1)}, \ldots, Y_{\sigma(n)}) \quad \text{and} \quad Y_{\sigma(i)} = X_{(\pi\sigma)(i)}.$$

The right-hand side of (2.2) is thus equal to

$$f(X_{(\pi\sigma)(1)}, \ldots, X_{(\pi\sigma)(n)}).$$

Applying the definition once again, we see that (2.2) can be rewritten as

$$(\pi \cdot (\sigma \cdot f))(X_1, \ldots, X_n) = ((\pi\sigma) \cdot f)(X_1, \ldots, X_n). \qquad (2.3)$$

This formula will be used in the proof of the next theorem.

Theorem 2.65. *If a permutation $\sigma \in S_n$ can be written as a product of an even (resp. odd) number of transpositions, then any product of transpositions equaling σ contains an even (resp. odd) number of factors.*

Proof. We first explain the idea of the proof. We will find a nonzero polynomial

$$f(X_1, \ldots, X_n) \in \mathbb{Z}[X_1, \ldots, X_n]$$

satisfying

$$(\tau \cdot f)(X_1, \ldots, X_n) = -f(X_1, \ldots, X_n) \tag{2.4}$$

for every transposition τ. This property, together with (2.3), implies that

$$((\tau_1 \cdots \tau_k) \cdot f)(X_1, \ldots, X_n) = (-1)^k f(X_1, \ldots, X_n)$$

for all transpositions $\tau_1, \ldots, \tau_k$. The desired conclusion that the product of an even number of transpositions is never equal to the product of an odd number of transpositions then readily follows.

Let us now find such a polynomial $f(X_1, \ldots, X_n)$. If $n = 2$, the choice is obvious: $f(X_1, X_2) = X_1 - X_2$. For $n = 3$, we take

$$f(X_1, X_2, X_3) := (X_1 - X_2)(X_1 - X_3)(X_2 - X_3).$$

Indeed, S_3 contains three transpositions, $(1\,2)$, $(1\,3)$, and $(2\,3)$, and the reader can verify that (2.4) holds for each of them. We can now guess that the polynomial we are searching for is

$$f(X_1, \ldots, X_n) := \prod_{i<j}(X_i - X_j) = (X_1 - X_2)(X_1 - X_3)\ldots(X_{n-1} - X_n).$$

Take a transposition $\tau = (i\,j)$, $i < j$, and let us check that (2.4) holds. The factors in the polynomial $(\tau \cdot f)(X_1, \ldots, X_n)$ can differ from the factors in the polynomial $f(X_1, \ldots, X_n)$ only in sign. Besides $X_j - X_i$, the only ones of different sign are $X_j - X_\ell$ and $X_\ell - X_i$ where $i < \ell < j$. Since they occur in pairs, the number of all factors of different sign is odd. Hence, (2.4) holds. $\square$

We will say that $\sigma \in S_n$ is an **even permutation** if it can be written as a product of an even number of transpositions, and is an **odd permutation** if it can be written as a product of an odd number of transpositions. Theorems 2.64 and 2.65 show that every permutation is either even or odd, but cannot be both. The number

$$\mathrm{sgn}(\sigma) := \begin{cases} 1 \; ; \; \sigma \text{ is an even permutation} \\ -1 \; ; \; \sigma \text{ is an odd permutation} \end{cases}$$

is called the **sign** of the permutation σ. The product of two even permutations is even, the product of an even permutation by an odd one (or an odd permutation by an even one) is an odd permutation, and the product of two odd permutations is even. These statements can be summarized in the following formula:

$$\text{sgn}(\sigma\pi) = \text{sgn}(\sigma)\text{sgn}(\pi) \quad \text{for all } \sigma, \pi \in S_n.$$

Taking $\pi = \sigma^{-1}$ we obtain

$$\text{sgn}(\sigma^{-1}) = \text{sgn}(\sigma)^{-1} \quad \text{for all } \sigma \in S_n.$$

This means that the inverse of an even permutation is an even permutation, and the inverse of an odd permutation is an odd permutation.

Let A_n denote the set of all *even* permutations in S_n. In the preceding paragraph, we noticed that A_n is closed under multiplication and contains inverses of its elements. This means that A_n is a subgroup of S_n. We call it the **alternating group of degree** n. This is an important group and will be frequently mentioned hereafter.

Theorem 2.64 shows that every permutation is "built" out of transpositions. We will now consider somewhat larger building blocks. A permutation σ is called a **cycle of length** k, or a k**-cycle**, if there exist distinct $i_1, \ldots, i_k \in \mathbb{N}_n$ such that

$$\sigma(i_1) = i_2, \ \sigma(i_2) = i_3, \ \ldots, \sigma(i_{k-1}) = i_k, \ \sigma(i_k) = i_1,$$

and σ leaves all other elements from $\mathbb{N}_n$ fixed. In this case, we write

$$\sigma = (i_1 \, i_2 \ldots i_k).$$

Of course, we can also write

$$\sigma = (i_k \, i_1 \ldots i_{k-1}), \quad \sigma = (i_{k-1} \, i_k \, i_1 \ldots i_{k-2}), \quad \text{etc.}$$

A 2-cycle is nothing but a transposition, and every 1-cycle (i_1) is the identity 1.

Every permutation is a product of cycles, since it is even a product of transpositions. However, we can say more about the factorization into cycles. The easiest way to explain this is by an example. Recall from Example 1.54 the two-line notation for permutations in S_n, and consider

$$\sigma = \begin{pmatrix} 1 \, 2 \, 3 \, 4 \, 5 \, 6 \, 7 \, 8 \, 9 \\ 7 \, 1 \, 3 \, 8 \, 9 \, 4 \, 2 \, 6 \, 5 \end{pmatrix}.$$

The element 3 is left fixed by σ, so we can, in some sense, forget about it. Consider the sequence of images:

$$1 \mapsto 7, \ 7 \mapsto 2, \ 2 \mapsto 1.$$

We have thus found the 3-cycle $(1\,7\,2)$ inside σ. Note that σ can be written as the product $\sigma = (1\,7\,2)\sigma_1$ where

$$\sigma_1 = \begin{pmatrix} 1 \, 2 \, 3 \, 4 \, 5 \, 6 \, 7 \, 8 \, 9 \\ 1 \, 2 \, 3 \, 8 \, 9 \, 4 \, 7 \, 6 \, 5 \end{pmatrix}.$$

Besides 3, σ_1 also preserves 1, 7, and 2. Let us focus on the other five elements. From the sequences of images

$$4 \mapsto 8, \quad 8 \mapsto 6, \quad 6 \mapsto 4 \quad \text{and} \quad 5 \mapsto 9, \quad 9 \mapsto 5$$

we infer that $\sigma_1 = (4\,8\,6)(5\,9)$. Hence, σ can be written as the product of three cycles,

$$\sigma = (1\,7\,2)(4\,8\,6)(5\,9),$$

which are independent of each other in the sense that they have no common elements. The element 3 is missing in this expression. To have a more clear picture, we can also write

$$\sigma = (1\,7\,2)(4\,8\,6)(5\,9)(3),$$

since the 1-cycle (3) equals the identity.

Any permutation σ can be handled in much the same way. We start with the element 1 and consider the sequence $1, \sigma(1), \sigma^2(1), \ldots$ Since $\mathbb{N}_n$ is a finite set, there exists a $k \geq 1$ such that $1, \sigma(1), \ldots, \sigma^{k-1}(1)$ are distinct, whereas $\sigma^k(1)$ is equal to one of them. In fact, the only possibility is that $\sigma^k(1) = 1$. Indeed, $\sigma^k(1) = \sigma^i(1)$ with $0 \leq i < k$ yields $\sigma^{k-i}(1) = 1$, so i can only be 0. We have thus encountered the k-cycle

$$(1\,\sigma(1)\ldots\sigma^{k-1}(1))$$

and, as in the above example, we have the decomposition

$$\sigma = (1\,\sigma(1)\ldots\sigma^{k-1}(1)) \cdot \sigma_1,$$

where σ_1 is a permutation preserving the elements $1, \sigma(1), \ldots, \sigma^{k-1}(1)$. We may now restrict our attention to σ_1. Repeating this procedure, we eventually arrive at expressing σ as a product of **disjoint cycles**. By this we mean that every element in $\mathbb{N}_n$ can appear in only one of these cycles. Using the standard convention that a product can contain only one factor, we thus have the following theorem.

Theorem 2.66. *Every permutation $\sigma \in S_n$ is a product of disjoint cycles.*

An expression of σ as a product of disjoint cycles is called the **cycle decomposition** of σ. It is a simple exercise to show that the factors in this decomposition commute with each other. We can even express σ as a product of transpositions, but they do not always commute. The cycle decomposition has another advantage: it is unique! This should be intuitively clear from the way this decomposition has been derived. The formal proof is not difficult, but we omit it. One actually has to be a little bit careful and say "unique up to the order of the factors". Anyway, the order is irrelevant as the factors commute.

Note a slight similarity between the cycle decomposition and the decomposition of integers into prime factors. They both serve as models for decomposing algebraic objects into simpler, indecomposable objects, in an essentially unique way.

Exercises

2.67. Let $\sigma = \left(\begin{smallmatrix} 1 & 2 & 3 & 4 \\ 4 & 2 & 1 & 3 \end{smallmatrix}\right)$ and $\pi = \left(\begin{smallmatrix} 1 & 2 & 3 & 4 \\ 3 & 4 & 1 & 2 \end{smallmatrix}\right)$. Determine $\sigma\pi$, $\pi\sigma$, σ^{-1}, and π^{-1}.

2.68. Find the cycle decomposition of

$$\sigma = \begin{pmatrix} 1\ 2\ 3\ 4\ 5\ 6\ 7\ 8\ 9 \\ 5\ 1\ 6\ 8\ 7\ 3\ 9\ 4\ 2 \end{pmatrix}.$$

What is the length of the longest cycle? Swap two elements in the lower row in such a way that the cycle decomposition of the obtained permutation will contain a cycle of longer length.

2.69. Find the cycle decomposition of $\sigma = (3\,5\,2)(5\,7\,1)(1\,3) \in S_7$. Consider the same problem for σ^{-1}.

2.70. Express $(1\,2)(2\,3)\cdots(k-1\,k)$ as a k-cycle.

2.71. Express $(1\,2)(1\,3)\cdots(1\,k)$ as a k-cycle.

2.72. Find the inverse of the k-cycle $(i_1\,i_2\ldots i_k)$.

2.73. Show that the sign of a k-cycle is $(-1)^{k-1}$.

2.74. Suppose $\sigma \in S_n$ is equal to the product of s disjoint cycles in which all elements of $\mathbb{N}_n$ are included (this can always be achieved by adding 1-cycles (i) if necessary). Using the preceding exercise, show that $\mathrm{sgn}(\sigma) = (-1)^{n-s}$.

2.75. Recall that the order of an element a of a group G is the smallest natural number (provided it exists) r such that $a^r = 1$. Prove that the order of a k-cycle is k.

2.76. Prove that the order of $\sigma \in S_n$ is the least common multiple of the lengths of the cycles in the cycle decomposition of σ.

2.77. Find a permutation of greatest possible order in S_{10}. Do groups S_{11} and S_{12} contain elements of higher order?

2.78. Let σ be a k-cycle, where $k \geq 3$. Prove that σ^2 is a cycle if and only if k is odd.

2.79. Prove that $\pi(i_1\,i_2\ldots i_k) = (\pi(i_1)\,\pi(i_2)\ldots\pi(i_k))\pi$ holds for every $\pi \in S_n$ and every k-cycle $(i_1\,i_2\ldots i_k) \in S_n$.

2.80. Let $\sigma, \sigma' \in S_n$. We say that σ and σ' have the **same cycle structure** if their cycle decompositions have the same number of cycles of the same lengths. For example, $(1\,2\,3)(4\,5\,6)(7\,8)$ and $(4\,2\,5)(8\,1)(6\,3\,7)$ have the same cycle structure. Next, recall from Section 1.6 that σ and σ' are conjugate if there exists a $\pi \in S_n$ such that $\sigma' = \pi\sigma\pi^{-1}$. Using the preceding exercise, prove the following:

(a) If σ and σ' are conjugate and σ is a k-cycle, then so is σ'.

(b) Any two k-cycles in S_n are conjugate.

(c) σ and σ' are conjugate if and only if they have the same cycle structure.

2.81. Prove that the group S_n is generated by the transpositions $(1\,2)$, $(2\,3)$, ..., $(n-1\,n)$.

2.82. Prove that the group S_n is generated by the transposition $(1\,2)$ and the n-cycle $(1\,2\dots n)$.

2.83. Prove that $Z(S_n)$, the center of S_n, equals $\{1\}$ for every $n \geq 3$.

2.84. Prove that $Z(A_n) = \{1\}$ for every $n \geq 4$.

2.85. Prove that $|A_n| = \frac{n!}{2}$ for every $n \geq 2$.

2.86. Prove that for every $n \geq 3$, the group A_n is generated by the 3-cycles.

2.6 Dihedral Groups

Imagine a square on a plane. Label its vertices 1, 2, 3, and 4, as in Figure 2.1. Remove the square from the plane, move it in 3-dimensional space in any possible way, and then put it back on the plane to cover the empty place. The plane again looks untouched, but the new position of the square may be different. We are interested in all possible new positions. This is all that matters to us, so we do not distinguish between the motions that lead to the same final position.

Obviously, the position of the square is determined by the location of the vertices, which in turn can be identified with a permutation of $\{1, 2, 3, 4\}$. For instance, the rotation of $90°$ counterclockwise about the center moves the square into the position shown in Figure 2.2, and this corresponds to the 4-cycle $(1\,2\,3\,4)$.

Note that performing one motion followed by another corresponds to a product of permutations. Accordingly, we will refer to the former as the product of motions. Thus, denoting the aforementioned rotation of $90°$ by r, r^2 is the rotation of $180°$, r^3 is the rotation of $270°$, and $r^4 = 1$ is the "motion" that leaves the square fixed.

We can also turn the square upside down. This can be done in several ways. One of them is to perform the rotation of $180°$ about a vertical axis (Figure 2.3). This motion, let us call it s, moves the square into the position shown in Figure 2.4. As a permutation, s can be described as the product of the transpositions $(1\,2)$ and $(3\,4)$. We can also rotate about a horizontal or one of the diagonal axes. However, there is no need to introduce additional notation since these motions can be described as rs, r^2s, and r^3s. The set

$$D_8 := \{1, r, r^2, r^3, s, rs, r^2s, r^3s\}$$

contains all the motions we are searching for. Indeed, the final position of the square is determined by the location of the vertex 1 and by whether or not the square has been turned upside down. Therefore, there cannot be more than $4 \times 2 = 8$ final positions, and the motions in D_8 lead to 8 different ones.

Fig. 2.1 Initial position

Fig. 2.2 Rotation of 90° counterclockwise about the center

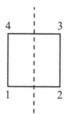

Fig. 2.3 Vertical axis

Fig. 2.4 Rotation of 180° about a vertical axis

Let us get to the point: D_8 is a group! Viewing elements in D_8 as permutations, one can easily check that D_8 is a subgroup of the symmetric group S_4. However, it is even easier to think geometrically, having in mind that the product of two motions simply means performing one motion after another. We call D_8 the **dihedral group of order** 8, and we call its elements **symmetries of a square**. Obviously, D_8 is generated by r and s. Note that $rs \neq sr$, so it is a non-Abelian group. We also record the following relations:

$$r^4 = 1, \quad s^2 = 1, \quad \text{and} \quad (rs)^2 = 1.$$

They can be derived either geometrically or by computing in S_4. There are other relations, but they can be deduced from these three.

In a similar fashion we consider **symmetries of a regular n-gon** for any $n \geq 3$. There is nothing special about the $n = 4$ case, it was chosen just to make the exposition simpler. The group of symmetries of a regular n-gon is called the **dihedral group of order** $2n$ and is denoted by D_{2n} (some actually denote it by D_n; unfortunately, there is no universal agreement about this). It is generated by r, the rotation of $\frac{360°}{n}$ about the center, and s, the rotation of $180°$ about an appropriate axis (one may also call it "reflection"). Note that r and s do not commute,

$$r^n = 1, \quad s^2 = 1, \quad \text{and} \quad (rs)^2 = 1,$$

and

$$D_{2n} = \{1, r, \ldots, r^{n-1}, s, rs, \ldots r^{n-1}s\}.$$

We can consider D_{2n} as a subgroup of S_n. It is a rather small subgroup for a large n, since $S_n = n!$ and $|D_{2n}| = 2n$. For $n = 3$, however, both groups have order 6, so $D_6 = S_3$.

What about if $n < 3$? For $n = 2$ we define D_4 as the group of symmetries of a rectangle that is not a square. It consists of four elements, $1, r, s$, and rs, that satisfy $r^2 = s^2 = 1$ and $rs = sr$. Finally, $D_2 = \{1, r\}$ with r satisfying $r^2 = 1$. The dihedral groups D_2 and D_4 cannot be considered as subgroups of the symmetric groups S_1 and S_2, respectively. After all, D_2 has two and S_1 has only one element, and D_4 has four and S_2 has only two elements.

The term "symmetry," which we use for an element of a dihedral group, has a much broader meaning in mathematics. In particular, it is often used for a map from an object onto itself that preserves certain properties, especially those of a geometric or topological nature. Different types of symmetries give rise to different examples of groups. We can reverse the viewpoint and consider symmetric objects, i.e., objects that are preserved under certain maps. In many different ways, symmetric objects, and symmetry in general, also appear in art. The consideration of symmetry is thus one of the meeting places of mathematics and art—or, better said, of mathematics and other kinds of art, since mathematics can be, among other things, also regarded as an art form.

Exercises

2.87. Find all $x \in D_{2n}$ such that $x^2 = 1$.

2.88. Find all $x \in D_{2n}$ such that $rxr = x$.

2.89. Find all $x \in D_{2n}$ such that $xsx = s$.

2.90. Determine $C_{D_{2n}}(r)$, the centralizer of the element $r \in D_{2n}$ (see Exercise 1.139).

2.91. Determine $C_{D_{2n}}(s)$, the centralizer of the element $s \in D_{2n}$.

2.92. Let $n \geq 3$. Prove that $Z(D_{2n})$, the center of D_{2n}, is equal to $\{1\}$ if n is odd, and to $\{1, r^{\frac{n}{2}}\}$ if n is even.

2.93. An element c of a group G is called a **commutator** if there exist $x, y \in G$ such that $c = xyx^{-1}y^{-1}$ (note that $c = 1$ if and only if x and y commute—hence the name). Show that the set of all commutators in D_{2n}, $n \geq 3$, is equal to the subgroup generated by the element r^2.

Comment. In general, the set of all commutators in a group need not be a subgroup.

2.94. The **infinite dihedral group** D_{∞} is defined as the subgroup of the symmetric group $\mathrm{Sym}(\mathbb{R})$ generated by the functions $r, s : \mathbb{R} \to \mathbb{R}$, $r(x) = x + 1$, $s(x) = -x$. Show that $r^n \neq 1$ for all $n \in \mathbb{Z} \setminus \{0\}$, $s^2 = 1$, $(rs)^2 = 1$, and $D_{\infty} = \{r^n, r^n s \mid n \in \mathbb{Z}\}$.

2.95. Prove that each of the groups D_{2n} in D_{∞} is generated by a pair of elements u in w satisfying $u^2 = w^2 = 1$.

Comment. Finding u and w should not be difficult. The message of this exercise, however, is interesting, especially in light of Exercise 3.34.

2.7 Rings of Matrices and Linear Groups

The meaning and significance of matrices should be well-understood after a course in linear algebra. They are also very important in abstract algebra, although not necessarily for the same reasons.

2.7.1 Rings of Matrices

The ring of 2×2 real matrices has already already been mentioned several times. Let now n be an arbitrary natural number and let R be an arbitrary ring. A square array of the form

$$
A = \begin{bmatrix}
a_{11} & a_{12} & \ldots & a_{1n} \\
a_{21} & a_{22} & \ldots & a_{2n} \\
\vdots & \vdots & \ddots & \vdots \\
a_{n1} & a_{n2} & \ldots & a_{nn}
\end{bmatrix},
$$

where $a_{ij} \in R$, is called an $n \times n$ **matrix over** R, and the elements a_{ij} are called the **entries** of A. We will usually write this matrix simply as

$$
A = (a_{ij}).
$$

The set of all $n \times n$ matrices over R will be denoted by $M_n(R)$. Addition in $M_n(R)$ is defined as follows:

$$
(a_{ij}) + (b_{ij}) := (a_{ij} + b_{ij}).
$$

It is easy to verify that this makes $M_n(R)$ an Abelian group. The zero element is the zero matrix, i.e., the matrix all of whose entries are 0. Defining multiplication by

$$(a_{ij})(b_{ij}) := (c_{ij}), \text{ where } c_{ij} = \sum_{k=1}^{n} a_{ik}b_{kj},$$

$M_n(R)$ becomes a ring. Checking the associative and distributive laws is straightforward, and is usually done in a linear algebra course. The fact that the entries now belong to an abstract ring rather than being real or complex numbers, does not affect the proof. The unity is the identity matrix I, i.e., the matrix with ones on the main diagonal and zeros elsewhere ($a_{ii} = 1$ and $a_{ij} = 0$ if $i \neq j$).

The ring $M_1(R)$ can be identified with R. When talking about matrices, we usually have in mind that $n \geq 2$. Then $M_n(R)$ is not commutative and has zero-divisors (see Example 1.73).

We call (a_{ij}) a **diagonal matrix** if $a_{ij} = 0$ for all $i \neq j$. In the case where $a_{ij} = 0$ for all $i > j$, we call (a_{ij}) an **upper triangular matrix**. An upper triangular matrix whose diagonal entries a_{ii} are also 0 is called a **strictly upper triangular matrix**. Analogously we define (strictly) lower triangular matrices. Every matrix is obviously equal to the sum of a strictly upper triangular, diagonal, and strictly lower triangular matrix.

As already mentioned in some of the exercises, a ring element e is called an **idempotent** if $e^2 = e$. In this case, $1 - e$ is also an idempotent; indeed,

$$(1 - e)^2 = 1 - 2e + e^2 = 1 - e.$$

Every ring R has at least two idempotents, 0 and 1. Rewriting $e^2 = e$ as $e(1 - e) = 0$ we see that these are also the only idempotents if R has no zero-divisors. An example of an idempotent in $M_n(R)$ is every diagonal matrix that has only zeros and ones on the diagonal.

A ring element a is said to be **nilpotent** if $a^n = 0$ for some $n \in \mathbb{N}$. In a ring without zero-divisors, 0 is the only nilpotent. Every strictly (upper or lower) triangular matrix in $M_n(R)$ is nilpotent. Showing this is an easy exercise in matrix multiplication.

If R is an algebra over a field F, the ring $M_n(R)$ also becomes an algebra over F by defining scalar multiplication in the following natural way:

$$\lambda(a_{ij}) := (\lambda a_{ij}).$$

As an important special case we have that $M_n(F)$ is an algebra over F. Observe that its dimension is n^2.

The reader may remember from linear algebra that the algebra $M_n(F)$ is intimately connected with the algebra of all linear maps from an n-dimensional vector space over F to itself. We will examine this connection in Example 3.77.

2.7.2 Linear Groups

From matrix rings we proceed to matrix groups. We will consider groups whose elements are matrices over a field F and the operation is multiplication. The set of all $n \times n$ matrices over F is, of course, only a monoid, not a group. Recall that the set of all invertible elements of an arbitrary monoid S forms a group, which we denote by S^* (Theorem 1.51). Thus,

$$\mathrm{GL}_n(F) := M_n(F)^*,$$

the set of all invertible $n \times n$ matrices over F, is a group. It is called the **general linear group**. It is easy to see that it is non-Abelian, as long as $n > 1$. The other **linear groups** are subgroups of $\mathrm{GL}_n(F)$. We will record some concrete examples, but first we have to recall the notion of the determinant of a matrix. It may be that some of the readers are familiar only with determinants of real and complex matrices, but the definition and basic properties are the same for any field F.

The **determinant** of a matrix $A = (a_{ij}) \in M_n(F)$ is defined as

$$\det(A) := \sum_{\sigma \in S_n} \mathrm{sgn}(\sigma) a_{1\sigma(1)} \cdots a_{n\sigma(n)}.$$

For $n = 2$ we thus have

$$\det\left(\begin{bmatrix} a_{11} & a_{12} \\ a_{21} & a_{22} \end{bmatrix}\right) = a_{11}a_{22} - a_{12}a_{21}.$$

This was short and simple, but for a large n the formula for $\det(A)$ looks intimidating. Computing the determinant of a concrete matrix by hand may be next to impossible, unless this matrix has some special form. For example, the determinant of an (upper or lower) triangular matrix is equal to the product of all diagonal entries (why?). However, calculating determinants is not our concern here. We only need the following well-known facts:

(a) $\det(I) = 1$.
(b) $\det(AB) = \det(A)\det(B)$ for all $A, B \in M_n(F)$.
(c) $A \in M_n(F)$ is invertible if and only if $\det(A) \neq 0$.

The first one, (a), is obvious. Proving (b) and (c) requires some effort, as the reader probably remembers from a linear algebra course. Those who feel a bit uncertain about these properties may wish to check (b) at least for upper triangular matrices, and (c) at least for $n = 2$. Concerning the latter, it is actually easy to see that if $A = \begin{bmatrix} a_{11} & a_{12} \\ a_{21} & a_{22} \end{bmatrix} \in M_2(F)$ has a nonzero determinant, then $A^{-1} = \frac{1}{\det(A)} \begin{bmatrix} a_{22} & -a_{12} \\ -a_{21} & a_{11} \end{bmatrix}$.

Let us make a small digression and touch upon a question that might have already crossed the reader's mind: may we replace the field F by a ring R in the above discussion? If R is commutative, then the answer is positive to some extent. The properties (a) and (b) still hold, while in (c) the condition that $\det(A) \neq 0$ must be replaced by the condition that $\det(A)$ is invertible in R. If, however, R is not commutative, then the definition of the determinant, as given above, does not make much sense.

We now return to linear groups. In light of (c), we have the following alternative description of the general linear group:

$$\mathrm{GL}_n(F) = \{A \in M_n(F) \mid \det(A) \neq 0\}.$$

Consider the set

$$\mathrm{SL}_n(F) := \{A \in M_n(F) \mid \det(A) = 1\}.$$

From (b), we see that it is closed under multiplication. Taking A^{-1} for B in (b) and using (a) we see that $\det(A^{-1}) = \det(A)^{-1}$ for every invertible matrix A. This implies that $\mathrm{SL}_n(F)$ contains inverses of its elements. Hence, $\mathrm{SL}_n(F)$ is a subgroup of $\mathrm{GL}_n(F)$. It is called the **special linear group**.

In the next paragraph, we will use some notions from linear algebra. The reader who is not familiar with them can skip this part.

We denote the transpose of a matrix A by A^t. The set of all orthogonal matrices,

$$O_n(F) := \{A \in M_n(F) \mid AA^t = I\},$$

is a subgroup of $\mathrm{GL}_n(F)$. It is called the **orthogonal group**. Similarly, we introduce the **unitary group**

$$U_n := \{A \in M_n(\mathbb{C}) \mid AA^* = I\}.$$

Here, A^* denotes the conjugate transpose of the matrix A. The groups

$$\mathrm{SO}_n(F) := O_n(F) \cap \mathrm{SL}_n(F) \quad \text{and} \quad \mathrm{SU}_n := U_n \cap \mathrm{SL}_n(\mathbb{C})$$

are called the **special orthogonal group** and the **special unitary group**, respectively. Finally, we mention the **symplectic group**

$$\mathrm{Sp}_{2n}(F) := \{A \in M_{2n}(F) \mid A^t J A = J\},$$

where $J = \begin{bmatrix} 0 & I_n \\ -I_n & 0 \end{bmatrix} \in M_{2n}(F)$. Here, I_n is the identity matrix of size $n \times n$.

One should of course check that all these sets are indeed subgroups of the general linear group. This is easy and left to the reader. We could think of many more examples of subgroups of $\mathrm{GL}_n(F)$. However, we have restricted ourselves to a few important, classical linear groups.

Exercises

2.96. How many elements does the group $\mathrm{GL}_2(\mathbb{Z}_2)$ have?

Comment. This group and the groups from the next two exercises will be described more thoroughly in the exercises of Section 3.1.

2.97. Describe the subgroup of $\mathrm{GL}_2(\mathbb{R})$ generated by the matrix $\begin{bmatrix} 1 & 1 \\ 0 & 1 \end{bmatrix}$.

2.98. Describe the subgroup of $\mathrm{GL}_2(\mathbb{C})$ generated by the matrices $\begin{bmatrix} i & 0 \\ 0 & -i \end{bmatrix}$ and $\begin{bmatrix} 0 & 1 \\ -1 & 0 \end{bmatrix}$.

2.99. Determine which of the following sets are subgroups of $\mathrm{GL}_n(\mathbb{R})$:

(a) $\{(a_{ij}) \in \mathrm{GL}_n(\mathbb{R}) \mid a_{ij} \in \mathbb{Z}\}$.
(b) $\{(a_{ij}) \in \mathrm{SL}_n(\mathbb{R}) \mid a_{ij} \in \mathbb{Z}\}$.
(c) $\{(a_{ij}) \in \mathrm{O}_n(\mathbb{R}) \mid a_{ij} \in \mathbb{Z}\}$.
(d) $\{A \in \mathrm{GL}_n(\mathbb{R}) \mid A = A^t\}$.
(e) $\{A \in \mathrm{GL}_n(\mathbb{R}) \mid |\det(A)| = 1\}$.
(f) $\{A \in \mathrm{GL}_n(\mathbb{R}) \mid A^2 = I\}$.

Advice. In this and some of the next exercises, you can restrict yourself to the case where $n = 2$. Computing with matrices of arbitrary size can be tedious and non-transparent, making it difficult to grasp the essence. As a matter of fact, even computations with 2×2 can be lengthy. When choosing an arbitrary matrix, first take one with many zeros.

2.100. Recall that elements a and b of a group G are *conjugate* in G if there exists a $p \in G$ such that $a = pbp^{-1}$ (see Section 1.6). Recall also from linear algebra that matrices $A, B \in M_n(F)$ are *similar* if there exists an invertible matrix $P \in M_n(F)$ such that $A = PBP^{-1}$. Thus, two matrices in $\mathrm{GL}_n(F)$ are similar if and only if they are conjugate in $\mathrm{GL}_n(F)$. Find two similar matrices in $\mathrm{SL}_2(\mathbb{R})$ that are not conjugate in $\mathrm{SL}_2(\mathbb{R})$.

2.101. Describe $Z(\mathrm{GL}_n(F))$, the center of the group $\mathrm{GL}_n(F)$.

2.102. Describe $Z(M_n(R))$, the center of the ring $M_n(R)$, in terms of the center of the ring R.

2.103. A ring R is said to be **prime** if, for every pair of nonzero elements $a, b \in R$, there exists an $x \in R$ such that $axb \neq 0$. Obvious examples are rings without zero-divisors, but there are others: prove that the ring $M_n(R)$ is prime if and only if R is prime.

2.104. Prove that a prime ring with zero-divisors contains a nonzero nilpotent.

Comment. The existence of a nonzero nilpotent is equivalent to the existence of a nonzero element whose square is zero. Indeed, $a^n = 0$ with $n \geq 2$ implies $(a^{n-1})^2 = 0$.

2.105. Let $A \in M_n(F)$ be a matrix such that any two of its rows are linearly dependent (a matrix of rank at most one). Show that $A^2 = \lambda A$ for some $\lambda \in F$. Now find a nilpotent matrix in $M_n(F)$ that is not strictly (upper or lower) triangular, and an idempotent matrix in $M_n(F)$ that is not diagonal.

2.106. Let e be an idempotent in a ring R. Show that for every $x \in R$, $e + ex(1 - e)$ is also an idempotent. Now find an idempotent matrix in $M_3(F)$ that is not diagonal and is not of rank one.

2.107. Let R be a ring in which the element $1 + 1$ is invertible. Prove that R contains an idempotent $e \neq 0, 1$ if and only if R contains an element $u \neq 1, -1$ such that $u^2 = 1$.

Hint. Considering the case where e and u are diagonal 2×2 matrices may give you an idea.

2.108. A matrix $A \in M_n(F)$ is called **diagonalizable** if it is similar to a diagonal matrix. Prove that such a matrix can be written as $A = \sum_{i=1}^{r} \lambda_i E_i$ where $\lambda_i \in F$ and $E_1, \ldots, E_r \in M_n(F)$ are idempotent matrices satisfying $E_i E_j = 0$ for all $i \neq j$.

2.109. If $A \in M_n(F)$ is strictly upper triangular, then $I - A$ has determinant 1 and is therefore invertible. Prove the following more general statement: if a ring element x is nilpotent, then $1 - x$ is invertible.

Hint. The geometric series can help you guess how to express $(1 - x)^{-1}$ (similarly as in Exercise 1.88).

2.110. Prove that in a ring without unity that has no zero-divisors, 0 is the only idempotent.

2.111. Let M be the set of all infinite matrices with real entries

$$\begin{bmatrix} a_{11} & a_{12} & a_{13} & \cdots \\ a_{21} & a_{22} & a_{23} & \cdots \\ a_{31} & a_{32} & a_{33} & \cdots \\ \vdots & \vdots & \vdots & \ddots \end{bmatrix}$$

such that each column has only finitely many nonzero entries. Show that M is a ring under the usual matrix operations. Show also that the matrix

$$\begin{bmatrix} 0 & 1 & 0 & 0 & \cdots \\ 0 & 0 & 1 & 0 & \cdots \\ 0 & 0 & 0 & 1 & \cdots \\ \vdots & \vdots & \vdots & \vdots & \ddots \end{bmatrix}$$

has a right inverse, but no left inverse.

2.112. Prove that a ring R contains an element that has a right inverse but no left inverse if and only if the ring $M_2(R)$ contains an invertible upper triangular matrix whose inverse is not upper triangular.

Hint. If $xy = 1$, then $e := 1 - yx$ is an idempotent and $xe = ey = 0$.

2.8 The Quaternions

An essential part of our mathematical education is learning, gradually and over a long period of time, about the number sets

$$\mathbb{N} \subset \mathbb{Z} \subset \mathbb{Q} \subset \mathbb{R} \subset \mathbb{C}.$$

Each term in this sequence is extremely important, yet, besides the last one, in some sense imperfect, which motivates the introduction of its successor. Is there some natural way to continue this sequence? Does everything that is interesting and useful really end with the complex numbers?

Let us pose the question more precisely. The complex numbers $\mathbb{C}$ form a 2-dimensional real algebra (see Examples 1.98 and 1.152). The first question one may ask is whether the multiplication on $\mathbb{C}$ can be extended to three dimensions. Let us try to do this. Thus, let us add a third vector, say denoted by j, to the standard basis $\{1, i\}$ of $\mathbb{C}$ over $\mathbb{R}$, and consider the 3-dimensional real vector space, say denoted by A, with basis $\{1, i, j\}$. We are interested in whether the multiplication on $\mathbb{C}$ can be extended to A in such a way that A will become a real algebra with unity 1. The element i satisfies $i^2 = -1$, while the products ij, ji, j^2 still have to be determined. As an element of A, ij must be of the form

$$ij = \lambda + \mu i + \nu j$$

for some $\lambda, \mu, \nu \in \mathbb{R}$ (just as in $\mathbb{C}$, λ stands for $\lambda 1$). Multiply this equality from the left by i. Since $i^2 = -1$ and the multiplication on A is assumed to be associative, we obtain

$$-j = \lambda i - \mu + \nu ij.$$

Again using $ij = \lambda + \mu i + \nu j$ and rearranging terms, we arrive at

$$(\nu\lambda - \mu) + (\nu\mu + \lambda)i + (\nu^2 + 1)j = 0.$$

Since $1, i, j$ are linearly independent and $\nu^2 + 1 \neq 0$ for every $\nu \in \mathbb{R}$, we have reached a contradiction. This means that the complex numbers cannot be reasonably extended to a 3-dimensional vector space.

We should not give up. The above calculation has shown that ij cannot lie in span $\{1, i, j\}$. Therefore, we need at least four dimensions. Denote a fourth vector by k and consider the 4-dimensional real vector space with basis $\{1, i, j, k\}$. Its elements can thus be written as

$$\lambda_0 + \lambda_1 i + \lambda_2 j + \lambda_3 k$$

where λ_i are uniquely determined real numbers, and they are added and multiplied by scalars as follows:

$$(\lambda_0 + \lambda_1 i + \lambda_2 j + \lambda_3 k) + (\mu_0 + \mu_1 i + \mu_2 j + \mu_3 k)$$
$$= (\lambda_0 + \mu_0) + (\lambda_1 + \mu_1)i + (\lambda_2 + \mu_2)j + (\lambda_3 + \mu_3)k$$

and

$$\lambda(\lambda_0 + \lambda_1 i + \lambda_2 j + \lambda_3 k) = (\lambda\lambda_0) + (\lambda\lambda_1)i + (\lambda\lambda_2)j + (\lambda\lambda_3)k.$$

For reasons that will be explained at the end of the section, we denote this vector space by $\mathbb{H}$. We will now endow $\mathbb{H}$ with a multiplication making it an algebra. The algebra axioms, specifically the distributive laws and the axiom $\lambda(xy) = (\lambda x)y = x(\lambda y)$, imply that the product of any two elements in $\mathbb{H}$ is determined by the product of basis elements. Since the element 1 of course plays the role of unity, the following rules contain all information about multiplication:

Fig. 2.5 Multiplication of quaternions

$$i^2 = j^2 = k^2 = -1,$$
$$ij = -ji = k,$$
$$jk = -kj = i, \tag{2.5}$$
$$ki = -ik = j.$$

A useful mnemonic is presented in Figure 2.5. The product of two consecutive elements is the third one if going around the circle clockwise. Going counterclockwise, the sign changes. So, for example,

$$(2 + i - k)(i + \frac{1}{2}j) = 2i + 2\frac{1}{2}j + i^2 + \frac{1}{2}ij - ki - \frac{1}{2}kj = -1 + \frac{5}{2}i + \frac{1}{2}k.$$

Why these multiplication rules? We could show that they occur naturally, but this would take some space. Let us just focus on the question of what these rules bring. First of all, we should check that they really make $\mathbb{H}$ an algebra. Only the associative law is not entirely obvious. It is enough to show that the multiplication of basis elements is associative (why?), so we have to verify formulas like $(ij)k = i(jk)$, $(i^2)j = i(ij)$, etc. This is easy, but slightly tedious. Let us omit the details—$\mathbb{H}$ is indeed a 4-*dimensional real algebra*. Elements of $\mathbb{H}$ are called **quaternions**. The quaternions of the form $\lambda_0 + \lambda_1 i$ can be identified with the complex numbers and in this way we can consider $\mathbb{C}$ as a subalgebra of $\mathbb{H}$. Accordingly, the quaternions of the form $\lambda_0 (= \lambda_0 1)$ are viewed as real numbers. Note that, unlike $\mathbb{R}$ and $\mathbb{C}$, the algebra $\mathbb{H}$ *is not commutative*.

The conjugation of complex numbers can be naturally extended to quaternions. Take $h = \lambda_0 + \lambda_1 i + \lambda_2 j + \lambda_3 k \in \mathbb{H}$. Its **conjugate** $\bar{h}$ is defined to be

$$\bar{h} := \lambda_0 - \lambda_1 i - \lambda_2 j - \lambda_3 k.$$

A straightforward computation shows that

$$h\bar{h} = \bar{h}h = \lambda_0^2 + \lambda_1^2 + \lambda_2^2 + \lambda_3^2.$$

Thus, $h\bar{h} = \bar{h}h$ is a real number (a scalar), which is 0 only when $h = 0$. This implies that $\mathbb{H}$ *is a division algebra*. That is, every quaternion $h \neq 0$ is invertible, with inverse

$$h^{-1} = \frac{1}{h\bar{h}}\bar{h}.$$

For example,

$$(3 - i - 3j + 2k)^{-1} = \frac{3}{23} + \frac{1}{23}i + \frac{3}{23}j - \frac{2}{23}k.$$

Let us summarize the most important properties of $\mathbb{H}$:

(a) $\mathbb{H}$ is a 4-dimensional real algebra.
(b) $\mathbb{H}$ contains $\mathbb{C}$ and $\mathbb{R}$ as subalgebras.
(c) $\mathbb{H}$ is a division algebra.
(d) $\mathbb{H}$ is not commutative.

Although the quaternions are not considered to be numbers, the first three properties make $\mathbb{H}$ a decent successor in the sequence $\mathbb{N}, \mathbb{Z}, \mathbb{Q}, \mathbb{R}, \mathbb{C}$. Being a division algebra is a particularly nice property, and also very exceptional! Indeed, it can be proved that $\mathbb{R}, \mathbb{C}$, and $\mathbb{H}$ are actually the *only* finite-dimensional real division algebras. There is also the last property, the noncommutativity, that makes $\mathbb{H}$ more intriguing than its predecessors.

Now, the reader might wonder whether $\mathbb{H}$ also has a "successor". The algebras $\mathbb{R}, \mathbb{C}$, and $\mathbb{H}$ are of dimension 1, 2, and 4, respectively. In dimension 8, we have the **octonions** that also have many nice properties, but their multiplication is not associative. Therefore we will not discuss them here.

We have already mentioned (in Example 1.111) that the complex numbers $1, -1, i, -i$ form a group under multiplication. We can now add them the quaternions $j, -j, k, -k$. The new set

$$Q := \{\pm 1, \pm i, \pm j, \pm k\}$$

is again a group under multiplication. We call it the **quaternion group**. With its 8 elements, it is one of the smallest non-Abelian groups. The symmetric group S_3 has even fewer, namely, 6 elements. The dihedral group D_8 is also non-Abelian and has 8 elements.

The reader might have noticed that the multiplication rules for the quaternions i, j, k coincide with the multiplication rules for the vectors $\mathbf{i}, \mathbf{j}, \mathbf{k}$ in $\mathbb{R}^3$ with respect to the cross product. Writing a quaternion $\lambda_0 + \lambda_1 i + \lambda_2 j + \lambda_3 k$ as the pair $(\lambda_0, \mathbf{u})$ where $\mathbf{u} = (\lambda_1, \lambda_2, \lambda_3) \in \mathbb{R}^3$, the formula for the product of quaternions reads as follows:

$$(\lambda_0, \mathbf{u})(\mu_0, \mathbf{v}) = (\lambda_0 \mu_0 - \mathbf{u} \cdot \mathbf{v}, \lambda_0 \mathbf{v} + \mu_0 \mathbf{u} + \mathbf{u} \times \mathbf{v}). \tag{2.6}$$

Here, $\mathbf{u} \cdot \mathbf{v}$ is the dot product and $\mathbf{u} \times \mathbf{v}$ is the cross product of the vectors $\mathbf{u}$ and $\mathbf{v}$. All basic vector operations are thus hidden in the multiplication of quaternions.

The letter $\mathbb{H}$ is in honor of the Irish mathematician *William Rowan Hamilton* (1805–1865), the inventor of the quaternions. After a long and unfruitful attempt to find a multiplication in 3-dimensional vector spaces that would be as meaningful as the multiplication of complex numbers, on October 16, 1843, while walking with his wife along the Royal Canal in Dublin, a sudden idea of 4-dimensional quaternions sparked in his mind. In a wave of enthusiasm, he carved the formula

$$i^2 = j^2 = k^2 = ijk = -1 \tag{2.7}$$

into the stone of Broom Bridge. This Eureka-like story is documented in Hamilton's letter to his son.

Exercises

2.113. Prove that the formula (2.7) implies the formulas (2.5) (and vice versa).

2.114. Solve the equations $(i + j)x = 1 + k$ and $y(i + j) = 1 + k$ in $\mathbb{H}$.

2.115. Find all solutions of the equation $ix - xi = j$ in $\mathbb{H}$.

2.116. Find all solutions of the equation $x^2 = -1$ in $\mathbb{H}$.

Advice. This can be solved in one way or another, but it will be easier and more instructive to use the notation $(\lambda_0, \mathbf{u})$ and the formula (2.6).

2.117. Determine $Z(Q)$, the center of the group Q.

2.118. Determine $Z(\mathbb{H})$, the center of the algebra $\mathbb{H}$.

2.119. Observe that the additive group $(\mathbb{H}, +)$ becomes a complex vector space if we define scalar multiplication λh as the product of the quaternions $\lambda \in \mathbb{C} \subseteq \mathbb{H}$ and $h \in \mathbb{H}$. What is the dimension of $\mathbb{H}$ over $\mathbb{C}$? Is $\mathbb{H}$ a complex algebra under this scalar multiplication?

2.120. Prove that, for all $h, h' \in \mathbb{H}$ and $\lambda \in \mathbb{R}$,

$$\overline{\lambda h} = \lambda \overline{h}, \quad \overline{h + h'} = \overline{h} + \overline{h'}, \quad \text{and} \quad \overline{hh'} = \overline{h'} \cdot \overline{h}.$$

2.121. The **norm of the quaternion** $h = \lambda_0 + \lambda_1 i + \lambda_2 j + \lambda_3 k$ is defined to be the nonnegative real number

$$\|h\| := \sqrt{\lambda_0^2 + \lambda_1^2 + \lambda_2^2 + \lambda_3^2}.$$

Prove that $\|hh'\| = \|h\| \|h'\|$ for all $h, h' \in \mathbb{H}$.

Hint. Use the preceding exercise and the identity $\|h\|^2 = h\overline{h}$.

2.122. Prove that for every $h \in \mathbb{H}$, there exist $\alpha, \beta \in \mathbb{R}$ such that $h^2 + \alpha h + \beta = 0$. Express α and β in terms of h and $\overline{h}$.

2.123. Let $h \in \mathbb{H} \setminus \mathbb{R}$. Prove that $h + \overline{h} = 0$ if and only if there exists an $x \in \mathbb{H}$ such that $hx \neq xh$ and $h^2 x = xh^2$.

2.124. In the definition of the algebra $\mathbb{H}$, we can replace the real numbers by elements of an arbitrary field F. The algebra elements are still of the form $\lambda_0 + \lambda_1 i + \lambda_2 j + \lambda_3 k$, only λ_i are now from F. Defining the operations in the same way as above, we obtain a 4-dimensional algebra over F. Let us denote it by $\mathbb{H}_F$. Show that $\mathbb{H}_{\mathbb{Q}}$ is a division algebra, whereas $\mathbb{H}_{\mathbb{C}}$ has zero-divisors.

2.125. Prove that a finite-dimensional complex algebra A without zero-divisors is necessarily 1-dimensional.

Hint. For each $a \in A$, $x \mapsto ax$ is a linear map from A to A. What do we know about linear maps from a finite-dimensional *complex* vector space to itself?

2.126. Let x and y be nonzero elements of a division ring D. Suppose that $x \neq y^{-1}$. Prove **Hua's identity**

$$xyx = x - \left(x^{-1} + (y^{-1} - x)^{-1}\right)^{-1}.$$

2.127. Prove that a division ring D is commutative provided that $(xy)^2 = (yx)^2$ for all $x, y \in D$.

Hint. The elements x and y can be replaced by a sum of two elements. Such substitutions bring you closer to the solution, but keep in mind that $x + x = 0$ does not necessarily imply $x = 0$ (just think of $\mathbb{Z}_2$).

2.128. Let G be a non-Abelian group satisfying $(xy)^2 = (yx)^2$ for all $x, y \in G$. Prove that G contains an element $a \neq 1$ such that $a^2 = 1$. Check also that Q is an example of such a group.

2.129. A step further from the ring of formal power series $R[[X]]$ is the **ring of formal Laurent series** $R((X))$. Its elements are formal series $\sum_{k \in \mathbb{Z}} a_k X^k$ where only finitely many coefficients a_k with $k < 0$ can be different from 0. Addition and multiplication are defined in a self-explanatory manner (so that $R[[X]]$ is a subring of $R((X))$). Prove that if R is a division ring, then so is $R((X))$.

Comment. We can thus construct new division rings from old ones. If R is a field, then $R((X))$ is a field too. If, however, R is not commutative, then neither is $R((X))$. Therefore, $\mathbb{H}$ is no longer the only example of a noncommutative division ring that we know of. There are also $\mathbb{H}((X))$, $\mathbb{H}((X))((Y))$, etc.

Chapter 3
Homomorphisms

It often happens in mathematics that we understand something intuitively, but do not know how to express it in words. We then need appropriate definitions that help us to clarify our thoughts and make discussion possible. The "right" definition should also reveal the essence of the issue and direct our way of thinking. One of the definitions that plays such a role in algebra is that of a homomorphism.

A homomorphism is a map from an algebraic structure to an algebraic structure of the same type that preserves the operation(s) of the structure. It is impossible to list all reasons why homomorphisms are important, they just arise naturally in various contexts. In particular, we need them for recognizing common features of algebraic structures.

3.1 Isomorphic Groups and Cyclic Groups

In different areas of mathematics, we think of objects that share the same key properties peculiar to the area as one and the same object. For example, in set theory we may be interested only in the cardinality of sets, and then the difference between, say, the sets $\{1, 2, 3\}$ and $\{a, b, c\}$ is irrelevant. In geometry, we very well understand that there is a difference between equal figures and congruent figures, but, on the other hand, it is clear to us that this difference is in many ways inessential. When are two groups essentially equal?

3.1.1 Isomorphic Groups

The following examples will lead us to an answer to the question we have just posed.

Example 3.1. All information about the addition in the group $\mathbb{Z}_4$ is given in Table 3.1. Its meaning can be easily inferred. For example, since $2 + 3 = 1$ in $\mathbb{Z}_4$, 1 is

© Springer Nature Switzerland AG 2019
M. Brešar, *Undergraduate Algebra*, Springer Undergraduate Mathematics Series,
https://doi.org/10.1007/978-3-030-14053-3_3

+	0	1	2	3
0	0	1	2	3
1	1	2	3	0
2	2	3	0	1
3	3	0	1	2

Table 3.1 Cayley table of $(\mathbb{Z}_4, +)$

·	1	i	-1	$-i$
1	1	i	-1	$-i$
i	i	-1	$-i$	1
-1	-1	$-i$	1	i
$-i$	$-i$	1	i	-1

Table 3.2 Cayley table of $(U_4, \cdot)$

placed at the intersection of the row headed by 2 with the column headed by 3. Every finite group G can be represented by such a table. We call it the **Cayley table** of G, named after the English mathematician *Arthur Cayley* (1821–1895).

Let us now consider the multiplicative group of complex numbers 1, i, -1, and $-i$. We denote it by U_4 (as in Example 1.111). Our first impression may be that U_4 does not have much in common with the group $\mathbb{Z}_4$. However, take a look at its Cayley table (Table 3.2). It has the same structure as the Cayley table of $\mathbb{Z}_4$, only the labeling of the elements is different. Indeed, the element 0 in Table 3.1 is placed at the same spots as the element 1 in Table 3.2, and the same is true for 1 in i, 2 and -1, and 3 and $-i$. The differences between the groups $(\mathbb{Z}_4, +)$ and $(U_4, \cdot)$ are thus inessential. From the point of view of group theory, they may be considered equal.

This example will be better understood by exploring a more general one.

Example 3.2. The elements of U_4 can be described as solutions of the equation $z^4 = 1$ in the set of complex numbers. Now take any $n \in \mathbb{N}$ and set

$$U_n := \{z \in \mathbb{C} \mid z^n = 1\}.$$

That is, U_n is the set of all nth roots of unity. As we know, $|U_n| = n$ and all elements in U_n are powers of the number $a := e^{\frac{2\pi i}{n}}$. More precisely, by setting

$$z_k := a^k, \quad k = 0, 1, \ldots, n - 1,$$

we have

$$U_n = \{z_0, z_1, \ldots, z_{n-1}\}.$$

One immediately checks that U_n is a group under multiplication. Note that

$$z_k z_\ell = \begin{cases} z_{k+\ell} & ; \quad k + \ell < n \\ z_{k+\ell-n} & ; \quad k + \ell \geq n \end{cases}$$

which can be more simply written as

$$z_k z_\ell = z_{k+\ell},$$

where $k + \ell$ is the sum of k and ℓ in the group $\mathbb{Z}_n$. The groups U_n and $\mathbb{Z}_n$ are thus essentially equal: multiplication of elements z_k in U_n corresponds to addition of their indices, viewed as elements in $\mathbb{Z}_n$. This can be more precisely described via the map

$$\varphi : \mathbb{Z}_n \to U_n, \quad \varphi(k) = z_k,$$

which transforms the group $\mathbb{Z}_n$ into the group U_n. That is, φ is bijective and connects addition in $\mathbb{Z}_n$ with multiplication in U_n through the formula

$$\varphi(k + \ell) = \varphi(k)\varphi(\ell) \quad \text{for all } k, \ell \in \mathbb{Z}_n. \tag{3.1}$$

Now let G and G' be arbitrary groups. When are they essentially equal? As the above example suggests, a convenient way to answer this question is through the existence of a certain map between G and G'. The term we use for what we informally called "essentially equal" is "isomorphic".

Definition 3.3. We say that the groups G and G' are **isomorphic**, and write $G \cong G'$, if there exists a bijective map $\varphi : G \to G'$ satisfying

$$\varphi(xy) = \varphi(x)\varphi(y) \quad \text{for all } x, y \in G. \tag{3.2}$$

Any such map φ is called a **group isomorphism**.

If one or both groups are additive, (3.2) must be changed accordingly. From (3.1) we thus see that $\mathbb{Z}_n \cong U_n$.

If G and G' are finite groups, then $G \cong G'$ if and only if the Cayley tables of G and G' are identical, only the elements in the two tables may be labeled differently. Proving this is merely a matter of understanding the definitions, so we leave it to the reader. One just has to be a bit careful when interpreting this statement, since the Cayley table of a group depends on the order in which we list the group elements.

At first glance, it may seem that the groups G and G' do not appear symmetrically in the definition of isomorphic groups. That is, the definition talks about a group isomorphism from G to G', and not from G' to G. The next theorem shows, however, that the existence of a group isomorphism φ from G to G' implies the existence of a group isomorphism from G' to G. By φ^{-1} we denote the inverse map of φ (which exists since φ is bijective).

Theorem 3.4. *If* $\varphi : G \to G'$ *is a group isomorphism, then so is* $\varphi^{-1} : G' \to G$.

Proof. We have to prove that

$$\varphi^{-1}(uv) = \varphi^{-1}(u)\varphi^{-1}(v)$$

for all $u, v \in G'$. As φ is injective, it is enough to show that

$$\varphi\big(\varphi^{-1}(uv)\big) = \varphi\big(\varphi^{-1}(u)\varphi^{-1}(v)\big).$$

This indeed holds: the left-hand side is equal to uv, and so is the right-hand side since φ satisfies (3.2).   □

Let us consider a few more examples. First a trivial one.

Example 3.5. Every group G satisfies $G \cong G$. The simplest example of an isomorphism from G to G is the identity map id_G.

Example 3.6. One of the first things we learn about the exponential function $x \mapsto e^x$ is that $e^{x+y} = e^x e^y$ for all $x, y \in \mathbb{R}$. In the language of algebra, we can interpret this as saying that the exponential function is an isomorphism from the group of real numbers under addition to the group of positive real numbers under multiplication. Thus, $(\mathbb{R}, +) \cong (\mathbb{R}^+, \cdot)$.

Example 3.7. The group $(\mathbb{Z}, +)$ is isomorphic to each of its nontrivial subgroups. Indeed, by Theorem 2.2, a nontrivial subgroup of $\mathbb{Z}$ is of the form $n\mathbb{Z}$ for some $n \in \mathbb{N}$, and $x \mapsto nx$ is an isomorphism from $\mathbb{Z}$ to $n\mathbb{Z}$.

A finite group, however, cannot be isomorphic to any of its proper subgroups. Indeed, isomorphic groups must have the same order since an isomorphism is a bijective map.

3.1.2   Cyclic Groups and the Order of Elements

The groups treated in Examples 3.2 and 3.7 have a common property described in the following definition.

Definition 3.8. A group generated by a single element is called a **cyclic group**.

A cyclic group generated by an element a is denoted by $\langle a \rangle$ (see Section 1.7). Every element in $\langle a \rangle$ is of the form a^k for some $k \in \mathbb{Z}$, and conversely, every element of the form a^k lies in $\langle a \rangle$. Thus,

$$\langle a \rangle = \{a^k \mid k \in \mathbb{Z}\}.$$

For example, $U_4 = \langle i \rangle$.

In an additive group, we of course write a^k as ka. The group $(\mathbb{Z}, +)$ is generated by the element 1 (as well as by -1) and is therefore cyclic. Similarly, for every $n \in \mathbb{N}$, $(\mathbb{Z}_n, +)$ is a cyclic group. It is generated by the element 1 if $n > 1$; if $n = 1$, then this is the trivial group, generated by 0. .

Every cyclic group is Abelian. This is because $a^k \cdot a^\ell = a^{k+\ell} = a^\ell \cdot a^k$ for all $k, \ell \in \mathbb{Z}$. The next theorem tells us much more: the cyclic groups mentioned in the preceding paragraph are essentially the only cyclic groups that exist. Some of the arguments from Example 3.2 will be repeated in its proof.

Theorem 3.9. *A cyclic group G is isomorphic either to the group $(\mathbb{Z}, +)$ or to the group $(\mathbb{Z}_n, +)$ for some $n \in \mathbb{N}$.*

Proof. Take $a \in G$ such that $G = \langle a \rangle$. Assume first that all the elements a^k are distinct, i.e., $a^k \neq a^\ell$ if $k \neq \ell$. The map

$$\varphi : \mathbb{Z} \to G, \quad \varphi(k) = a^k,$$

is then injective. Since $G = \langle a \rangle$, φ is also surjective. Next we have

$$\varphi(k + \ell) = a^{k+\ell} = a^k a^\ell = \varphi(k)\varphi(\ell)$$

for all $k, \ell \in \mathbb{Z}$. Thus, φ is a group isomorphism.

Consider now the second possibility where there exist integers k and ℓ such that $k < \ell$ and $a^k = a^\ell$. The latter can be written as $a^{\ell-k} = 1$. This means that there exist natural numbers s such that $a^s = 1$. Let n be the smallest of these. We claim that

$$|G| = n \quad \text{and} \quad G = \{1, a, \ldots, a^{n-1}\}. \tag{3.3}$$

If $0 \leq p < q < n$, then a^p are a^q distinct, for otherwise $q - p$ would be a natural number smaller than n that satisfies $a^{q-p} = 1$. Consequently, $|G| \geq n$. Now take any $a^k \in G$. By the division algorithm for integers (Theorem 2.1), there exist $q, r \in \mathbb{Z}$ such that $k = qn + r$ and $0 \leq r < n$. Hence,

$$a^k = a^{qn+r} = (a^n)^q a^r = a^r,$$

which proves (3.3). As in Example 3.2, we set $z_k := a^k$, $k = 0, 1, \ldots, n - 1$. Observe that $z_k z_\ell = z_{k+\ell}$, where $k + \ell$ is the sum of k and ℓ in the group $\mathbb{Z}_n$. This implies that

$$\varphi : \mathbb{Z}_n \to G, \quad \varphi(k) = z_k,$$

is a group isomorphism. $\qquad\square$

A more elegant proof, based on the theory developed in Chapter 4, will be given in Example 4.112.

Since isomorphic groups have the same order, Theorem 3.9 implies that any finite cyclic group of order n is isomorphic to $(\mathbb{Z}_n, +)$. Therefore, we informally do not distinguish between "cyclic group of order n" and "additive group $\mathbb{Z}_n$", as well as between "infinite cyclic group" and "additive group $\mathbb{Z}$". Another standard notation for a cyclic group of order n is C_n, but we will not use it in this book.

One of the reasons why cyclic groups are important is that they can be found inside any group G. Indeed, every element a in G generates the **cyclic subgroup** $\langle a \rangle$. This subgroup is connected with the following notion, which we have already encountered in some of the exercises.

Definition 3.10. Let a be a group element. If $a^n = 1$ for some $n \in \mathbb{N}$, then we say that a has **finite order**, and we call the smallest such n the **order** of a. If $a^n \neq 1$ for all $n \in \mathbb{N}$, then we say that a has **infinite order**.

An element a has finite order if and only if the cyclic subgroup $\langle a \rangle$ is finite. If a has order n, then $\langle a \rangle = \{1, a, \ldots, a^{n-1}\}$ and $|\langle a \rangle| = n$. Hence,

$$\text{order of } a = |\langle a \rangle|. \tag{3.4}$$

In words: the order of an element equals the order of the cyclic subgroup generated by this element. So, the apparently entirely different notions "order of an element" and "order of a group" are related after all.

The order of an element a in an additive group is, of course, the smallest natural number n such that $na = 0$. All nonzero elements in the group $(\mathbb{Z}, +)$ are thus of infinite order. A finite group, on the other hand, cannot contain an infinite cyclic subgroup, so all of its elements have finite order.

Example 3.11. In the group $(\mathbb{Z}_4, +)$, 0 has order 1, 2 has order 2, and 1 and 3 have order 4. The dihedral group D_4 (see Section 2.6) also has four elements, but, besides the identity element, they all have order 2. Since isomorphisms preserve orders of elements (why?), the groups $\mathbb{Z}_4$ and D_4 therefore are not isomorphic. The reader can verify that D_4 is isomorphic to the group $(\mathbb{Z}_2 \oplus \mathbb{Z}_2, +)$.

Exercises

3.12. Prove that in any group, the elements x and x^{-1} have the same order.

3.13. Prove that in any group, the elements x and yxy^{-1} have the same order (in order words, conjugate elements have the same order).

3.14. Prove that in any group, the elements xy and yx have the same order.

3.15. Suppose that a cyclic group G has only one generator, i.e., $G = \langle a \rangle$ holds for exactly one element $a \in G$. Prove that G is either a trivial group or $G \cong \mathbb{Z}_2$.

3.16. The group $(\mathbb{Z}_2 \oplus \mathbb{Z}_2, +)$ has order 4 but contains no element of order 4 and so is not cyclic. Show that, on the other hand, the group $(\mathbb{Z}_3 \oplus \mathbb{Z}_2, +)$ is cyclic (and is hence isomorphic to $(\mathbb{Z}_6, +)$).

Comment. The answer to the general question of determining all $m, n \in \mathbb{N}$ such that the group $(\mathbb{Z}_m \oplus \mathbb{Z}_n, +)$ is cyclic will be given in Exercise 3.20. The next three exercises serve as a bridge to it.

3.17. Let a be a group element. Prove:

(a) If $a^m = 1$ for some $m \in \mathbb{Z}$, then the order of a divides m.
(b) If $a \neq 1$ and $a^p = 1$ for some prime p, then p is the order of a.

Hint. Division algorithm.

3.18. Let a and b be group elements of orders m and n, respectively. Suppose that a and b commute. Prove that the order of ab divides the least common multiple of m and n. Find an example showing that it is not necessarily equal to it.

3.19. Let G and H be groups. Suppose that $a \in G$ has order m and $b \in H$ has order n. Prove that the order of $(a, b) \in G \times H$ is the least common multiple of m and n.

3.20. Let G and H be finite cyclic groups. Prove that the group $G \times H$ is cyclic if and only if the orders of G and H are relatively prime. (Therefore, the additive groups $\mathbb{Z}_m \oplus \mathbb{Z}_n$ and $\mathbb{Z}_{mn}$ are isomorphic if and only if m and n are relatively prime.)

Hint. A finite group K is cyclic if and only if K contains an element of order $|K|$.

3.21. Prove that a subgroup of a cyclic group is cyclic.

Hint. For the group $(\mathbb{Z}, +)$, this follows from Theorem 2.2. For a general cyclic group, one only has to modify the proof.

3.22. Let $G = \langle a \rangle$ be a finite cyclic group of order n. Prove that a^k generates G if and only if k and n are relatively prime.

Comment. Combining this with Exercise 2.35, we see that the number of all elements that generate the group $(\mathbb{Z}_n, +)$ is equal to the number of all invertible elements of the ring $(\mathbb{Z}_n, +, \cdot)$, which in turn is equal to $\varphi(n)$, where φ is Euler's phi function.

3.23. Prove that a cyclic group G of order n contains a subgroup of order d if and only if $d \mid n$. Show also that this subgroup is unique.

Comment. From the comment to the previous exercise we see that this subgroup has $\varphi(d)$ different generators. Each of them has order d, and conversely, each element of order d generates this (unique) subgroup of order d. Therefore, the group G has exactly $\varphi(d)$ elements of order d. Partitioning G into subsets consisting of elements of the same order, we thus get the formula $n = \sum_{d \mid n} \varphi(d)$. (For example, $6 = \varphi(1) + \varphi(2) + \varphi(3) + \varphi(6) = 1 + 1 + 2 + 2$.)

3.24. Suppose that an element a of a group G has order 5. What is the order of $b \in G$ if $b \neq 1$ and $bab = a$? What is the order of $c \in G$ if $c \neq 1$ and $ac = c^2a$?

Hint. Consider $a^j ba^{-j}$ and $a^j ca^{-j}$, $j = 1, 2, \ldots$, and use the result of Exercise 3.17 (b).

3.25. Prove that an infinite group has infinitely many subgroups.

Hint. Consider separately the case where all group elements have finite order and the case where there exists an element of infinite order.

3.26. Suppose a finite Abelian group G contains an element of order 2. Prove that the product of all elements in G coincides with the product of all elements of order 2.

Comment. Along with Exercise 1.64, this implies that the product of all elements in any finite Abelian group either equals 1 or has order 2.

3.27. Using the previous exercise for the case where $F = \mathbb{Z}_p^*$, the multiplicative group of all nonzero elements of the field $\mathbb{Z}_p$, deduce **Wilson's Theorem**, which states that $p \in \mathbb{N}$ is a prime number if and only if $(p - 1)! \equiv -1 \pmod{p}$.

Hint. What are the solutions of the equation $x^2 = 1$ in a field (or, more generally, in any ring without zero-divisors)?

3.28. Prove that the cyclic subgroup of the group $GL_2(\mathbb{R})$ generated by the matrix $\left[\begin{smallmatrix} 1 & 1 \\ 0 & 1 \end{smallmatrix}\right]$ is isomorphic to $(\mathbb{Z}, +)$.

3.29. Prove that the subgroup of the group $GL_2(\mathbb{C})$ generated by the matrices $\left[\begin{smallmatrix} i & 0 \\ 0 & -i \end{smallmatrix}\right]$ and $\left[\begin{smallmatrix} 0 & 1 \\ -1 & 0 \end{smallmatrix}\right]$ is isomorphic to the quaternion group Q.

Comment. In principle, one has some freedom when searching for an isomorphism, there may be several possibilities. Indeed, sometimes some isomorphism seems natural and we think of it immediately (this should be the case in the previous exercise). In the quaternion group Q, however, several elements have identical properties, and hence this exercise has several "natural" solutions.

3.30. Prove that the dihedral group D_8 is isomorphic to the subgroup of the orthogonal group $O_2(\mathbb{R})$ generated by the matrices $R = \left[\begin{smallmatrix} 0 & -1 \\ 1 & 0 \end{smallmatrix}\right]$ and $S = \left[\begin{smallmatrix} -1 & 0 \\ 0 & 1 \end{smallmatrix}\right]$.

Comment. The reader familiar with the geometric meaning of orthogonal matrices should be able to interpret the solution geometrically.

3.31. Prove that the group $GL_2(\mathbb{Z}_2)$ is isomorphic to the symmetric group S_3.

Hint. An isomorphism preserves the orders of elements.

3.32. The dihedral group D_8 and the quaternion group Q both have order 8 and are non-Abelian. However, they are not isomorphic. Why not?

3.33. Prove that the groups $(\mathbb{R}^+, \cdot)$, $(\mathbb{R}^*, \cdot)$, and $(\mathbb{C}^*, \cdot)$ are not isomorphic to each other.

Comment. The first group is a proper subgroup of the second group, and the second group is a proper subgroup of the third group. However, as Example 3.7 shows, from this we cannot yet conclude that the groups are mutually non-isomorphic.

3.34. Suppose a group G is generated by two elements u and w, both of which have order 2. Set $t := uw$. Prove:

(a) If t has finite order n, then G is isomorphic to the dihedral group D_{2n}.
(b) If t has infinite order, then G is isomorphic to the dihedral group D_∞.

Hint. Show first that every element in G can be written as t^i or $t^j w$ for some integers i and j, and that an element of the form t^i cannot be equal to an element of the form $t^j w$.

3.2 Isomorphic Vector Spaces

Other isomorphic algebraic structures can be introduced and studied similarly to isomorphic groups. The goal of this short section—which the reader with a good linear algebra background may skip—is to establish a fundamental result concerning isomorphic vector spaces.

Let V and V' be vector spaces over a field F. We say that a map $\varphi : V \to V'$ is **linear** if

$$\varphi(x + y) = \varphi(x) + \varphi(y) \quad \text{and} \quad \varphi(\lambda x) = \lambda\varphi(x) \quad \text{for all } x, y \in V, \lambda \in F.$$

It is easy to see that this is equivalent to the condition that

$$\varphi(\lambda x + \mu y) = \lambda\varphi(x) + \mu\varphi(y) \quad \text{for all } x, y \in V, \lambda, \mu \in F.$$

A bijective linear map from V to V' is called a **vector space isomorphism**. If such a map exists, we say that V and V' are **isomorphic vector spaces**. The intuition behind this concept can be described just as in the case of isomorphic groups.

We restrict ourselves to finite-dimensional vector spaces. Let us just mention that the next theorem also makes sense, and is true, if V and V' are infinite-dimensional. One defines the dimension of an arbitrary vector space as the cardinal number of any of its bases. Every vector space has a basis (see Exercise 1.174) and it turns out that they all have the same cardinality. However, we will not go into this here.

Theorem 3.35. *Finite-dimensional vector spaces V and V' over a field F are isomorphic if and only if they have the same dimension.*

Proof. Let $\varphi : V \to V'$ be an isomorphism. Take a basis $\{b_1, \ldots, b_n\}$ of V. Let us prove that $\{\varphi(b_1), \ldots, \varphi(b_n)\}$ is a basis V'. This will show that both vector spaces have dimension n. For any $\lambda_1, \ldots, \lambda_n \in F$, we have

$$\varphi(\lambda_1 b_1 + \cdots + \lambda_n b_n) = \lambda_1\varphi(b_1) + \cdots + \lambda_n\varphi(b_n). \tag{3.5}$$

Since φ is surjective and $V = \operatorname{span}\{b_1, \ldots, b_n\}$, (3.5) implies that V' is equal to $\operatorname{span}\{\varphi(b_1), \ldots, \varphi(b_n)\}$. It remains to show that the vectors $\varphi(b_1), \ldots, \varphi(b_n)$ are linearly independent. Suppose there exist $\lambda_i \in F$ such that

$$\lambda_1\varphi(b_1) + \cdots + \lambda_n\varphi(b_n) = 0.$$

By (3.5), we can write this as

$$\varphi(\lambda_1 b_1 + \cdots + \lambda_n b_n) = 0.$$

Since $\varphi(0) = 0$ (this follows from $\varphi(0) = \varphi(0 + 0) = 2\varphi(0)$) and φ is injective, we may conclude that

$$\lambda_1 b_1 + \cdots + \lambda_n b_n = 0.$$

But then each $\lambda_i = 0$ since the vectors $b_1, \ldots, b_n$ are linearly independent. This completes the proof of the "only if" part.

To prove the converse, assume that both V and V' have dimension n. Take a basis $\{b_1, \ldots, b_n\}$ of V and a basis $\{b'_1, \ldots, b'_n\}$ of V'. Every vector in V can be uniquely written in the form $\lambda_1 b_1 + \cdots + \lambda_n b_n$ for some $\lambda_i \in F$. Define $\varphi : V \to V'$ by

$$\varphi(\lambda_1 b_1 + \cdots + \lambda_n b_n) = \lambda_1 b'_1 + \cdots + \lambda_n b'_n$$

for any $\lambda_i \in F$. We leave it to the reader to check that φ is a vector space isomorphism. $\qquad\qquad\Box$

Corollary 3.36. *A nontrivial finite-dimensional vector space over a field F is isomorphic to the vector space F^n for some $n \in \mathbb{N}$.*

The ultimate goal when studying some algebraic objects is to classify them "up to isomorphism". By this we mean that we would like to describe all of them, but without distinguishing among isomorphic ones. From this point of view, the study of vector spaces (over a given field F) is extremely simple. Theorem 3.35 gives a definitive description of their structure. For each $n \in \mathbb{N}$, there exists a vector space of dimension n (for example, F^n) and arbitrary vector spaces of the same dimension are isomorphic. The study of other algebraic structures that are considered in this book is much more involved. In the previous section, we have indeed derived a simple classification of cyclic groups, but these groups are very special. To give a small indication of the complexity of the structure of groups, we recall from Example 3.11 that there exist two nonisomorphic groups with four elements.

Exercises

3.37. Write down explicitly the isomorphism φ between the 2-dimensional real vector spaces $V = \{(x, y, z) \in \mathbb{R}^3 \mid x+y+z = 0\}$ and $\mathbb{C}$ that satisfies $\varphi((1, -2, 1))=1+i$ and $\varphi^{-1}(i) = (0, 1, -1)$.

3.38. In the context of Exercise 3.37, describe all isomorphisms from V to $\mathbb{C}$.

3.39. Let V and V' be 2-dimensional vector spaces over the field $\mathbb{Z}_2$. How many isomorphisms from V to V' are there?

3.40. Let p be a prime number and let V and V' be finite-dimensional vector spaces over the field $\mathbb{Z}_p$. Show that V and V' are isomorphic if and only if $|V| = |V'|$. *Hint.* How many elements does an n-dimensional vector space over $\mathbb{Z}_p$ have?

3.41. The sets of square matrices over the complex numbers $M_m(\mathbb{C})$ and square matrices over the quaternions $M_n(\mathbb{H})$ can be treated as real vector spaces (see Section 2.7). Are they isomorphic for some pair of natural numbers m and n?

3.42. The real algebras $\mathbb{H}$ and $M_2(\mathbb{R})$ both have dimension 4, so they are isomorphic as vector spaces. Does there exist a vector space isomorphism $\varphi : \mathbb{H} \to M_2(\mathbb{R})$ such that $\varphi(h^2) = \varphi(h)^2$ for all $h \in \mathbb{H}$?

3.3 The Concept of a Homomorphism

A homomorphism can be briefly described as a map preserving operations that are peculiar to the algebraic structure under consideration. In the preceding two sections, we considered bijective homomorphisms of groups and vector spaces. These special cases were treated first only to provide some motivation for the concept of a homomorphism. In this section, we will establish a solid foundation for its general treatment.

First, a warning about notation. Unlike in the previous sections, we will denote groups, rings, vector spaces, and algebras by the same letters A and A'. Since we will consider properties that are common to all algebraic structures, such a uniform notation will be convenient.

Definition 3.43. A map $\varphi : A \to A'$ is

- a **group homomorphism** if A and A' are groups and

$$\varphi(xy) = \varphi(x)\varphi(y)$$

for all $x, y \in A$;
- a **ring homomorphism** if A and A' are rings and

$$\varphi(1) = 1, \quad \varphi(x + y) = \varphi(x) + \varphi(y), \quad \text{and} \quad \varphi(xy) = \varphi(x)\varphi(y)$$

for all $x, y \in A$;
- a **vector space homomorphism** (or a **linear map**) if A and A' are vector spaces over a field F and

$$\varphi(x + y) = \varphi(x) + \varphi(y) \quad \text{and} \quad \varphi(\lambda x) = \lambda\varphi(x)$$

for all $x, y \in A$ and $\lambda \in F$;
- an **algebra homomorphism** if A and A' are algebras over a field F and

$$\varphi(1) = 1, \quad \varphi(x + y) = \varphi(x) + \varphi(y),$$

$$\varphi(xy) = \varphi(x)\varphi(y), \quad \text{and} \quad \varphi(\lambda x) = \lambda\varphi(x)$$

for all $x, y \in A$ and $\lambda \in F$.

Usually it is clear from the context whether we have in mind a group, a ring, a vector space, or an algebra homomorphism. In such a case (i.e., almost always) we talk simply about a **homomorphism**. We will also use the term "homomorphism" to mean a homomorphism of any possible type.

A bijective homomorphism is called an **isomorphism**, a surjective homomorphism is called an **epimorphism**, and an injective homomorphism is called an **embedding** or a **monomorphism** (we will use the former term). A homomorphism from A to itself is called an **endomorphism**, and a bijective endomorphism is called an **automorphism**; in other words, an automorphism is an endomorphism which is an isomorphism.

We add some comments to these definitions.

Remark 3.44. It goes without saying that the equality defining a group homomorphism must be appropriately adjusted if one of the groups A and A' is additive (see

Section 3.1). An **additive group homomorphism**, i.e., a map φ from an additive group A to an additive group A' satisfying

$$\varphi(x + y) = \varphi(x) + \varphi(y) \quad \text{for all } x, y \in A,$$

is also called an **additive map**. Ring, vector space, and algebra homomorphisms are thus additive maps.

Remark 3.45. For ring (and algebra) homomorphisms we have required that they send the unity of A into the unity of A'. Therefore, for example, the map $\varphi : \mathbb{R} \to M_2(\mathbb{R})$, $\varphi(x) = \begin{bmatrix} x & 0 \\ 0 & 0 \end{bmatrix}$, is not a ring homomorphism, although it preserves addition and multiplication.

Remark 3.46. **Field homomorphisms** are not defined separately. These are just ring homomorphisms between fields. Indeed fields (and, more generally, division rings) are quite different from "ordinary" rings because of the invertibility of their nonzero elements. However, ring homomorphisms send inverses of invertible elements to inverses of their images (see Theorem 3.53 below). Therefore, there is simply no need to define a field homomorphism differently than a ring homomorphism.

We will now prove several basic facts about homomorphisms. The readers can try to find the proofs themselves immediately after reading the statements. This will be more instructive (and more fun) than just reading proofs line-by-line.

Theorem 3.47. *The composition of two homomorphisms is a homomorphism.*

Proof. Let $\varphi : A \to A'$ and $\psi : A' \to A''$ be group homomorphisms. Then

$$\begin{aligned}
(\psi \circ \varphi)(xy) &= \psi(\varphi(xy)) = \psi(\varphi(x) \cdot \varphi(y)) \\
&= \psi(\varphi(x)) \cdot \psi(\varphi(y)) = (\psi \circ \varphi)(x) \cdot (\psi \circ \varphi)(y),
\end{aligned}$$

meaning that $\psi \circ \varphi : A \to A''$ is a group homomorphism. A similar argument works for homomorphisms of other structures. $\qquad\square$

The next theorem has already been established for groups (Theorem 3.4). The proof for homomorphisms of other structures requires only obvious modifications, so we omit it. However, the readers are strongly advised to go through the details.

Theorem 3.48. *The inverse of an isomorphism is an isomorphism.*

We say that A and A' are **isomorphic**, and write

$$A \cong A',$$

if there exists an isomorphism from A to A'. Here, A and A' may be groups, rings, vector spaces, or algebras.

Remark 3.49. The relation of being isomorphic, $\cong$, is an *equivalence relation* on the class of all groups (rings etc.). Indeed, Theorem 3.48 shows that it is symmetric, i.e., $A \cong A'$ implies $A' \cong A$. Since the composition of bijective maps is bijective, Theorem 3.47 tells us that the composition of isomorphisms is an isomorphism. The relation $\cong$ is therefore transitive: $A \cong A'$ and $A' \cong A''$ implies $A \cong A''$. Finally, $\cong$ is reflexive, $A \cong A$, since the identity map id_A is an automorphism of A, no matter which algebraic structure we consider.

The meaning of being isomorphic was explained in Section 3.1 for the case of groups. A similar explanation can be given for all other cases. Take, for example, isomorphic rings A and A'. They may appear quite different at first glance, but they share all of the ring-theoretic properties, i.e., properties that depend on addition, subtraction and multiplication. Just think of something concrete, like commutativity, being a division ring, having zero-divisors, containing nonzero nilpotents, etc., and check that A has this property if and only if A' has it. From the ring-theoretical point of view, A and A' can therefore be considered identical. In general, we do not distinguish between isomorphic algebraic objects. Remark 3.49 actually gives some formal justification for this: isomorphic objects belong to the same equivalence class. However, an intuitive understanding, which usually comes with experience, is more important.

As a corollary to the last two theorems we obtain new interesting examples of groups.

Corollary 3.50. *Let A be a group, a ring, a vector space, or an algebra. The set of all automorphisms of A is a group under composition.*

Proof. Since the composition of bijective maps is bijective, Theorem 3.47 shows that the composition of automorphisms is an automorphism. Composition is an associative operation, the identity map id_A is an automorphism and is the identity element for composition, and the inverse of an automorphism is an automorphism by Theorem 3.48. $\square$

Theorem 3.51. *If $\varphi : A \to A'$ is a group homomorphism, then $\varphi(1)=1$ and $\varphi(x^{-1}) = \varphi(x)^{-1}$ for every $x \in A$.*

Proof. We have
$$\varphi(1) = \varphi(1^2) = \varphi(1)^2.$$
By the cancellation law, it follows that $\varphi(1) = 1$. Hence,
$$\varphi(x)\varphi(x^{-1}) = \varphi(xx^{-1}) = \varphi(1) = 1,$$
showing that $\varphi(x^{-1}) = \varphi(x)^{-1}$. $\square$

If A and A' are additive groups, Theorem 3.51 reads as follows.

Theorem 3.52. *Let A and A' be additive groups. If $\varphi : A \to A'$ is a group homomorphism, then $\varphi(0) = 0$ and $\varphi(-x) = -\varphi(x)$ for all $x \in A$.*

In particular, Theorem 3.52 holds for ring, vector space, and algebra homomorphisms. Ring and algebra homomorphisms thus satisfy $\varphi(1) = 1$ as well as $\varphi(0) = 0$. The first equality is part of the definition, whereas the second equality follows from the definition.

A common generalization of both equalities from Theorem 3.51 is

$$\varphi(x^n) = \varphi(x)^n,$$

where n can be any (possibly negative) integer. We leave the proof to the reader. If the groups are additive, this reads as

$$\varphi(nx) = n\varphi(x).$$

A result similar to Theorem 3.51 holds for ring homomorphisms.

Theorem 3.53. *Let $\varphi : A \to A'$ be a ring homomorphism. If $x \in A$ is invertible, then so is $\varphi(x)$ and $\varphi(x^{-1}) = \varphi(x)^{-1}$.*

Proof. The proof differs from the proof of Theorem 3.51 in two details. The first one is that the equality $\varphi(1) = 1$ is now an assumption, so we do not need to prove it. The second one is that, besides $\varphi(x)\varphi(x^{-1}) = 1$, we also have to show that $\varphi(x^{-1})\varphi(x) = 1$ (which is, of course, very easy). This is because in rings, $ab = 1$ in general does not imply $ba = 1$ (see, for example, Exercise 2.111). $\qquad\square$

When dealing with homomorphisms that are not isomorphisms, the notions of kernel and image play an important role. Let us first discuss the latter. The **image of the homomorphism** $\varphi : A \to A'$ will be denoted by im φ. Thus,

$$\operatorname{im} \varphi = \{\varphi(x) \mid x \in A\}.$$

Theorem 3.54. *The image of a group, a ring, a vector space, and an algebra homomorphism is a subgroup, a subring, a subspace, and a subalgebra, respectively.*

Proof. Let $\varphi : A \to A'$ be a group homomorphism. From $\varphi(x)\varphi(y) = \varphi(xy)$ it follows that im φ is closed under multiplication, and from $\varphi(x)^{-1} = \varphi(x^{-1})$ it follows that it contains inverses of its elements. Hence, it is a subgroup of A'. Similarly, we see that the image of a ring homomorphism is a subring, the image of a vector space homomorphism is a subspace, and the image of an algebra homomorphism is a subalgebra. The details are left to the reader. $\qquad\square$

Remark 3.55. Every homomorphism $\varphi : A \to A'$ therefore becomes an epimorphism if we shrink the codomain A' and consider it as a homomorphism from A to im φ. An embedding in this way becomes an isomorphism. When dealing with an embedding $\varphi : A \to A'$, we therefore often identify A with im φ and consider A as a subgroup (subring, etc.) of A'. For example, in Section 2.4 we wrote that by identifying elements from a ring R with constant polynomials, we can consider R as a subring of the polynomial ring $R[X]$. We can now make this precise: the map

$$a \mapsto a + 0X + 0X^2 + \dots$$

is an embedding of the ring R into the ring $R[X]$. An even simpler example is the following: the field of real numbers $\mathbb{R}$ embeds into the field of complex numbers $\mathbb{C}$ via

$$a \mapsto a + i0.$$

Similarly, we can embed the fields $\mathbb{R}$ and $\mathbb{C}$ into the division ring of quaternions $\mathbb{H}$.

The **kernel of a group homomorphism** $\varphi : A \to A'$ is the set

$$\ker \varphi := \{x \in A \mid \varphi(x) = 1\}.$$

If A' is an additive group, this definition reads as follows:

$$\ker \varphi := \{x \in A \mid \varphi(x) = 0\}. \tag{3.6}$$

Rings, vector spaces, and algebras are, in particular, additive groups. Accordingly, (3.6) also defines the **kernel of a ring homomorphism, a vector space homomorphism**, and **an algebra homomorphism**.

Theorem 3.51 shows that the kernel of a group homomorphism φ contains the identity element 1. If 1 is the only element in the kernel, then we say that φ has **trivial kernel**. For an additive group, a ring, a vector space, and an algebra homomorphism we of course say that its kernel is trivial if it contains only the zero element 0.

Theorem 3.56. *A homomorphism* $\varphi : A \to A'$ *is injective if and only if* φ *has trivial kernel.*

Proof. It is enough to consider the case where A and A' are groups under multiplication. In other cases, we only change the notation. If φ is injective, then $\ker \varphi$ cannot contain an element x different from 1, for otherwise we would have $\varphi(x) = \varphi(1)$. Conversely, if $\ker \varphi$ is trivial and $\varphi(x) = \varphi(y)$ holds for some $x, y \in A$, then

$$\varphi(xy^{-1}) = \varphi(x)\varphi(y)^{-1} = 1,$$

and hence

$$xy^{-1} \in \ker \varphi = \{1\},$$

i.e., $x = y$. This means that φ is injective. $\qquad\square$

Let us reword this theorem in a less condensed form. If $\varphi : A \to A'$ is a group homomorphism and the operation in A is written as multiplication, then

$$\varphi \text{ is injective} \iff \ker \varphi = \{1\}.$$

If the operation in A is written as addition, i.e., if A is an additive group, then

$$\varphi \text{ is injective} \iff \ker \varphi = \{0\}.$$

The latter therefore holds for ring, vector space, and algebra homomorphisms, since they are (in particular) additive group homomorphisms. We advise the reader to also write down the proof of Theorem 3.56 for the case where the groups are additive. A full understanding of such basic arguments is necessary in order to follow the text in subsequent sections.

The image of the homomorphism φ measures the surjectivity—the bigger it is, the closer φ is to being surjective. From Theorem 3.56 (and even more clearly from Exercise 3.59) we see that, analogously, the kernel of φ measures the injectivity— the smaller it is, the closer φ is to being injective. The most "perfect" homomorphisms are isomorphisms which have the smallest (trivial) kernel and the largest image. Their opposites, i.e., homomorphisms with the largest kernel and the smallest image, are called **trivial homomorphisms**. Trivial group homomorphisms map all elements from the group A into the identity element 1 of the group A'. In the case of homomorphisms of other types, one must replace 1 by 0. (For formal reasons, trivial ring (algebra) homomorphisms may map only to the zero ring $\{0\}$, since homomorphisms send 1 into 1 and $1 = 0$ holds only in the zero ring.)

The image of a homomorphism $\varphi : A \to A'$ is a subset of A', and the kernel is a subset of A. Theorem 3.54 describes the structure of the image. What can be said about the structure of the kernel? This question is more interesting than one might think. It is straightforward to verify that the kernel of a group homomorphism is a subgroup, the kernel of a vector space homomorphism is a subspace, and that the kernel of a ring (algebra) homomorphism has all the properties of a subring (subalgebra) except that it does not (necessarily) contain the unity element. But more can be said. However, let us wait until Sections 4.2 and 4.3.

Exercises

3.57. Prove that $A_1 \cong B_1$ and $A_2 \cong B_2$ implies $A_1 \times A_2 \cong B_1 \times B_2$. Here, A_1, A_2, B_1 and B_2 can be either groups, rings, or algebras.

3.58. Prove that $A_1 \times A_2 \cong A_2 \times A_1$. Here as well, A_1 and A_2 can be either groups, rings, or algebras.

3.59. Let $\varphi : A \to A'$ be a group homomorphism and let $a \in A$, $b \in A'$ be such that $\varphi(a) = b$. Show that the set of all solutions of the equation $\varphi(x) = b$ is equal to $a \ker \varphi := \{au \mid u \in \ker \varphi\}$.

Comment. If φ is an additive group homomorphism, in particular if φ is a ring, a vector space, or an algebra homomorphism, then we write this set as $a + \ker \varphi := \{a + u \mid u \in \ker \varphi\}$.

3.60. Let $\varphi : A \to A'$ be a group homomorphism and let $a \in A$ have finite order. Prove that the order of $\varphi(a)$ divides the order of a. Moreover, if φ is an embedding, then a and $\varphi(a)$ have the same order.

Comment. In general, the orders are not the same. After all, a can lie in ker φ. See also Exercise 3.83.

3.61. A semigroup homomorphism is defined in the same way as a group homomorphism, whereas for a monoid homomorphism we additionally require that it preserves the identity element. The set $\mathbb{Z} \times \mathbb{Z}$ is a monoid under componentwise multiplication. Find a semigroup endomorphism φ of $\mathbb{Z} \times \mathbb{Z}$ which is not a monoid endomorphism.

3.62. Let A and A' be vector spaces over $\mathbb{Q}$. Show that every additive group homomorphism $\varphi : A \to A'$ is also a vector space homomorphism. In other words, show that $\varphi(x + y) = \varphi(x) + \varphi(y)$ for all $x, y \in A$ implies $\varphi(qx) = q\varphi(x)$ for all $q \in \mathbb{Q}$ and $x \in A$.
Comment. This is a peculiarity of the field $\mathbb{Q}$. Additive maps on real or complex vector spaces may not be linear.

3.63. Show that the kernel of a nontrivial ring homomorphism does not contain invertible elements.

3.64. Determine which of the following statements are true for every nontrivial ring homomorphism $\varphi : A \to A'$:

(a) If A is commutative, then im φ is commutative.
(b) If A has no zero-divisors, then im φ has no zero-divisors.
(c) If A has zero-divisors, then im φ has zero-divisors.
(d) If A is a division ring, then im φ is a division ring.
(e) If A is a field, then im φ is a field.

Find counterexamples to false statements.

Comment. Finding counterexamples might be quite challenging at this stage, before seeing concrete examples of homomorphisms. After reading the next section this should be a much easier task. Nevertheless, give it a try! Do not worry too much if you do not succeed. Sometimes we learn more from unsuccessful attempts.

3.4 Examples of Homomorphisms

The purpose of this section is to present various examples of homomorphisms. Some examples have already been encountered in the previous sections. For instance, in Example 3.6 we saw that the exponential function can be viewed as a group isomorphism. Its inverse, the logarithm function, is therefore also a group isomorphism. Generally speaking, various important maps that appear throughout mathematics often preserve some algebraic property and are therefore homomorphisms with respect to some algebraic structure.

3.4.1   Examples of Group Homomorphisms

We begin with three examples in Abelian groups.

Example 3.65. Let G be an Abelian group. The map that assigns to each $x \in G$ its inverse x^{-1} is an automorphism of G. More generally, the map

$$x \mapsto x^m$$

is an endomorphism of G for every integer m. (If G is an additive group, we write this as $x \mapsto mx$.)

Example 3.66. The map

$$z \mapsto |z|$$

is an epimorphism from the group of nonzero complex numbers $(\mathbb{C}^*, \cdot)$ to the group of positive real numbers $(\mathbb{R}^+, \cdot)$. Its kernel is the circle group $\mathbb{T} = \{z \in \mathbb{C} \mid |z| = 1\}$.

Example 3.67. The map

$$x \mapsto e^{ix}$$

is an epimorphism from the group $(\mathbb{R}, +)$ to the group $(\mathbb{T}, \cdot)$. Its kernel is the set $\{2k\pi \mid k \in \mathbb{Z}\}$.

Example 3.68. The identity $\text{sgn}(\sigma\pi) = \text{sgn}(\sigma)\text{sgn}(\pi)$, which holds for all elements σ and π of the symmetric group S_n, shows that the map

$$\sigma \mapsto \text{sgn}(\sigma)$$

is an epimorphism from S_n to the group $(\{1, -1\}, \cdot)$. Its kernel is the alternating group A_n.

Example 3.69. The determinant of a matrix has a similar property as the sign of a permutation, namely, the determinant of the product of two matrices if equal to the product of their determinants. This means that the map

$$A \mapsto \det(A)$$

is an epimorphism from the general linear group $\text{GL}_n(F)$ to the group of all nonzero scalars $(F^*, \cdot)$. Its kernel is the special linear group $\text{SL}_n(F)$.

Example 3.70. Let G_1 and G_2 be arbitrary groups. The map

$$\pi : G_1 \times G_2 \to G_1, \quad \pi\big((x_1, x_2)\big) = x_1,$$

is an epimorphism with kernel $\{1\} \times G_2$, whereas the map

$$\iota : G_1 \to G_1 \times G_2, \quad \iota(x_1) = (x_1, 1)$$

is an embedding of G_1 into the direct product $G_1 \times G_2$. Note that $\pi \circ \iota = \text{id}_{G_1}$ and $\iota \circ \pi\big((x_1, x_2)\big) = (x_1, 1)$.

Example 3.71. Let G be an arbitrary group. For each $a \in G$, define $\varphi_a : G \to G$ by

$$\varphi_a(x) = axa^{-1}.$$

It is easy to check that φ_a is bijective. Since

$$\varphi_a(xy) = axya^{-1} = (axa^{-1})(aya^{-1}) = \varphi_a(x)\varphi_a(y),$$

φ_a is an automorphism of G. Such an automorphism is called an **inner automorphism of the group** G. If G is Abelian, then the identity id_G is the only inner automorphism of G. This notion is therefore only of interest in non-Abelian groups.

The elements x and $\varphi_a(x)$ are, of course, conjugate in G. Further, φ_a maps a subgroup H of G onto the conjugate subgroup aHa^{-1}. Any two conjugate subgroups are therefore isomorphic.

Example 3.72. Let G be a group. We write

$$\mathrm{Aut}(G) \quad \text{and} \quad \mathrm{Inn}(G)$$

for the set of all automorphisms of G and the set of all inner automorphisms of G, respectively. Recall that $\mathrm{Aut}(G)$ is a group under composition (Corollary 3.50). From

$$\varphi_a\varphi_b = \varphi_{ab} \quad \text{and} \quad \varphi_a^{-1} = \varphi_{a^{-1}}$$

we see that $\mathrm{Inn}(G)$ is a subgroup of $\mathrm{Aut}(G)$. Moreover, the equality $\varphi_a\varphi_b = \varphi_{ab}$ tells us that the map

$$a \mapsto \varphi_a$$

is an epimorphism from G to $\mathrm{Inn}(G)$. Its kernel consists of all elements $c \in G$ satisfying $\varphi_c = \mathrm{id}_G$, that is, the elements with the property that $cxc^{-1} = x$ for all $x \in G$. Multiplying from the right by c we see that these are exactly the elements of the center $Z(G)$.

Vector space homomorphisms, i.e., linear maps, are studied in detail in linear algebra. Therefore, we will not discuss them here. We proceed to ring and algebra homomorphisms.

3.4.2 Examples of Ring and Algebra Homomorphisms

Let us start where we ended our consideration of group homomorphisms.

Example 3.73. An **inner automorphism of the ring** (or **algebra**) R is defined in the same way as an inner automorphism of a group. The only difference is that we cannot take just any element $a \in R$—it has to be invertible. Then the map $\varphi_a(x) = axa^{-1}$ is indeed a ring (algebra) automorphism. Moreover, the inner automorphisms of R form a subgroup of the group of all automorphisms of R.

Example 3.74. Is the map $f : \mathbb{Z}_2 \to \mathbb{Z}$, defined by $f(0) = 0$ in $f(1) = 1$, a ring homomorphism? All of the requirements of the definition are met except one:

$$f(1 + 1) \neq f(1) + f(1).$$

Therefore, f is not even an additive group homomorphism. On the other hand, the map $\varphi : \mathbb{Z} \to \mathbb{Z}_2$ given by

$$\varphi(a) = \begin{cases} 1 ; & a \text{ is odd} \\ 0 ; & a \text{ is even} \end{cases}$$

is a ring epimorphism. More generally, so is the map

$$\varphi : \mathbb{Z} \to \mathbb{Z}_n, \quad \varphi(a) = [a],$$

where n is an arbitrary positive integer (here we used the original notation for elements of $\mathbb{Z}_n$). Note that $\ker \varphi = n\mathbb{Z}$.

Example 3.75. In Remark 3.55, we mentioned that every ring R embeds into the polynomial ring $R[X]$ by assigning the corresponding constant polynomial to every element in R. A sort of converse to this embedding is the epimorphism

$$\varphi : R[X] \to R, \quad \varphi(a_0 + a_1 X + \cdots + a_n X^n) = a_0.$$

Its kernel is the set of all polynomials with constant term 0. If R is commutative, we can describe φ as the map that assigns to each polynomial $f(X)$ the element $f(0) \in R$, i.e., the evaluation of $f(X)$ at 0. Here we can replace 0 by any element x in R and thus obtain the epimorphism

$$f(X) \mapsto f(x)$$

from $R[X]$ to R (incidentally, Exercise 2.52 asks to show that this map preserves multiplication). We call it the **evaluation homomorphism at** x. Its kernel is the set of all polynomials having x as a root.

Example 3.76. The concept from the previous example can be used in rings of functions. For example, fixing any $x \in [a, b]$, the map

$$f \mapsto f(x),$$

which assigns to every continuous function $f \in C[a, b]$ the function value at x, is an epimorphism from the ring $C[a, b]$ to the ring $\mathbb{R}$. It is actually an algebra homomorphism, not only a ring homomorphism.

Example 3.77. Let V be a vector space over a field F. By $\text{End}_F(V)$ we denote the set of all endomorphisms of V, that is, the set of all linear maps from V to V. We denote the elements of $\text{End}_F(V)$ by capital letters S, T, etc., as is common in linear algebra. As the reader presumably knows, we define addition, scalar multiplication, and multiplication in $\text{End}_F(V)$ as follows. The sum of endomorphisms $S, T \in \text{End}_F(V)$ is the map $S + T$ from V to V defined by

$$(S + T)(v) := S(v) + T(v),$$

the product λT, where $\lambda \in F$ and $T \in \text{End}_F(V)$, is the map defined by

$$(\lambda T)(v) := \lambda T(v),$$

and the product ST is the map defined by

$$(ST)(v) := S(T(v));$$

ST is thus the composition $S \circ T$. It is easy to see that $S + T$, λS, and ST are endomorphisms. Moreover, $\text{End}_F(V)$ is an algebra over F under these operations. The proof is a simple verification of the algebra axioms, which the reader should be able to supply.

Assume now that $\dim_F V = n < \infty$. We claim that, in this case, the endomorphism algebra is isomorphic to the algebra of $n \times n$ matrices over F, that is,

$$\text{End}_F(V) \cong M_n(F).$$

Let us sketch the proof. For simplicity, we restrict ourselves to the case where $n = 2$. The general case is only notationally heavier. Pick a basis $B = \{b_1, b_2\}$ of V. For every $T \in \text{End}_F(V)$, there exist unique $t_{ij} \in F$ such that

$$T(b_1) = t_{11}b_1 + t_{21}b_2 \quad \text{and} \quad T(b_2) = t_{12}b_1 + t_{22}b_2.$$

Set

$$T_B := \begin{bmatrix} t_{11} & t_{12} \\ t_{21} & t_{22} \end{bmatrix} \in M_2(F).$$

As the reader hopefully remembers from a linear algebra course, T_B is nothing but the matrix of the linear map T with respect to the basis B. By a somewhat lengthy but straightforward calculation one shows that

$$(S + T)_B = S_B + T_B, \quad (\lambda T)_B = \lambda T_B,$$
$$(\text{id}_V)_B = I, \quad (ST)_B = S_B T_B$$

for all $S, T \in \text{End}_F(V)$ and $\lambda \in F$. This means that the map $T \mapsto T_B$ is an algebra homomorphism from $\text{End}_F(V)$ to $M_2(F)$. It is easy to check that it is bijective and thus an isomorphism.

We conclude with a longer example in which we tackle a typical algebraic problem. It is a kind of detective work: we are given some objects, in our case these will be rings, which seem rather special or unusual at first glance, and our task is to "uncover" them. That is, we try to show that they are actually some well-known objects in disguise.

Example 3.78. Consider the following four subrings of the ring of 2×2 real and, in the last case, complex matrices:

$$R_1 := \left\{ \begin{bmatrix} x & 0 \\ 0 & x \end{bmatrix} \mid x \in \mathbb{R} \right\},$$

$$R_2 := \left\{ \begin{bmatrix} x & 0 \\ 0 & y \end{bmatrix} \mid x, y \in \mathbb{R} \right\},$$

$$R_3 := \left\{ \begin{bmatrix} x & y \\ -y & x \end{bmatrix} \mid x, y \in \mathbb{R} \right\},$$

$$R_4 := \left\{ \begin{bmatrix} z & w \\ -\overline{w} & \overline{z} \end{bmatrix} \mid z, w \in \mathbb{C} \right\}.$$

Our goal is to recognize them. That is, we would like to find rings that are already familiar to us and are isomorphic to the rings R_i.

1. The equalities

$$\begin{bmatrix} x & 0 \\ 0 & x \end{bmatrix} \pm \begin{bmatrix} x' & 0 \\ 0 & x' \end{bmatrix} = \begin{bmatrix} x \pm x' & 0 \\ 0 & x \pm x' \end{bmatrix} \quad \text{and} \quad \begin{bmatrix} x & 0 \\ 0 & x \end{bmatrix} \begin{bmatrix} x' & 0 \\ 0 & x' \end{bmatrix} = \begin{bmatrix} xx' & 0 \\ 0 & xx' \end{bmatrix}$$

show that R_1 is indeed a subring of $M_2(\mathbb{R})$. Moreover, they show that the operations in R_1 essentially coincide with the operations in $\mathbb{R}$. It is therefore no surprise that

$$R_1 \cong \mathbb{R}.$$

The obvious isomorphism is given by

$$\begin{bmatrix} x & 0 \\ 0 & x \end{bmatrix} \mapsto x.$$

2. Matrices in R_2 are added and multiplied by adding and multiplying diagonal entries. The operations in R_2 thus essentially coincide with the operations in the direct product of two copies of $\mathbb{R}$. Therefore,

$$R_2 \cong \mathbb{R} \times \mathbb{R}$$

with the isomorphism given by

$$\begin{bmatrix} x & 0 \\ 0 & y \end{bmatrix} \mapsto (x, y).$$

3. The addition in R_3 is just as simple as in R_1 and R_2. The multiplication is more intriguing:

$$\begin{bmatrix} x & y \\ -y & x \end{bmatrix} \begin{bmatrix} x' & y' \\ -y' & x' \end{bmatrix} = \begin{bmatrix} xx' - yy' & xy' + x'y \\ -(xy' + x'y) & xx' - yy' \end{bmatrix}.$$

First of all, it should be noted that the result of the multiplication lies in R_3, implying that R_3 is a subring of $M_2(\mathbb{R})$. Now, does this result remind us of something familiar? The multiplication of complex numbers is defined by a similar formula:

$$(x + yi)(x' + y'i) = (xx' - yy') + (xy' + x'y)i.$$

Hence,

$$R_3 \cong \mathbb{C}$$

via the isomorphism

$$\begin{bmatrix} x & y \\ -y & x \end{bmatrix} \mapsto x + yi.$$

4. Checking that R_4 is a subring of $M_2(\mathbb{C})$ is an easy computational exercise. However, unlike in the previous cases, it does not seem to help us much in finding a ring that we know and is isomorphic to R_4. What are the properties of matrices in R_4? Observe that a nonzero matrix

$$\begin{bmatrix} z & w \\ -\overline{w} & \overline{z} \end{bmatrix} \in R_4$$

has a nonzero determinant $z\overline{z} + w\overline{w}$, and is therefore invertible in $M_2(\mathbb{C})$. Its inverse actually lies in R_4, namely

$$\begin{bmatrix} z & w \\ -\overline{w} & \overline{z} \end{bmatrix}^{-1} = \frac{1}{z\overline{z} + w\overline{w}} \begin{bmatrix} \overline{z} & -w \\ \overline{w} & z \end{bmatrix} \in R_4.$$

This means that R_4 is a division ring! Therefore, it should not be surprising that R_4 is isomorphic to the division ring of quaternions:

$$R_4 \cong \mathbb{H}.$$

An isomorphism is given by

$$\begin{bmatrix} z & w \\ -\overline{w} & \overline{z} \end{bmatrix} \mapsto x + yi + uj + vk,$$

where $z = x + yi$ and $w = u + vi$. Verifying this is a bit tedious but straightforward. However, the readers should ask themselves how to find this or some other isomorphism (hint: what are the properties of matrices that correspond to the quaternions i, j, and k?).

The four rings from this example are also real algebras, and all four ring isomorphisms are actually algebra isomorphisms. The algebra of matrices R_4 is thus isomorphic to the algebra of quaternions $\mathbb{H}$. In Section 2.8, we wrote that $\mathbb{R}$, $\mathbb{C}$, and $\mathbb{H}$ are the only finite-dimensional real division algebras. Obviously, this was not entirely accurate. In precise terms, every finite-dimensional real division algebra is *isomorphic* to $\mathbb{R}$, $\mathbb{C}$, or $\mathbb{H}$. However, in algebra we often neglect the difference between "isomorphic" and "equal". This may be distracting for a novice. It often happens that we view isomorphic objects as different in one sentence, and as equal in the next one. Eventually one gets used to this, and after some time one does not even notice.

Exercises

3.79. Suppose that group (ring, vector space, algebra) homomorphisms $\varphi, \psi : A \to A'$ coincide on a set of generators X of the group (ring, vector space, algebra) A. Show that then $\varphi = \psi$.

Comment. Homomorphisms are thus uniquely determined on *any* set of generators. Several of the following exercises consider whether there exists a homomorphism having prescribed values on *some* set of generators.

3.80. Let G be an arbitrary group. Show that for every $a \in G$, there exists a group homomorphism $\varphi : \mathbb{Z} \to G$ such that $\varphi(1) = a$.

3.81. Describe all endomorphisms of the additive group $\mathbb{Z}$. Which among them are automorphisms? Show that the group $\mathrm{Aut}(\mathbb{Z})$ is isomorphic to the group $\mathbb{Z}_2$.

3.82. Describe all endomorphisms of the ring $\mathbb{Z}$.

3.83. Let G be an arbitrary group and let $a \in G$. Prove that there exists a homomorphism $\varphi : \mathbb{Z}_n \to G$ such that $\varphi(1) = a$ if and only if $a^n = 1$.

Comment. Here, n is not necessarily equal to the order of a, it may be any of its multiples. The order of $\varphi(1)$ can therefore be smaller than the order of 1.

3.84. Describe all automorphisms of the additive group $\mathbb{Z}_n$ and show that the group $\mathrm{Aut}(\mathbb{Z}_n)$ is isomorphic to $\mathbb{Z}_n^*$, the group of all invertible elements of the ring $\mathbb{Z}_n$.

Hint. An endomorphism of $\mathbb{Z}_n$ is uniquely determined by its value at the generator 1. An automorphism sends a generator to a generator.

3.85. Let G be a group with trivial center. Show that the group $\mathrm{Aut}(G)$ also has trivial center.

3.86. Let G be a nontrivial group. Find three different embeddings of G into the group $G \times G$.

Hint. Example 3.70 indicates two of them.

3.87. Find all homomorphisms from the group $G = \mathbb{Z}_3$ to the group $\mathbb{C}^*$. Consider the same question for $G = \mathbb{Z}_2 \oplus \mathbb{Z}_2$.

3.88. Find an embedding of the group $G = \mathbb{Z}_3$ into the group $\mathrm{GL}_2(\mathbb{C})$. Consider the same question for $G = \mathbb{Z}_2 \oplus \mathbb{Z}_2$, $G = D_8$, and $G = Q$.

Hint. Exercise 3.30, Example 3.78.

Comment. A homomorphism from a group G to the group $\mathrm{GL}_n(F)$ is called a **representation of G** on an n-dimensional vector space over F. Group representations are important not only in mathematics, but also in theoretical physics and other fields.

3.89. Let R be an arbitrary ring. Show that for every $a \in R$, there exists a ring homomorphism $\varphi : \mathbb{Z}[X] \to R$ such that $\varphi(X) = a$.

3.90. Does there exist an endomorphism φ of the ring $\mathbb{Z}[X]$ such that $\varphi(X^2) = X^3$?

3.91. Describe all automorphisms of the ring $\mathbb{Z}[X]$.

3.92. Let A be an arbitrary algebra over a field F. Show that for every $a \in A$, there exists an algebra homomorphism $\varphi : F[X] \to A$ such that $\varphi(X) = a$.

3.93. Describe all automorphisms of the algebra $F[X]$.

3.94. Let A be an arbitrary real algebra and let $a, b \in A$. Prove that there exists an algebra homomorphism $\varphi : \mathbb{H} \to A$ such that $\varphi(i) = a$ and $\varphi(j) = b$ if and only if $a^2 = b^2 = -1$ and $ab = -ba$.

3.95. Explain why an endomorphism of the algebra $M_2(\mathbb{R})$ cannot send the matrix E_{11} into the matrix E_{12}. Find an endomorphism that sends E_{11} into E_{22}.

3.96. Let V and V' be vector spaces over a field F. Prove that for an arbitrary basis $\{b_i \,|\, i \in I\}$ of V and an arbitrary set of vectors $\{a_i \,|\, i \in I\}$ in V', there exists a vector space homomorphism $\varphi : V \to V'$ such that $\varphi(b_i) = a_i$ for all $i \in I$.

Comment. Note the similarity with Exercises 3.80, 3.89, and 3.92. They all ask to prove the existence of a homomorphism having prescribed values on a certain set of generators. However, the vector space V in this exercise is arbitrary, whereas Exercises 3.80, 3.89, and 3.92 deal with a very special group, ring, and algebra, respectively, which is, in particular, generated by a single element. There actually exist groups, rings, and algebras with larger sets of generators that have the same property. The interested reader may look at Subsection 4.4.4 where some of them will be touched upon (although from a different perspective). However, they are all quite special examples of groups, rings, and algebras. Thus we see again that vector spaces are different, or we could say simpler, than other algebraic structures.

One more comment. Some of the readers may feel uncertain about working with infinite bases. They should be advised to first consider the case where the set I is finite and then ask themselves what changes are to be made to cover the general case. Dealing with infinite bases is not as hard as it may seem. No matter whether a basis is finite or infinite, every vector is a finite linear combination of some basis vectors.

3.97. Show that the real algebra $\mathbb{C}$ has only two automorphisms: the identity map $\mathrm{id}_{\mathbb{C}}$ and complex conjugation $z \mapsto \bar{z}$.

3.98. By identifying the quaternions of the form $\lambda_0 + \lambda_1 i$ with complex numbers, we can consider $\mathbb{C}$ as a subalgebra of the real algebra $\mathbb{H}$. Find an extension of the automorphism $z \mapsto \bar{z}$ of $\mathbb{C}$ to an automorphism of $\mathbb{H}$. Is this automorphism inner?

Comment. The conjugation of quaternions $h \mapsto \bar{h}$ is not a right answer. This map is linear and bijective, but reverses the order of multiplication, i.e., $\overline{hh'} = \bar{h'} \cdot \bar{h}$. A map with such properties is called an antiautomorphism. The most well-known example is transposition of matrices in the algebra $M_n(F)$.

3.99. Find a subring of the ring $M_2(\mathbb{Z})$ which is isomorphic to the ring of Gaussian integers $\mathbb{Z}[i]$.

Hint. Example 3.78.

3.100. Find a nonzero map $\delta : \mathbb{R}[X] \to \mathbb{R}[X]$ such that

$$f(X) \mapsto \begin{bmatrix} f(X) & \delta(f(X)) \\ 0 & f(X) \end{bmatrix}$$

defines an embedding of the algebra $\mathbb{R}[X]$ into the algebra $M_2(\mathbb{R}[X])$.

3.101. Prove that the algebra $\mathbb{H}_\mathbb{C}$ from Exercise 2.124 is isomorphic to the algebra $M_2(\mathbb{C})$.

3.102. Let A be the set of all matrices of the form $\begin{bmatrix} a & b & d \\ 0 & a & c \\ 0 & 0 & a \end{bmatrix}$, and let A' be the set of all matrices of the form $\begin{bmatrix} a & 0 & 0 \\ b & a & 0 \\ d & c & a \end{bmatrix}$, where $a, b, c, d \in \mathbb{R}$. Verify that A and A' are 4-dimensional real algebras under the usual matrix operations. Prove that A and A' are isomorphic, but are not isomorphic to the algebra $M_2(\mathbb{R})$.

3.5 Cayley's Theorem and Other Embedding Theorems

The message of the first theorem below might surprise the reader: the "only" groups are symmetric groups and their subgroups. Similar theorems will be proved for rings and algebras, with the ring (resp. algebra) of endomorphisms of an additive group (resp. vector space) playing the role of the symmetric group.

To be honest, these theorems are not as applicable to concrete problems as one might first think. However, they give an important insight into the understanding of algebraic structures.

3.5.1 Embedding a Group into a Symmetric Group

Recall that a symmetric group is the group of all permutations of a set. The following theorem was established, in some form, by *Arthur Cayley* in 1854.

Theorem 3.103. (**Cayley's Theorem**) *Every group G can be embedded into a symmetric group.*

Proof. For every $a \in G$, define the map

$$\ell_a : G \to G, \quad \ell_a(x) = ax.$$

If $\ell_a(x) = \ell_a(y)$, then $ax = ay$ and hence $x = y$. This means that ℓ_a is injective. Since $\ell_a(a^{-1}x) = x$ for every $x \in G$, ℓ_a is also surjective. Therefore, ℓ_a lies in $\mathrm{Sym}(G)$, the symmetric group of the set G.

Now, consider the map

$$\varphi : G \to \mathrm{Sym}(G), \quad \varphi(a) = \ell_a.$$

Let us show that φ is a group homomorphism, i.e., that $\varphi(ab) = \ell_{ab}$ is equal to $\varphi(a) \circ \varphi(b) = \ell_a \circ \ell_b$. This follows from the associative law for multiplication in G. Indeed,

$$\ell_{ab}(x) = (ab)x = a(bx) = \ell_a(\ell_b(x)) = (\ell_a \circ \ell_b)(x)$$

for each $x \in G$. If $a \in \ker \varphi$, then $\ell_a = \mathrm{id}_G$ and hence $a = \ell_a(1) = \mathrm{id}_G(1) = 1$. The kernel of φ is thus trivial, so φ is an embedding. □

Subgroups of symmetric groups are called **permutation groups**. Cayley's Theorem thus tells us that every group is isomorphic to a permutation group. This is illuminating, but does not essentially affect the way we study abstract groups. The meaning of the theorem can be better understood from a historical perspective: the first definition of a group, given by Galois around 1830, essentially coincides with what is now called a (finite) permutation group, and it was Cayley who defined (finite) abstract groups (although not in entirely the same way as we do now) and showed that they are isomorphic to permutation groups. The development was thus natural, from concrete to abstract, and not vice versa as in our exposition.

The proof we gave actually shows that every group G can be embedded into the symmetric group $\mathrm{Sym}(G)$. We have deliberately stated the theorem loosely, saying only that G can be embedded into $\mathrm{Sym}(X)$ for some set X. Indeed, sometimes there are more convenient choices for X than G. Still, knowing that we may always choose $X = G$ is useful. In particular, if G is finite, then $\mathrm{Sym}(G)$ is the symmetric group of degree $n := |G|$. Hence, we have the following corollary.

Corollary 3.104. *Every finite group G can be embedded into the symmetric group S_n for some $n \in \mathbb{N}$.*

It turns out that general homomorphisms, not necessarily embeddings, from arbitrary groups to symmetric groups are also important. However, usually they appear under a different name, along with a different (and simpler) notation.

Definition 3.105. A group G is said to **act on a set** X if there exists a map $(a, x) \mapsto a \cdot x$ from $G \times X$ to X which satisfies the following two conditions:

(a) $(ab) \cdot x = a \cdot (b \cdot x)$ for all $a, b \in G$, $x \in X$.
(b) $1 \cdot x = x$ for all $x \in X$.

We call this map a **group action** of G on X.

The notion of a group action is equivalent to the notion of a homomorphism from a group to a symmetric group, in the following sense: if φ is a homomorphism from a group G to a symmetric group $\mathrm{Sym}(X)$, then $a \cdot x := \varphi(a)(x)$ defines an action of

G on X. Conversely, given an action of G on X, we can define the homomorphism $\varphi : G \to \text{Sym}(X)$ as follows: $\varphi(a)$ is the permutation of X given by $\varphi(a)(x) := a \cdot x$. Checking both statements should be an easy exercise for the reader. Let us only mention that in order to prove that $\varphi(a)$ is a permutation, it is helpful to put a^{-1} for b in (a).

Every group G acts on itself by **left multiplication**: $a \cdot x = ax$. This is the action used in the proof of Cayley's Theorem. We could list many other examples, but let us put this topic aside for now. Group actions together with their applications will be considered in Section 6.3.

3.5.2 Embedding a Ring into an Endomorphism Ring

Let M be an additive (and hence, by our convention, Abelian) group. Denote by $\text{End}(M)$ the set of all of its endomorphisms, i.e., the set of all additive maps from M to M. We define the sum $f + g$ and the product fg of endomorphisms $f, g \in \text{End}(M)$ just as in Example 3.77, i.e.,

$$(f + g)(v) := f(v) + g(v),$$

$$(fg)(v) := f(g(v)).$$

By routine verification, we see that $\text{End}(M)$ is a ring under these operations. The following theorem shows that its role in ring theory is similar to the role of the symmetric group in group theory.

Theorem 3.106. *Every ring R can be embedded into the ring of endomorphisms of an additive group.*

Proof. The proof is similar to that of Cayley's Theorem. Given any $a \in R$, we define

$$\ell_a : R \to R, \quad \ell_a(x) = ax.$$

It is clear that ℓ_a is an additive map. That is, ℓ_a belongs to $\text{End}(R)$, the ring of endomorphisms of the additive group of R. As in the proof of Cayley's Theorem, we see that $\ell_{ab} = \ell_a \circ \ell_b$. One also immediately checks that $\ell_{a+b} = \ell_a + \ell_b$ and $\ell_1 = \text{id}_R$. The map

$$\varphi : R \to \text{End}(R), \quad \varphi(a) = \ell_a,$$

is thus a ring homomorphism. If $a \in R$ is such that $\ell_a = 0$, then $a = \ell_a(1) = 0$. This means that φ is an embedding. $\square$

When treating the ring $\text{End}(R)$, we "forget" that R is a ring and consider it only as an additive group. Similarly, when treating the group $\text{Sym}(G)$, we consider G only as a set.

Following a similar path that led us to the notion of a group action, we arrive at the notion of a module M over a ring R.

Definition 3.107. An additive group M together with an external binary operation $(a, v) \mapsto av$, called **module multiplication**, from $R \times M$ to M is called a **module over the ring** R if the following properties are satisfied:

(a) $a(u + v) = au + av$ for all $a \in R$ and $u, v \in M$.
(b) $(a + b)v = av + bv$ for all $a, b \in R$ and $v \in M$.
(c) $(ab)v = a(bv)$ for all $a, b \in R$ and $v \in M$.
(d) $1v = v$ for all $v \in M$.

The notion of a module M over R is equivalent to the notion of a homomorphism from a ring to the endomorphism ring of an additive group. Indeed, if $\varphi : R \to \text{End}(M)$ is a homomorphism, then M becomes a module over R under the operation $av := \varphi(a)(v)$. Conversely, if M is a module over R, then we can define a homomorphism $\varphi : R \to \text{End}(M)$ by $\varphi(a)(v) := av$. Proving these two facts is a good exercise in understanding the definitions.

We have actually defined a **left module** over R. A **right module** is defined analogously, only the elements from a ring are written on the right-hand side; that is, we write va instead of av and the axioms (a)–(d) are changed accordingly (in particular, (c) becomes $v(ab) = (va)b$). Since we will consider only left modules, we have taken the liberty to omit the adjective "left" in the definition. A similar remark applies to groups actions; we have defined **left group actions**, but there are also analogously defined **right group actions**.

The module axioms are identical to the vector space axioms, except that scalars (i.e., elements of a field) are replaced by ring elements. Thus, a module over a field F is nothing but a vector space over F. Another example that immediately presents itself is that every ring R is a module over itself if we interpret multiplication in R as the module multiplication (as we indirectly did in the proof of Theorem 3.106). Finally, the discussion in Remark 1.94 shows that every additive group can be viewed as a module over the ring $\mathbb{Z}$.

Let that be all for now. We will return to modules in Section 5.4.

3.5.3 Embedding an Algebra into an Endomorphism Algebra

Let A be an algebra over a field F. It is rather obvious that one can modify the proof of Theorem 3.106 so that the role of the ring of endomorphisms of an additive group is played by the algebra of endomorphisms of a vector space over F. We thus state the following theorem without further explanation.

Theorem 3.108. *Every algebra A can be embedded into the algebra of endomorphisms of a vector space.*

A (left) **module over an algebra** A is defined as a vector space M that, besides conditions (a)–(d), also satisfies

(e) $\lambda(av) = (\lambda a)v = a(\lambda v)$ for all $\lambda \in F$, $a \in A$, and $v \in M$.

The notion of a module over an algebra is equivalent to the notion of a homomorphism from an algebra to an algebra of endomorphisms of a vector space.

In Theorem 3.108, we may take A for the vector space. If A is finite-dimensional, we can therefore embed A into the algebra of endomorphisms of a finite-dimensional vector space. This algebra, on the other hand, is isomorphic to a matrix algebra $M_n(F)$ (see Example 3.77). We thus have the following corollary.

Corollary 3.109. *Every finite-dimensional algebra A over a field F can be embedded into the matrix algebra $M_n(F)$ for some $n \in \mathbb{N}$.*

An informal résumé of our discussion in this section is that we can always imagine that group elements are permutations and that ring (resp. algebra) elements are endomorphisms of an additive group (resp. vector space). Maybe not too often, but sometimes this is helpful. Let us illustrate this with an example.

Example 3.110. Let A be a finite-dimensional algebra over, say, the field of real numbers. Can A contain elements s and t such that

$$st - ts = 1?$$

Since A is arbitrary, our first thought may be that we do not know where to start attacking this question. However, using Corollary 3.109, it can be transformed into the following more concrete form: do there exist matrices $S, T \in M_n(\mathbb{R})$ such that

$$ST - TS = I?$$

This no longer seems out of reach since we know how to calculate with matrices. But do not just rush into computations. Using the notion of the **trace** of a matrix, the solution is at hand. Recall that the trace of $(a_{ij}) \in M_n(\mathbb{R})$ is defined to be the number $\sum_{i=1}^{n} a_{ii}$. It is well-known (and easy to see) that the matrices ST and TS have equal trace. Consequently, the trace of the matrix $ST - TS$ is 0. As the identity matrix $I \in M_n(\mathbb{R})$ has trace n, it follows that $ST - TS$ cannot be equal to I. The answer to our initial question is therefore negative: $st - ts \neq 1$ for all $s, t \in A$.

We did not consider vector spaces in this section. Their structure is already simple, so there is no need to embed them in some other vector spaces. Every nonzero finite-dimensional vector space over a field F is isomorphic to F^n (Corollary 3.36), and there is nothing to add here.

Exercises

3.111. Every element in the symmetric group S_n can be written as a product of elements a in S_n satisfying $a^2 = 1$. This follows from the fact that every permutation is a product of transpositions. Since every finite group G can be embedded into S_n for some $n \in \mathbb{N}$, it follows that we can write every element in G as a product of elements a in G satisfying $a^2 = 1$. Is this true? It cannot be, since, for example, a cyclic group of odd order does not contain elements $a \neq 1$ such that $a^2 = 1$ (why not?). Where did we make a mistake?

3.112. Test your understanding of the proof of Cayley's Theorem and find a subgroup of the symmetric group S_4 isomorphic to the group $\mathbb{Z}_4$. Consider the same question for $\mathbb{Z}_2 \oplus \mathbb{Z}_2$.

3.113. Test your understanding of the proof of Theorem 3.106 and find a subring of the ring $\text{End}(\mathbb{Z})$ (i.e., the ring of endomorphisms of the additive group $\mathbb{Z}$) isomorphic to the ring $\mathbb{Z}$.

3.114. Test your understanding of the proof of Corollary 3.109 and find a subalgebra of the real algebra $M_2(\mathbb{R})$ isomorphic to the algebra of complex numbers $\mathbb{C}$.

3.115. Test your understanding of the proof of Corollary 3.109 and find a subalgebra of the real algebra $M_4(\mathbb{R})$ isomorphic to the algebra of quaternions $\mathbb{H}$.

3.116. Let V be the vector space of all polynomials with real coefficients, and let $T \in \text{End}_{\mathbb{R}}(V)$ be defined by $T(f(X)) = Xf(X)$. Find $S \in \text{End}_{\mathbb{R}}(V)$ such that $ST - TS = I$.

Comment. The finite-dimensionality assumption in Example 3.110 is thus necessary.

3.117. Let s be an invertible element of a finite-dimensional real algebra A. Prove that $sts^{-1} \neq t + 1$ for every $t \in A$.

3.118. The preceding exercise states that if φ is an inner automorphism of a finite-dimensional real algebra A, then $\varphi(t) \neq t + 1$ for every $t \in A$. Prove that this holds for all automorphisms φ of A, not only for the inner ones.

Hint. Apply the preceding exercise with $\text{End}_{\mathbb{R}}(A)$ playing the role of A, φ playing the role of s, and ℓ_t playing the role of t.

3.6 Embedding an Integral Domain into a Field

Having only two invertible elements (1 and -1), the ring of integers $\mathbb{Z}$ seems to be far from being a field. On the other hand, one can say that all nonzero elements in $\mathbb{Z}$ are invertible, it is just that their inverses do not necessarily lie in $\mathbb{Z}$ but in a larger ring $\mathbb{Q}$. From this perspective, $\mathbb{Z}$ is close to a field. More specifically, we can "enlarge" it to a field. For which rings R can this be done? Or, to pose the question more precisely, when can R be embedded into a field? There are two obvious restrictions: R should be commutative and should be without zero-divisors (see Theorem 1.77). In short, R should be an integral domain. We will show that this condition is not only necessary, but also sufficient for the existence of an embedding of R into a field.

Constructing this field is the main theme of the section. The construction is essentially no harder than the construction of the field of rational numbers. But how exactly do we construct $(\mathbb{Q}, +, \cdot)$? We are so familiar with rational numbers that, perhaps, we have never felt the need to think about the details concerning their formal definition. It is clear that this definition must take into account that different fractions, like $\frac{1}{2}$ and $\frac{-2}{-4}$, may represent the same rational number. This kind of problem also occurs in the abstract situation which we are about to consider.

Throughout the section, R denotes an integral domain. Our goal is to construct a field F_R into which R can be embedded, in such a way that $F_R = \mathbb{Q}$ if $R = \mathbb{Z}$.

Lemma 3.119. *The following rule*

$$(a, b) \sim (a', b') \iff ab' = a'b$$

defines an equivalence relation on the set $R \times (R \setminus \{0\})$.

Proof. The reflexivity and symmetry are obvious. To verify the transitivity, assume that $(a, b) \sim (a', b')$ and $(a', b') \sim (a'', b'')$, i.e., $ab' = a'b$ and $a'b'' = a''b'$. Multiplying the first equality by b'' and the second equality by b we get $ab'b'' = a'bb''$ and $a'bb'' = a''bb'$. Comparing the new equalities gives $ab'b'' = a''bb'$. We can rewrite this as $(ab'' - a''b)b' = 0$. Since $b' \in R \setminus \{0\}$ and, by assumption, R has no zero-divisors, it follows that $ab'' - a''b = 0$. That is, $(a, b) \sim (a'', b'')$. $\square$

For any $a \in R$ and $b \in R \setminus \{0\}$, we set

$$\frac{a}{b} := \text{the equivalence class of the element } (a, b).$$

Of course, the equivalence classes $\frac{a}{b}$ and $\frac{a'}{b'}$ are equal if and only if $(a, b) \sim (a', b')$. That is,

$$\frac{a}{b} = \frac{a'}{b'} \iff ab' = a'b.$$

Lemma 3.120. *Suppose that $a, a', c, c' \in R$ and $b, b', d, d' \in R \setminus \{0\}$ are such that*

$$\frac{a}{b} = \frac{a'}{b'} \quad \text{and} \quad \frac{c}{d} = \frac{c'}{d'}.$$

Then

Then
$$\frac{ad + bc}{bd} = \frac{a'd' + b'c'}{b'd'} \quad \text{and} \quad \frac{ac}{bd} = \frac{a'c'}{b'd'}.$$

Proof. Since R has no zero-divisors, both bd and $b'd'$ are different from 0. All expressions appearing in the statement of the lemma thus make sense. We have to show that the conditions

$$ab' = a'b \quad \text{and} \quad cd' = c'd$$

imply that

$$(ad + bc)b'd' = (a'd' + b'c')bd \quad \text{and} \quad (ac)(b'd') = (a'c')(bd).$$

Rearranging the terms, we see that both equalities indeed hold. □

The reader has probably guessed the meaning of the expressions in the lemma. Let us come to the point.

Theorem 3.121. *Let R be an integral domain. Endowing the set of all equivalence classes*

$$F_R := \left\{ \frac{a}{b} \mid a, b \in R, b \neq 0 \right\}$$

with addition and multiplication given by

$$\frac{a}{b} + \frac{c}{d} := \frac{ad + bc}{bd} \quad and \quad \frac{a}{b} \cdot \frac{c}{d} := \frac{ac}{bd},$$

F_R becomes a field. The map

$$\varphi : R \to F_R, \quad \varphi(a) = \frac{a}{1},$$

is an embedding of the integral domain R into the field F_R.

Proof. Lemma 3.120 tells us that addition and multiplication are well-defined. It is a matter of direct computation to check the associativity and commutativity of both operations, as well as the distributive law. The zero element is $\frac{0}{1}$, and the unity is $\frac{1}{1}$. The additive inverse of $\frac{a}{b}$ is $\frac{-a}{b}$. Note that $\frac{a}{b}$ is different from $0 = \frac{0}{1}$ if and only if $a \neq 0$, and that $\left(\frac{a}{b}\right)^{-1} = \frac{b}{a}$ holds in this case. All this shows that F_R is a field. From

$$\frac{a+b}{1} = \frac{a}{1} + \frac{b}{1} \quad and \quad \frac{ab}{1} = \frac{a}{1} \cdot \frac{b}{1}$$

it follows that φ is a homomorphism. It is clear that $\varphi(a) = 0$ implies $a = 0$. Hence, φ is an embedding. □

Definition 3.122. The field F_R is called the **field of fractions** of the integral domain R.

One usually writes a instead of $\frac{a}{1}$ and, accordingly, considers R as a subring of F_R. If F_0 is a subfield of F_R that contains R, then F_0 also contains the inverse $\frac{1}{b}$ of every nonzero element $b \in R$, from which it clearly follows that $F_0 = F_R$. The field of fractions F_R of R is thus generated by R. Putting this in less formal terms, F_R is the smallest field containing R.

Let us look at a few special cases.

Example 3.123. If R is itself a field, then $F_R = R$.

Example 3.124. The field $\mathbb{Q}$ is the field of fractions of the ring $\mathbb{Z}$.

Example 3.125. The polynomial ring $F[X]$ is an integral domain for every field F (see Theorem 2.46). Its field of fractions is called the **field of rational functions** in X and is denoted by $F(X)$. Every $q(X) \in F(X)$ can be written as $q(X) = \frac{f(X)}{g(X)}$

for some $f(X), g(X) \in F[X]$. More generally, the field of fractions of the ring of polynomials in several variables $F[X_1, \ldots, X_n]$ is called the field of rational functions in $X_1, \ldots, X_n$ and is denoted by $F(X_1, \ldots, X_n)$.

A rough summary of this section is that the notion of an integral domain is equivalent to the notion of a subring of a field. More precisely, the following holds.

Corollary 3.126. *Every integral domain can be embedded into a field.*

Let us remark that the commutativity of R was heavily used in our arguments. Similar constructions for noncommutative rings are much more demanding and work under more severe restrictions. In particular, it turns out that there are noncommutative rings without zero-divisors that cannot be embedded into division rings.

Exercises

3.127. Let $f(X), g(X) \in \mathbb{R}[X]$ be non-constant polynomials. Suppose that

$$f(X) + \frac{1}{f(X)} = g(X) + \frac{1}{g(X)}$$

holds in $\mathbb{R}(X)$. Prove that $f(X) = g(X)$.

3.128. Let $D : \mathbb{R}(X) \to \mathbb{R}(X)$ be an additive map that sends constant polynomials into 0 and satisfies

$$D\big(q(X)r(X)\big) = D\big(q(X)\big)r(X) + q(X)D\big(r(X)\big)$$

for all $q(X), r(X) \in \mathbb{R}(X)$. Prove that

$$D\left(\frac{f(X)}{g(X)}\right) = u(X)\frac{f'(X)g(X) - f(X)g'(X)}{g(X)^2}$$

for all $f(X), g(X) \in \mathbb{R}[X]$, where $u(X) = D(X)$ and $f'(X)$ is the derivative of $f(X)$ (which is, of course, defined as follows: if $f(X) = \sum_{k=0}^{n} a_k X^k$ then $f'(X) = \sum_{k=1}^{n} k a_k X^{k-1}$).

Hint. First compute $D(f(X))$.

3.129. Show that an automorphism of an integral domain R can be extended to an automorphism of its field of fractions F_R.

Comment. In this and the next exercise, do not forget to check that the maps are well-defined.

3.130. Let R be a subring of a field K. Prove that the subfield of K generated by R is isomorphic to the field of fractions F_R of R.

Comment. A nice example of an integral domain is the ring of Gaussian integers $\mathbb{Z}[i]$. Its field of fractions is thus isomorphic to the field $\mathbb{Q}(i) = \{p + qi \mid p, q \in \mathbb{Q}\}$ (see Example 1.149).

3.131. If F is a field, then the ring of formal power series $F[[X]]$ is an integral domain (why?). Prove that the field of fractions of $F[[X]]$ is isomorphic to the field of formal Laurent series $F((X))$ (see Exercise 2.129).

3.7 The Characteristic of a Ring and Prime Subfields

One of the goals of this section is to show that a certain concrete field sits inside an abstract field. We will begin somewhat informally. The next paragraph should be read with some caution.

Let F be a field. Since F contains the unity 1, it also contains the elements

$$2 \cdot 1 = 1 + 1, \quad 3 \cdot 1 = 1 + 1 + 1, \quad 4 \cdot 1 = 1 + 1 + 1 + 1, \text{ etc.}$$

For simplicity, let us write these elements as 2, 3, 4, etc. Of course, F also contains their additive inverses and the zero element. We have thus found a copy of the integers inside F. But F is a field, so it also contains elements of the form mn^{-1} where m and n are integers with $n \neq 0$. This means that F contains a copy of the rational numbers, or, in more precise terms, F has a subfield isomorphic to $\mathbb{Q}$.

Is the last statement really true? Unfortunately, not always. Recall that the ring $\mathbb{Z}_p$ is a field for any prime p (see Theorem 2.24), and it is obvious that a field isomorphic to $\mathbb{Q}$ cannot be contained in $\mathbb{Z}_p$ or in any other finite field. In $\mathbb{Z}_p$, $p \cdot 1 = 0$ and so $\mathbb{Z}_p$ does not contain a copy of $\mathbb{Z}$. We now see where the mistake was made. But we can remedy the situation by adding the assumption that $n \cdot 1 \neq 0$ for every $n \in \mathbb{N}$. The arguments in the preceding paragraph are then indeed loose, but essentially correct.

We need the following definition for further discussion.

Definition 3.132. We say that the **characteristic** of a ring R is 0 if $n \cdot 1 \neq 0$ for every $n \in \mathbb{N}$. Otherwise, the characteristic of R is the smallest natural number n such that $n \cdot 1 = 0$.

We can phrase the definition differently. A ring is a group under addition and its characteristic is equal to the order of 1 (in the sense of Definition 3.10), provided it is finite. If 1 has infinite order, then the characteristic is 0.

Example 3.133. The most basic examples of rings, such as $\mathbb{Z}$, $\mathbb{Q}$, $\mathbb{R}$, and $\mathbb{C}$, have characteristic 0.

Note that the characteristic of a finite ring cannot be 0.

Example 3.134. The ring $\mathbb{Z}_n$ has characteristic n.

Example 3.135. If a ring R has characteristic n, so do the rings $R \times R$, $M_k(R)$, $R[X]$, and many other rings constructed from R. The characteristic of an infinite ring therefore is not always 0. For example, the field of rational functions $\mathbb{Z}_p(X)$ is an infinite field with characteristic p.

In the next theorem, we gather together some basic facts concerning nonzero characteristic.

Theorem 3.136. *Let R be a ring of characteristic $n > 0$. Then:*

(a) $n \cdot x = 0$ *for all $x \in R$.*
(b) *For every $m \in \mathbb{Z}$, $m \cdot 1 = 0$ if and only if $n \mid m$.*
(c) *If $R \neq \{0\}$ and has no zero-divisors, then n is a prime number.*

Proof. (a) We have

$$n \cdot x = x + \cdots + x = (1 + \cdots + 1)x = (n \cdot 1)x = 0.$$

(b) By the division algorithm, we can write $m = qn + r$ with $0 \leq r < n$. This implies that $m \cdot 1 = r \cdot 1$. Since n is the smallest natural number satisfying $n \cdot 1 = 0$, $r \cdot 1$ can be 0 only when $r = 0$. Hence, $m \cdot 1 = 0$ if and only if $n \mid m$.

(c) Suppose that $n = k\ell$ for some $k, \ell \in \mathbb{N}$. Then

$$(k \cdot 1)(\ell \cdot 1) = n \cdot 1 = 0.$$

By our assumption, one of the elements $k \cdot 1$ and $\ell \cdot 1$ must be 0. Since $k \leq n$ and $\ell \leq n$, this is possible only if $k = n$ or $\ell = n$. Thus, n is a prime. $\qquad\square$

We are now in a position to clarify the statements from the beginning of the section. Take an arbitrary field F. The subfield F_0 of F generated by the unity 1 is called the **prime subfield** of F. Since every subfield of F contains 1, it also contains F_0. Thus, F_0 is the smallest subfield of F. Note that F and F_0 have the same characteristic, which can only be 0 or a prime by Theorem 3.136 (c).

Theorem 3.137. *Let F be a field and let F_0 be its prime subfield.*

(a) *If the characteristic of F is 0, then $F_0 \cong \mathbb{Q}$.*
(b) *If the characteristic of F is a prime number p, then $F_0 \cong \mathbb{Z}_p$.*

Proof. (a) The proof for this case was outlined in the second paragraph of the section. We only have to formalize it. Since F has characteristic 0, $n \cdot 1$ is a nonzero element in F_0 for every nonzero integer n. Define

$$\varphi : \mathbb{Q} \to F_0, \quad \varphi\left(\frac{m}{n}\right) = (m \cdot 1)(n \cdot 1)^{-1}$$

for all $m, n \in \mathbb{Z}$, $n \neq 0$. We have to verify that φ is well-defined. Assuming $\frac{m}{n} = \frac{m'}{n'}$ it follows that $mn' = nm'$, and hence

$$(m \cdot 1)(n' \cdot 1) = (n \cdot 1)(m' \cdot 1).$$

Multiplying by the inverses of $n \cdot 1$ and $n' \cdot 1$, we obtain

$$(m \cdot 1)(n \cdot 1)^{-1} = (m' \cdot 1)(n' \cdot 1)^{-1}.$$

This proves that φ is well-defined. It is straightforward to check that φ is a ring homomorphism. We claim that φ is actually an isomorphism. It is clear that $\ker \varphi = \{0\}$. Since $\mathbb{Q}$ is a field, so is $\operatorname{im} \varphi$. Now, since $\operatorname{im} \varphi \subseteq F_0$ and F_0 is, by definition, the smallest subfield of F, it follows that $\operatorname{im} \varphi = F_0$.

(b) To avoid confusion, we will use the original notation $[a]$, where $a \in \mathbb{Z}$, for elements in $\mathbb{Z}_p$. Define

$$\varphi : \mathbb{Z}_p \to F_0, \quad \varphi([a]) = a \cdot 1.$$

Note that $[a] = [b]$ implies $a - b \in p\mathbb{Z}$ and hence $a \cdot 1 = b \cdot 1$. This shows that φ is well-defined. One easily checks that φ is a ring homomorphism and that $\ker \varphi = \{0\}$. Since $\mathbb{Z}_p$ is a field, we can repeat the argument from the end of the proof of (a) to conclude that $\operatorname{im} \varphi = F_0$. Thus, φ is an isomorphism. $\qquad\square$

For a more conceptual proof of (b), see Exercise 4.136.

Exercises

3.138. Prove that a nonzero commutative ring R has characteristic 2 if and only if $(x + y)^2 = x^2 + y^2$ for all $x, y \in R$.

3.139. What is the characteristic of the ring $\mathbb{Z}_3 \times \mathbb{Z}_2$? More generally, what is the characteristic of $\mathbb{Z}_m \times \mathbb{Z}_n$?

3.140. Let R be a ring whose characteristic is a prime p. Prove that $a \in R$ satisfies $a^p = 1$ if and only if $(a - 1)^p = 0$.
Hint. If you cannot find a solution yourself, look at the proof of Lemma 7.108.

3.141. Prove that the field $\mathbb{Q}$ can be embedded into a field of characteristic 0 in exactly one way.

3.142. Prove that the field $\mathbb{Z}_p$ can be embedded into a field of characteristic p in exactly one way.

3.143. Let R and R' be rings. Which of the following two statements is true:

(a) If there exists an embedding from R into R', then R and R' have the same characteristic.
(b) If there exists an epimorphism from R to R', then R and R' have the same characteristic.

Provide a counterexample to the false statement.

3.144. Let $n \in \mathbb{N}$. Find all $k \in \mathbb{N}$ for which there exists a ring homomorphism from $\mathbb{Z}_n$ to $\mathbb{Z}_k$.

3.145. Provide an example of a ring of characteristic 0 that contains a nonzero element a such that $n \cdot a = 0$ for some $n \in \mathbb{N}$.

3.146. Prove that an algebra A over a field F has the same characteristic as F.

3.147. Let A be an algebra over a field of characteristic different from 2, and let $a \in A$ be such that $a^2 = 1$. Prove that A is the direct sum of the subspaces

$$A_0 = \{x \in A \,|\, ax = xa\} \quad \text{and} \quad A_1 = \{x \in A \,|\, ax = -xa\}.$$

Comment. If the characteristic is 2, then $A_0 = A_1$. Excluding this case is thus really necessary. Generally speaking, the characteristic 2 is something special. Addition and subtraction then coincide, and this has various consequences. Incidentally, look again at the last question of Exercise 1.101: did you notice before that fields of characteristic 2 require special attention?

Chapter 4
Quotient Structures

In this chapter, we will get acquainted with special types of subgroups called normal subgroups. With the help of a normal subgroup we can construct a new group called a quotient group. In a similar fashion, we construct quotient rings from certain subsets of rings called ideals. We have already encountered a particular case of these constructions in Section 2.2 on integers modulo n, but at that time we avoided the terms "quotient group" and "quotient ring".

Quotient structures are intimately connected with homomorphisms. Every homomorphism gives rise to a quotient structure, and every quotient structure gives rise to a homomorphism. Through this connection, we will better understand the meaning of homomorphisms that are not injective. However, the real meaning and importance of both homomorphisms and quotient structures will become evident in Part II, where we will use them as tools for solving various problems.

4.1 Cosets and Lagrange's Theorem

This chapter is centered around the following notion.

Definition 4.1. Let H be a subgroup of a group G, and let $a \in G$. The set

$$aH := \{ah \mid h \in H\}$$

is called a **coset** of H in G.

More precisely, we call aH a **left coset**. A **right coset** is defined as

$$Ha := \{ha \mid h \in H\}.$$

Some books give preference to left and some to right cosets. We decided to choose the former option. By "coset" we will therefore mean "left coset". On those few occasions where right cosets will be considered, we will always use the adjective

© Springer Nature Switzerland AG 2019

M. Brešar, *Undergraduate Algebra*, Springer Undergraduate Mathematics Series,

https://doi.org/10.1007/978-3-030-14053-3_4

"right". Sometimes, in particular when G is Abelian, aH and Ha are equal for every $a \in G$. As we will see in the next sections, this case is of special importance. In general, however, aH and Ha may be different.

Let us emphasize that a is an element of G, not necessarily of H. The following basic fact

$$aH = H \iff a \in H \tag{4.1}$$

is a special case of Lemma 4.6 below, but we advise the reader to prove it independently right now in order to gain some familiarity with cosets. Note also that aH does not contain the identity element if $a \notin H$, and hence is not a subgroup.

If G is an additive group, a coset is written as $a + H$. Thus,

$$a + H = \{a + h \mid h \in H\}.$$

Let us look at a few examples.

Example 4.2. Let $G = \mathbb{Z}$ and $H = n\mathbb{Z}$, $n \in \mathbb{N}$. The cosets generated by $0, 1, \ldots, n - 1$ are

$$n\mathbb{Z}, \ 1 + n\mathbb{Z}, \ \ldots, (n - 1) + n\mathbb{Z}.$$

The coset $n\mathbb{Z}$ contains all integers divisible by n, $1 + n\mathbb{Z}$ contains all integers that have remainder 1 when divided by n, etc. These cosets are the only ones that exist, since

$$n + n\mathbb{Z} = n\mathbb{Z},$$
$$(n + 1) + n\mathbb{Z} = 1 + n\mathbb{Z}, \text{ etc.},$$

and similarly,

$$-1 + n\mathbb{Z} = (n - 1) + n\mathbb{Z},$$
$$-2 + n\mathbb{Z} = (n - 2) + n\mathbb{Z}, \text{ etc.}$$

We have already dealt with these sets when defining the set $\mathbb{Z}_n$. However, the coset $a + n\mathbb{Z}$ was then denoted by $[a]$ (and later simply by a where $0 \leq a < n$).

Example 4.3. Let G be the additive group $\mathbb{R}^2$ and let H be the x-axis. Given $a = (a_1, a_2) \in \mathbb{R}^2$, the coset $a + H$ is the set $\{(x, a_2) \mid x \in \mathbb{R}\}$. The cosets are thus the horizontal lines, i.e., the lines parallel to H.

Example 4.4. Let G be the multiplicative group $\mathbb{C}^*$ of all nonzero complex numbers and let H be the circle group $\mathbb{T}$. The coset zH consists of all complex numbers having the same absolute value as z. Geometrically, the cosets are concentric circles centered at the origin.

Example 4.5. Let G be the symmetric group S_n and let H be the alternating group A_n. If $\sigma \in H$, then $\sigma H = H$. If $\sigma \notin H$, i.e., if σ is an odd permutation, then σH is readily seen to be the set of all odd permutations. Thus, there are only two cosets: the one consisting of all even permutations, and the one consisting of all odd permutations.

Until further notice, we assume that G is an arbitrary group and H is an arbitrary subgroup of G. The next lemma answers the natural question of when two elements generate the same coset.

Lemma 4.6. *For all* $a, b \in G$,

$$aH = bH \iff a^{-1}b \in H.$$

Proof. If $aH = bH$, then $b = b \cdot 1 \in bH = aH$. Therefore, $b = ah_0$ for some $h_0 \in H$, which yields $a^{-1}b = h_0 \in H$.

To prove the converse, assume that $h_0 := a^{-1}b \in H$. Then $b = ah_0$ and hence $bh = a(h_0 h) \in aH$ for every $h \in H$. Thus, $bH \subseteq aH$. Since

$$b^{-1}a = (a^{-1}b)^{-1} = h_0^{-1} \in H,$$

the same argument shows that $aH \subseteq bH$.   □

The condition $a^{-1}b \in H$ might remind one of the statement that a nonempty subset H of a group G is a subgroup if and only if $a^{-1}b \in H$ whenever $a, b \in H$. However, the latter concerns all elements of H, whereas Lemma 4.6 treats the situation where a and b are two fixed elements of G. Subgroups can be also characterized by the condition that $ab^{-1} \in H$ for all $a, b \in H$. If a and b are elements of G, the condition $ab^{-1} \in H$ is equivalent to the equality of the *right* cosets Ha and Hb. Thus, some caution is necessary when treating the equality of cosets. The following alternative proof of Lemma 4.6 may help one not to confuse which of the two similar conditions, $a^{-1}b \in H$ and $ab^{-1} \in H$, corresponds to "our," i.e., left cosets. Consider the equality $aH = bH$. Multiplying it from the left by a^{-1} we see that it can be equivalently written as $H = a^{-1}bH$. By (4.1), this is further equivalent to $a^{-1}b \in H$.

The last remarks are, of course, irrelevant for Abelian groups. In an additive group, the statement of the lemma reads as follows:

$$a + H = b + H \iff b - a \in H \,(\iff a - b \in H).$$

In each of the above examples, two different cosets have no common element. This was no coincidence.

Lemma 4.7. *For all* $a, b \in G$, *the cosets* aH *and* bH *are either equal or disjoint.*

Proof. Suppose $aH \cap bH \neq \emptyset$. Let $h_1, h_2 \in H$ be such that $ah_1 = bh_2$. Multiplying from the left by a^{-1} and from the right by h_2^{-1}, we obtain $a^{-1}b = h_1 h_2^{-1} \in H$. Hence, $aH = bH$ by Lemma 4.6.   □

This lemma implies that G is a disjoint union of all the cosets of H. Indeed, every element in G lies in the coset it generates ($a \in aH$), and different cosets are disjoint. The cosets thus form a partition of G. Therefore, there is an equivalence relation on G whose equivalence classes are exactly the cosets. Observe that we can define it as follows:

$$a \sim b \iff a^{-1}b \in H.$$

This was mentioned just to give the reader a more complete picture, but will not be needed later.

The number of distinct cosets of H in G is called the **index of H in G** and is denoted by $[G : H]$. If G is a finite group, we clearly have $[G : H] < \infty$ for every subgroup H. If G is infinite, then the index may be infinite, but it may also be finite.

Example 4.8. From Example 4.2, we see that $[\mathbb{Z} : n\mathbb{Z}] = n$ for every $n \in \mathbb{N}$.

The next theorem is named after *Joseph-Louis Lagrange* (1736–1813) who proved a special case of it in 1771, even before the notion of a group existed.

Theorem 4.9. (Lagrange's Theorem) *If H is a subgroup of a finite group G, then*

$$|G| = [G : H] \cdot |H|.$$

Proof. Let us write r for $[G : H]$. The set of all cosets of H in G can thus be written as $\{a_1 H, \ldots, a_r H\}$ for some $a_i \in G$. By Lemma 4.7, G is the disjoint union of the sets $a_1 H, \ldots, a_r H$. Hence,

$$|G| = |a_1 H| + \cdots + |a_r H|.$$

However, each of the numbers $|a_i H|$ is equal to $|H|$. This is because the map $h \mapsto a_i h$ from H to $a_i H$ is bijective. It is obviously surjective, and it is injective since $a_i h = a_i h'$ implies $h = h'$. Therefore, $|G| = r|H|$. $\square$

Example 4.10. Since $|S_n| = n!$ and $[S_n : A_n] = 2$ by Example 4.5, $|A_n| = \frac{n!}{2}$.

Lagrange's Theorem is one of the cornerstones of group theory. Its main point is that the order of a subgroup divides the order of a (finite) group. This has important consequences, some of which can be derived quite easily. They will be discussed in Section 6.1. We thus have to wait for a while before exploring the true meaning of Lagrange's Theorem. The theme of this chapter is cosets, so right now we are interested only in the following aspect of this far-reaching theorem: if H is a subgroup of a finite group G, then there are exactly $\frac{|G|}{|H|}$ distinct cosets of H in G.

Exercises

4.11. As just mentioned, we will consider various consequences of Lagrange's Theorem in Section 6.1. However, it will be instructive for the reader to derive the simplest and the most important ones independently already now. Show that if G is a finite group, then (a) the order of each element divides $|G|$, (b) $a^{|G|} = 1$ for every $a \in G$, and (c) G is cyclic if $|G|$ is a prime number.

4.12. Let G be as in Example 4.3, and let H be any line through the origin. What are the cosets of H in G?

4.13. Describe the cosets of the subgroup $\{1, -1, i, -i\}$ in the quaternion group Q.

4.14. Describe the cosets of the subgroup $\{-1, 1\}$ in the quaternion group Q.

4.15. Describe the cosets of the subgroup $\{0, 3, 6, 9\}$ in the group $(\mathbb{Z}_{12}, +)$.

4.16. Describe the cosets of the subgroup $\{0, 4, 8\}$ in the group $(\mathbb{Z}_{12}, +)$.

4.17. Describe the cosets of the subgroup $\{(x, 0, 0) \mid x \in \mathbb{R}\}$ in the group $(\mathbb{R}^3, +)$.

4.18. Describe the cosets of the subgroup $\{(x, y, 0) \mid x, y \in \mathbb{R}\}$ in the group $(\mathbb{R}^3, +)$.

4.19. Describe the cosets of the subgroup $\{1, -1\}$ in the group $(\mathbb{C}^*, \cdot)$.

4.20. Describe the cosets of the subgroup $\mathbb{R}^+$ in the group $(\mathbb{C}^*, \cdot)$.

4.21. Describe the cosets of the subgroup $\mathbb{R}^*$ in the group $(\mathbb{C}^*, \cdot)$.

4.22. Let H be a finite subgroup of an infinite group G. Prove that there are infinitely many cosets of H in G.

4.23. Let G be a finite group and let $H \leq G$. Prove that G contains elements a and b such that $a \notin H$, $b \notin H$, and $ab \notin H$ if and only if $|G| > 2|H|$.

4.24. Let G be an arbitrary finite group and let $H \leq G$. Verify that the right cosets Ha and Hb are equal if and only if $ab^{-1} \in H$, and hence derive versions of Lemma 4.7 and Theorem 4.9 for right cosets.

4.25. Let $H = \{1, (1\,2)\} \leq S_3$. Describe all left cosets aH and all right cosets Ha. Show that $aH = bH$ does not imply $Ha = Hb$.

4.26. Let $H \leq G$. Exercise 4.24 implies that the number of all left cosets of H in G is equal to the number of all right cosets of H in G, provided that G is finite. Extend this to general (possibly infinite) groups G by finding a bijective map from the set of all left cosets $\{aH \mid a \in G\}$ to the set of all right cosets $\{Ha \mid a \in G\}$.

Hint. Do not forget to check that your map is well-defined. What does the preceding exercise tell you about the "obvious" choice $aH \mapsto Ha$?

4.2 Normal Subgroups and Quotient Groups

As pointed out in Example 4.2, the elements of the group $(\mathbb{Z}_n, +)$ are the cosets $a + n\mathbb{Z}$ of the subgroup $n\mathbb{Z}$ in the group $\mathbb{Z}$. In this notation, the definition of addition reads as follows:

$$(a + n\mathbb{Z}) + (b + n\mathbb{Z}) = (a + b) + n\mathbb{Z}. \tag{4.2}$$

The set of all cosets of $n\mathbb{Z}$ in $\mathbb{Z}$ is thus a group under the operation defined in a simple and natural manner.

Now take an arbitrary group G and a subgroup N of G (the reason for this notation will become clear shortly). As usual, we write the operation in G as multiplication. In light of (4.2), one can ask whether the set of all cosets

$$G/N := \{aN \mid a \in G\}$$

becomes a group under the operation defined by

$$aN \cdot bN := (ab)N.$$

The main point of this question is whether this operation is well-defined. As we will see, this is the case only when N is a *normal subgroup*, the notion we are about to introduce.

4.2.1 Normal Subgroups

Let N be a subgroup of a group G, and let $a \in G$. Recall that

$$aNa^{-1} = \{ana^{-1} \mid a \in G\}$$

is again a subgroup of G, called a conjugate of N (see Example 1.114).

Definition 4.27. A subgroup N of a group G is called a **normal subgroup** if $aNa^{-1} \subseteq N$ for all $a \in G$. In this case, we write $N \triangleleft G$.

Thus, a subgroup N is normal if every conjugate of N is contained in N. We will usually denote normal subgroups by N and sometimes M, and use H and K for subgroups that are not necessarily normal.

The central theme of this section is the group structure of the set of all cosets G/N. But first we discuss some general facts about normal subgroups. It is worth mentioning that different texts define normal subgroups in different, but equivalent, ways. The following theorem reveals some of the conditions equivalent to the one from our definition.

Theorem 4.28. *Let N be a subgroup of a group G. The following conditions are equivalent:*

(i) $N \triangleleft G$.
(ii) $aN \subseteq Na$ *for all $a \in G$.*
(iii) $aN = Na$ *for all $a \in G$.*
(vi) $aNa^{-1} = N$ *for all $a \in G$.*

Proof. (i) $\Longrightarrow$ (ii). Multiplying $aNa^{-1} \subseteq N$ from the right by a gives $aN \subseteq Na$ (do check that set inclusion is preserved under such multiplication!).

(ii) $\Longrightarrow$ (iii). Since $aN \subseteq Na$ holds for all $a \in G$, we also have $a^{-1}N \subseteq Na^{-1}$. By multiplying this inclusion from both sides by a we get $Na \subseteq aN$. Hence, $aN = Na$.

(iii) $\implies$ (iv). Multiply $aN = Na$ from the right by a^{-1}.

(iv) $\implies$ (i). Trivial. $\qquad\square$

Condition (iv) states that the only conjugate of N is N itself. This does not necessarily mean that $ana^{-1} = n$ for all $a \in G$ and $n \in N$, but only that the sets aNa^{-1} and N are equal for every $a \in G$. Similarly, condition (iii), stating that the left coset aN is equal to the right coset Na for every $a \in G$, should not be interpreted as that $an = na$ for all $a \in G$ and $n \in N$. The latter holds only if N is contained in the center $Z(G)$ of G. A subgroup contained in $Z(G)$ is thus automatically normal. In particular, all subgroups of Abelian groups are normal. Later we will also see that many important subgroups of non-Abelian subgroups are normal. However, it would be inappropriate to say that "most" subgroups are normal. Let us give just one simple non-example: $H = \{1, (1\,2)\}$ is not a normal subgroup of the symmetric group S_3, since, for example, $(1\,3)(1\,2)(1\,3)^{-1} = (2\,3) \notin H$.

Every nontrivial group G has at least two normal subgroups, $\{1\}$ and G. It may happen that these two are also the only ones.

Definition 4.29. A nontrivial group is said to be **simple** if its only normal subgroups are $\{1\}$ and G.

These groups will be discussed in Section 6.5. Here we only mention without proof that the cyclic groups of prime order and the alternating groups A_n with $n \geq 5$ are simple.

Admittedly, it is not clear from the definition in what way normal subgroups are better than ordinary subgroups. The following discussion will indicate some of their advantages.

Let $H, K \leq G$. The set

$$HK := \{hk \mid h \in H, k \in K\}$$

is called the **product of subgroups** H and K. If G is an additive group, then we of course talk about the **sum of subgroups** H and K, and write

$$H + K := \{h + k \mid h \in H, k \in K\}.$$

As noticed already in Section 2.1, $H + K$ is again a subgroup. The same proof, only written in a different notation, shows that HK is a subgroup of G if G is Abelian. In general, however, HK may not be a subgroup (see Exercise 4.55). The following theorem is therefore of interest.

Theorem 4.30. *Let G be a group.*

(a) *If $H, K \leq G$ and $HK = KH$, then $HK \leq G$.*

(b) *If $H \leq G$ and $N \triangleleft G$, then $HN = NH \leq G$.*

(c) *If $M, N \triangleleft G$, then $MN = NM \triangleleft G$.*

(d) *If $M, N \triangleleft G$ and $M \cap N = \{1\}$, then $mn = nm$ for all $m \in M$ and $n \in N$.*

Proof. (a) Take $h_1, h_2 \in H$ and $k_1, k_2 \in K$. We must show that

$$(h_1 k_1)(h_2 k_2)^{-1} = h_1 k_1 k_2^{-1} h_2^{-1}$$

lies in HK. Since $k_1 k_2^{-1} \in K$, $h_2^{-1} \in H$, and $KH = HK$, we can write $(k_1 k_2^{-1}) h_2^{-1}$ as $h_3 k_3$ for some $h_3 \in H$ and $k_3 \in K$. Consequently,

$$h_1 k_1 k_2^{-1} h_2^{-1} = (h_1 h_3) k_3 \in HK.$$

(b) By Theorem 4.28, $hN = Nh$ for all $h \in H$, which implies that $HN = NH$. By (a), $HN \leq G$.

(c) We only have to show that $aMNa^{-1} \subseteq MN$ for every $a \in G$. This follows from $a(mn)a^{-1} = (ama^{-1})(ana^{-1})$.

(d) Recall that an element of the form $aba^{-1}b^{-1}$ is called a *commutator*. We denote it by $[a, b]$. Clearly, $[a, b] = 1$ if and only if a and b commute.

Take $m \in M$ and $n \in N$. Our goal is to show that $[m, n] = 1$. Since N is normal, $[m, n] = (mnm^{-1})n^{-1}$ implies that $[m, n] \in N$. Similarly, since M is normal, $[m, n] = m(nm^{-1}n^{-1})$ implies that $[m, n] \in M$. Hence, $[m, n] \in M \cap N = \{1\}$. $\square$

Remark 4.31. We can, of course, consider the product of more than two subgroups. By induction, we derive from Theorem 4.30(c) that

$$N_1 N_2 \cdots N_s := \{n_1 n_2 \cdots n_s \mid n_i \in N_i, i = 1, \ldots, s\}$$

is a normal subgroup, provided that all the subgroups $N_1, N_2, \ldots, N_s$ are normal.

One easily checks that the **intersection of normal subgroups** M and N is again a normal subgroup. Note that $M \cap N$ is the largest normal subgroup contained in both M and N, and MN is the smallest normal subgroup containing both M and N.

4.2.2 Quotient Groups

We return to the problem of defining multiplication on the set of all cosets G/N. First we settle the question of well-definedness.

Lemma 4.32. *Let $N \lhd G$. If $a, a', b, b' \in G$ are such that $aN = a'N$ and $bN = b'N$, then $(ab)N = (a'b')N$.*

Proof. Set $n_1 := a^{-1}a'$ and $n_2 := b^{-1}b'$. In view of Lemma 4.6, n_1 and n_2 lie in N, and we have to prove that $(ab)^{-1}(a'b')$ also lies in N. Since

$$(ab)^{-1}(a'b') = b^{-1}a^{-1}a'b' = b^{-1}n_1(bb^{-1})b' = (b^{-1}n_1 b)n_2$$

and N is a normal subgroup, this indeed holds. $\square$

We remark that the converse of Lemma 4.32 is also true, in the sense that any subgroup for which this lemma holds must be normal (see Exercise 4.51). Although not so important by itself, this provides some insight into the meaning of normal subgroups. These are exactly the subgroups for which $aN \cdot bN := (ab)N$ is a well-defined operation on the set of all cosets G/N.

We are now in a position to answer the question posed at the beginning of the section.

Theorem 4.33. *Let N be a normal subgroup of a group G. The set G/N of all cosets of N in G is a group under the multiplication defined by*

$$aN \cdot bN = (ab)N.$$

The map

$$\pi : G \to G/N, \quad \pi(a) = aN,$$

is a group epimorphism with $\ker \pi = N$.

Proof. Lemma 4.32 shows that multiplication is well-defined. The associative law follows from the associative law for multiplication in G:

$$(aN \cdot bN) \cdot cN = (ab)N \cdot cN = ((ab)c)N = (a(bc))N = aN \cdot (bN \cdot cN).$$

The identity element is $N(= 1N)$. The inverse of the coset aN is the coset $a^{-1}N$. This proves that G/N is a group.

We can rewrite $aN \cdot bN = (ab)N$ as $\pi(a)\pi(b) = \pi(ab)$. This means that π is a homomorphism. It is obviously surjective. An element $a \in G$ lies in the kernel of π if and only if $aN = N$, which is by (4.1) equivalent to $a \in N$. $\qquad\square$

Definition 4.34. The group G/N from Theorem 4.33 is called the **quotient group** (or **factor group**) of G by N, and the map π is called the **canonical epimorphism** (or **projection**) of G onto G/N.

If G is a finite group and $N \triangleleft G$, then

$$|G/N| = \frac{|G|}{|N|}$$

by Lagrange's Theorem. In particular, the order of G/N divides the order of G.

Theorem 4.33 indicates a connection between cosets and homomorphisms. The formula $\pi(ab) = \pi(a)\pi(b)$ is just another way of writing $aN \cdot bN = (ab)N$, and $\ker \pi = N$ is just another way of describing the equivalence between the conditions $aN = N$ and $a \in N$. Every normal subgroup is thus the kernel of a homomorphism. The next theorem shows that the converse is also true. The two seemingly unrelated notions, "normal subgroup" and "kernel of a group homomorphism," thus coincide.

Theorem 4.35. *A subset N of a group G is a normal subgroup if and only if N is the kernel of a homomorphism from G to some group G'.*

Proof. Assume that there exists a homomorphism $\varphi : G \to G'$ such that $N = \ker \varphi$. If m and n lie in N, then so does mn^{-1}, since

$$\varphi(mn^{-1}) = \varphi(m)\varphi(n)^{-1} = 1 \cdot 1^{-1} = 1.$$

Thus, N is a subgroup. For any $a \in G$ and $n \in N$, we have

$$\varphi(ana^{-1}) = \varphi(a)\varphi(n)\varphi(a^{-1}) = \varphi(a)1\varphi(a)^{-1} = 1.$$

That is, $ana^{-1} \in N$ and so N is normal.

Conversely, a normal subgroup N of G is the kernel of the canonical epimorphism $\pi : G \to G/N$. $\square$

If G is an additive (and hence Abelian) group, every subgroup N of G is normal. The operation in G/N is, of course, written as addition. Thus,

$$G/N = \{a + N \mid a \in G\}$$

is an additive group under the operation

$$(a + N) + (b + N) = (a + b) + N.$$

The quotient group $\mathbb{Z}/n\mathbb{Z}$ is nothing but $\mathbb{Z}_n$, the group of integers modulo n (see (4.2)). This example is both simple and illustrative, and for some time it will be our only example of a quotient group. We will consider a variety of examples of normal subgroups and quotient groups in Section 4.4, when the *Isomorphism Theorem* will be available and will help us to get a better insight into the matter.

4.2.3 Quotient Vector Spaces

If U is a subspace of a vector space V, then U is in particular a subgroup of the additive group $(V, +)$, so we can form the additive quotient group V/U. This is obviously not entirely satisfactory in the context of vector spaces. Let us show that we can make a vector space out of this additive group.

Theorem 4.36. *Let U be a subspace of a vector space V. The set V/U of all cosets of U in V is a vector space under the addition and scalar multiplication defined by*

$$(v + U) + (w + U) = (v + w) + U,$$
$$\lambda(v + U) = \lambda v + U.$$

The map

$$\pi : V \to V/U, \quad \pi(v) = v + U,$$

is a vector space epimorphism with $\ker \pi = U$.

Proof. We already know that V/U is an additive group and that π is a group homomorphism with $\ker \pi = U$. Let us show that the scalar multiplication is well-defined. Suppose $v + U = v' + U$. Then $v - v' \in U$. Since U is a subspace, it follows that

$$\lambda v - \lambda v' = \lambda(v - v') \in U$$

for every scalar λ. This means that $\lambda v + U = \lambda v' + U$, which is the desired conclusion. A routine verification shows that V/U satisfies all the vector space axioms. It is also clear that π satisfies $\pi(\lambda v) = \lambda \pi(v)$ for every scalar λ and every vector v. $\square$

We call V/U the **quotient** (or **factor**) **vector space** of V by U, and, as above, we call π the canonical epimorphism.

4.2.4 Subgroups of a Quotient Group

If we wish to understand the structure of a group, we are often forced to examine its subgroups, particularly normal subgroups. What can be said about subgroups of a quotient group G/N? In what way are they connected with subgroups of G? The following lemma will lead us to answers to these questions. We first recall some notation. Given a map $f : A \to A'$ and subsets $B \subseteq A$, $B' \subseteq A'$, we write $f(B) = \{f(b) \mid b \in B\}$ and $f^{-1}(B') = \{a \in A \mid f(a) \in B'\}$.

Lemma 4.37. *Let $\varphi : G \to G'$ be a group homomorphism.*

(a) *If $H' \leq G'$, then $\varphi^{-1}(H') \leq G$.*
(b) *If $N' \lhd G'$, then $\varphi^{-1}(N') \lhd G$.*
(c) *If $H \leq G$, then $\varphi(H) \leq G'$.*
(d) *If $N \lhd G$ and φ is an epimorphism, then $\varphi(N) \lhd G'$.*

Proof. (a) Take $h, k \in \varphi^{-1}(H')$. Then $\varphi(h), \varphi(k) \in H'$. We have to show that $hk^{-1} \in \varphi^{-1}(H')$, i.e., that $\varphi(hk^{-1}) \in H'$. Since

$$\varphi(hk^{-1}) = \varphi(h)\varphi(k)^{-1}$$

and H' is a subgroup, this indeed holds.

(b) By (a), we know that $\varphi^{-1}(N')$ is a subgroup. It remains to show that $ana^{-1} \in \varphi^{-1}(N')$ for all $a \in G$ and $n \in \varphi^{-1}(N')$. This is immediate from

$$\varphi(ana^{-1}) = \varphi(a)\varphi(n)\varphi(a)^{-1},$$

since $\varphi(n) \in N'$ and N' is a normal subgroup.

(c) Since a subgroup H is itself a group and the restriction of φ to H is a homomorphism from H to G', this follows from the fact that the image of a group homomorphism is a subgroup (Theorem 3.54).

(d) By (c), $\varphi(N)$ is a subgroup. Since $ana^{-1} \in N$ for all $a \in G$ and $n \in N$, we have

$$\varphi(a)\varphi(n)\varphi(a)^{-1} = \varphi(ana^{-1}) \in \varphi(N).$$

As φ is surjective, this means that $\varphi(N)$ is normal. $\square$

We can now address the question posed above. Let $N \triangleleft G$. If a subgroup H of G contains N, then N is obviously a normal subgroup of H. We can thus form the quotient group H/N, which is readily seen to be a subgroup of G/N; as a matter of fact, this also follows from Lemma 4.37 (c) if we take the canonical epimorphism $\pi : G \to G/N$ for φ. Similarly, using Lemma 4.37 (d) we see that $M/N \triangleleft G/N$ whenever $M \triangleleft G$ and $M \supseteq N$. The next theorem tells us that G/N has no other subgroups, ordinary and normal, than those just described.

Theorem 4.38. *Let N be a normal subgroup of a group G.*

(a) *Every subgroup of G/N is of the form H/N, where H is a subgroup of G that contains N.*
(b) *Every normal subgroup of G/N is of the form M/N, where M is a normal subgroup of G that contains N.*

Proof. Let π denote the canonical epimorphism from G to G/N.

(a) Take a subgroup H' of G/N. Lemma 4.37 (a) shows that $H := \pi^{-1}(H')$ is a subgroup of G. It consists of those elements h in G such that the coset $\pi(h) = hN$ lies in H'. Therefore $N \subseteq H$; indeed, $nN = N$ for every $n \in N$, and N is the identity element of H'. Since π is surjective, we have

$$\pi(\pi^{-1}(H')) = H'.$$

That is,

$$H' = \pi(H) = H/N.$$

(b) The proof of the second statement follows the same pattern. Take a normal subgroup N' of G/N. By Lemma 4.37 (b), $M := \pi^{-1}(N')$ is a normal subgroup of G. Note that $N \subseteq M$ and $N' = \pi(M) = M/N$. $\square$

Remark 4.39. If $H_0 \le G$ and $N \triangleleft G$, then the set of all cosets of the form $h_0 N$ with $h_0 \in H_0$ forms a subgroup of G/N, regardless of whether H_0 contains N or not. As is evident from the proof of Theorem 4.38, this subgroup is equal to H/N where $H = H_0 N$ (if $N \subseteq H_0$, then $H = H_0$).

Remark 4.40. An almost identical proof shows that every subspace of the vector space V/U is of the form W/U, where W is a subspace of V that contains U.

Theorem 4.38 is sometimes called the **Correspondence Theorem**; it shows that (normal) subgroups of G that contain N correspond to (normal) subgroups of G/N. The quotient group G/N thus has at most as many (normal) subgroups as the original group G. This indicates that G/N might have a simpler structure than G.

To illustrate the applicability of Theorem 4.38, we describe the subgroups of finite cyclic groups $\mathbb{Z}_n$ (this description was already given in a slightly different form in Exercise 3.23).

Example 4.41. First observe that if G is any additive group and k is any integer, $kG := \{kx \mid x \in G\}$ is a subgroup of G. For example, $2\mathbb{Z}_6 = \{0, 2, 4\}$ and is equal to $4\mathbb{Z}_6$, $3\mathbb{Z}_6 = \{0, 3\}$, $6\mathbb{Z}_6 = 0\mathbb{Z}_6 = \{0\}$, $1\mathbb{Z}_6 = 7\mathbb{Z}_6 = \mathbb{Z}_6$, etc. We claim that every subgroup K of $\mathbb{Z}_n = \mathbb{Z}/n\mathbb{Z}$ can be written as

$$K = k\mathbb{Z}_n, \quad \text{where } k \in \mathbb{N} \text{ and } k \mid n$$

(so $\mathbb{Z}_6$ has only the aforementioned four subgroups). Indeed, since every subgroup of $\mathbb{Z}$ can be written as $k\mathbb{Z}$ with $k \geq 0$ (Theorem 2.2), Theorem 4.38 implies that K is of the form $K = k\mathbb{Z}/n\mathbb{Z}$, where $n\mathbb{Z} \subseteq k\mathbb{Z}$. The latter condition is fulfilled if and only if k divides n. Since $kx + n\mathbb{Z} = k(x + n\mathbb{Z})$ for every $x \in \mathbb{Z}$, $k\mathbb{Z}/n\mathbb{Z}$ is equal to $k\mathbb{Z}_n$.

Exercises

4.42. Explain why a nontrivial homomorphism from a simple group to another group is always injective.

4.43. Let a be an element of a group G such that $a^2 = 1$. Show that the subgroup $\langle a \rangle = \{1, a\}$ is normal if and only if a lies in the center of G.

4.44. Show that all subgroups of the quaternion group Q are normal.

Comment. Thus, there do exist non-Abelian groups having only normal subgroups.

4.45. Determine which subgroups of the symmetric group S_3 are normal.

4.46. Which of the subgroups $\langle r \rangle$ and $\langle s \rangle$ of the dihedral group D_{2n} is normal?

4.47. Find subgroups H and N of D_8 such that H is a normal subgroup of N and N is a normal subgroup of D_8, but H is not a normal subgroup of D_8.

4.48. Let N be a finite subgroup of a group G. Suppose that G has no other subgroup of the same order as N. Prove that N is normal.

4.49. Let X be a nonempty subset of a group G. Prove that the subgroup generated by the set $\{gxg^{-1} \mid g \in G, x \in X\}$ is equal to the normal subgroup generated by X (by this we mean the normal subgroup that contains X and is contained in any other normal subgroup that contains X).

4.50. Let G be a finite non-Abelian simple group, and let x be an element of G different from 1. Prove that every element in G can be written as a product of conjugates of x.

4.51. Let N be a subgroup of a group G. Suppose that for all $a, a', b, b' \in G$, the conditions $aN = a'N$ and $bN = b'N$ imply that $(ab)N = (a'b')N$. Prove that N is normal.

4.52. The product of nonempty subsets X and Y of a group G is defined in the same way as the product of subgroups, that is, $XY := \{xy \mid x \in X, y \in Y\}$. Prove that the product of cosets aN and bN is equal to $(ab)N$ by this definition as well, provided of course that $N \lhd G$.

4.53. Let G be an arbitrary group. Show that $\mathrm{Inn}(G)$, the group of inner automorphisms of G, is a normal subgroup of the group of all automorphisms $\mathrm{Aut}(G)$.

4.54. A normal subgroup of a group G can be described as a subgroup N with the property that $\varphi(N) \subseteq N$ for every $\varphi \in \mathrm{Inn}(G)$. We call $K \leq G$ a **characteristic subgroup** if $\varphi(K) \subseteq K$ for every $\varphi \in \mathrm{Aut}(G)$. Every characteristic subgroup is thus normal. Prove:

(a) The subgroup $\{1, -1, i, -i\}$ of the quaternion group Q is normal but not characteristic.
(b) For any nontrivial group G, the subgroup $G \times \{1\}$ of the direct product $G \times G$ is normal but not characteristic.
(c) For any group G, the center $Z(G)$ is a characteristic subgroup of G.

4.55. Let a and b be elements of a group G such that $a^2 = b^2 = 1$ and $ab \neq ba$. Show that the product of $H = \{1, a\}$ and $K = \{1, b\}$ is not a subgroup. Provide a concrete example of a group containing such a pair of elements.

4.56. Let H and K be subgroups of a group G such that $HK = H \cup K$. Prove that $H \subseteq K$ or $K \subseteq H$.

Comment. This is similar to Exercise 1.137. Both exercises can be solved in the same way.

4.57. Show that Lemma 4.37 (d) does not hold in general for homomorphisms that are not surjective.

Hint. If H is not a normal subgroup of G, it is still a normal subgroup of itself.

4.58. Let $\varphi : G \to G'$ be a group epimorphism. Prove that $H \mapsto \varphi(H)$ defines a bijective correspondence between the set of all subgroups of G that contain $\ker \varphi$ and the set of all subgroups of G'. Moreover, normal subgroups correspond to normal subgroups under this correspondence.

Comment. Taking φ to be the canonical epimorphism $\pi : G \to G/N$, we see that this is a generalization of Theorem 4.38.

4.59. Let $N \triangleleft G$. We say that N is a **maximal normal subgroup** of G if $N \neq G$ and there does not exist a normal subgroup M of G such that $N \subsetneq M \subsetneq G$. Show that N is a maximal subgroup of G if and only if the group G/N is simple.

4.60. Let $N \leq G$. Prove that $[G : N] = 2$ implies $N \triangleleft G$ and $G/N \cong \mathbb{Z}_2$.

Hint. If $a \notin N$, then G is a disjoint union of the left cosets N and aN, as well as of the right cosets N and Na.

4.61. As mentioned above, the center $Z(G)$ of a group G is a normal subgroup. Prove that if G is non-Abelian, then $G/Z(G)$ is not a cyclic group.

4.62. Let G be an Abelian group. Denote by N the set of all elements in G of finite order. Prove that N is a subgroup of G and that G/N has no elements of finite order different from the identity element.

4.63. Prove that $\mathbb{Q}/\mathbb{Z}$ is an infinite group in which every element has finite order.

Comment. A group in which every element has finite order is called a **torsion group**. Finite groups are obviously torsion. The group $\mathbb{Q}/\mathbb{Z}$ is thus an example of an infinite torsion group. Can you think of more examples?

4.64. Let N be a normal subgroup of a group G. Which of the following statements are true:

(a) If N and G/N are finite, then so is G.
(b) If N and G/N are finitely generated, then so is G.
(c) If N and G/N are Abelian, then so is G.
(d) If N and G/N are torsion, then so is G.

Provide a counterexample to the false statement.

4.3   Ideals and Quotient Rings

This section is a ring-theoretic analogue of the preceding one.

Let R be a ring and let I be an additive subgroup of R (i.e., a subgroup of R under addition). The set

$$R/I = \{a + I \mid a \in R\}$$

becomes an additive group if we define addition of cosets by

$$(a + I) + (b + I) = (a + b) + I.$$

This is what we learned in the preceding section. Our goal now is to make R/I a ring. The natural, self-evident, way to define multiplication is the following:

$$(a + I)(b + I) := ab + I.$$

As we know from Section 2.2, this works in the special case where $R = \mathbb{Z}$ and $I = n\mathbb{Z}$. In general, however, this multiplication is not well-defined without imposing certain constraints on I. We call additive subgroups of rings satisfying these constraints *ideals*. They play a similar role in ring theory as normal subgroups do in group theory. The first part of the section is devoted to gaining some familiarity with ideals, and after that we will proceed to studying the set of cosets R/I as a ring.

4.3.1   Ideals

The role of ideals is indeed similar to that of normal subgroups, but their definition, and hence also various properties, are quite different.

Definition 4.65. An additive subgroup I of a ring R is called an **ideal of** R if for all $x \in R$ and all $u \in I$, both xu and ux lie in I. In this case, we write $I \triangleleft R$.

To show that a nonempty subset I of a ring R is an ideal, one thus has to check three conditions:

(a) $u - v \in I$ for all $u, v \in I$.
(b) $xu \in I$ for all $x \in R$ and $u \in I$.
(c) $ux \in I$ for all $x \in R$ and $u \in I$.

We often write (b) as $RI \subseteq I$ and (c) as $IR \subseteq I$. Each of these two conditions implies that an ideal is closed under multiplication. However, in general it is not a subring since it does not necessarily contain the unity 1. In fact, if it does contain 1, then, by (b), it also contains $x = x1$ for every $x \in R$, and so it is equal to the whole ring R. Of course, R is indeed an ideal of R. Also, $\{0\}$ is an ideal of every ring R. The next example is more illustrative.

Example 4.66. Let R be a *commutative* ring. For any $a \in R$, the set

$$aR := \{ax \mid x \in R\}$$

is readily seen to be an ideal of R. Such ideals are called **principal** and are usually denoted by (a). A principal ideal can also be described as an *ideal generated by a single element*; specifically, (a) is the ideal generated by a, i.e., it is the smallest ideal of R that contains a. Indeed, (a) is an ideal containing a and every other ideal I of R that contains a must also contain all elements of the form ax with $x \in R$, so $I \supseteq (a)$.

Principal ideals will play an important role in Chapter 5. If $R = \mathbb{Z}$, then every ideal of R is principal. This follows from Theorem 2.2, which tells us even more, namely that every additive subgroup of $\mathbb{Z}$ is of the form $n\mathbb{Z}$ where $n \in \mathbb{N} \cup \{0\}$. Hence these are also the only ideals of $\mathbb{Z}$.

There are several ways of forming new ideals from old ones. One can easily check that the **intersection of ideals** I and J is again an ideal, and so is the **sum of ideals**

$$I + J := \{u + v \mid u \in I, v \in J\}.$$

Note also that $I \cap J$ is the largest ideal contained in both I and J, and $I + J$ is the smallest ideal that contains both I and J. The **product of ideals** IJ is defined as the additive subgroup generated by all products uv where $u \in I$ and $v \in J$. Thus, IJ consists of all elements that can be written as

$$u_1 v_1 + \cdots + u_n v_n, \quad \text{where } u_i \in I \text{ and } v_i \in J.$$

It is clear that $IJ \triangleleft R$. From (c) we see that $IJ \subseteq I$, and from (b) that $IJ \subseteq J$. Thus,

$$IJ \subseteq I \cap J.$$

Of course, JI is also an ideal contained in $I \cap J$. If R is commutative, then it is equal to IJ.

Example 4.67. Let $I = 4\mathbb{Z} \lhd \mathbb{Z}$ and $J = 6\mathbb{Z} \lhd \mathbb{Z}$. Then $IJ = 24\mathbb{Z}$, $I \cap J = 12\mathbb{Z}$, and $I + J = 2\mathbb{Z}$. What do we get if we replace 4 and 6 by arbitrary m and n? We leave this question as an exercise for the reader.

Ideals are also called **two-sided ideals**. If we require only one of the conditions (b) and (c), then we talk about **one-sided ideals**. More precisely, an additive subgroup L of R is called a **left ideal** of R if $RL \subseteq L$. A **right ideal** is defined analogously via the condition $LR \subseteq L$. Thus, I is a two-sided ideal if it is simultaneously a left and right ideal. If R is commutative, then there is obviously no distinction between left, right, and two-sided ideals. Noncommutative rings, however, often contain many one-sided ideals that are not two-sided.

Example 4.68. Let R be an arbitrary ring, and let $a \in R$. The set

$$Ra := \{xa \mid x \in R\}$$

is a left ideal of R which is not necessarily a right ideal. Similarly, aR, considered in Example 4.66 under the assumption that R is commutative, is a right ideal which may not be left. For example, if $R = M_2(\mathbb{R})$ and $a = \left[\begin{smallmatrix} 1 & 0 \\ 0 & 0 \end{smallmatrix}\right] \in R$, then Ra consists of all matrices of the form $\left[\begin{smallmatrix} s & 0 \\ t & 0 \end{smallmatrix}\right]$, $s, t \in \mathbb{R}$. It is easy to check that it is a left, but not a right ideal of R. Similarly, aR consists of all matrices of the form $\left[\begin{smallmatrix} s & t \\ 0 & 0 \end{smallmatrix}\right]$ and is a right, but not a left ideal.

One-sided ideals are important, but in this book they will play only a marginal role simply because noncommutative rings will not be studied in greater depth. Still, the next two results concern one-sided ideals. They are formulated for left ideals, but from their proofs it should be clear that they also hold for right ideals. Later we will actually need these two results only for two-sided ideals.

Lemma 4.69. *If a left ideal L of a ring R contains an invertible element, then $L = R$.*

Proof. Let $\ell \in L$ be invertible. Then $1 = \ell^{-1}\ell \in RL \subseteq L$, and hence also $a = a1 \in RL \subseteq L$ for every $a \in R$. $\square$

Theorem 4.70. *A nonzero ring R is a division ring if and only if $\{0\}$ and R are its only left ideals.*

Proof. Let $L \neq \{0\}$ be a left ideal of a division ring R. Since nonzero elements in R are invertible, Lemma 4.69 implies that $L = R$.

For the converse, assume that $\{0\}$ and R are the only left ideals of R. Take $a \neq 0$ in R. We must show that a is invertible. Since the left ideal Ra contains a and is thus different from $\{0\}$, we must have $Ra = R$. Hence, there exists a $b \in R$ such that $ba = 1$. Now, b is also nonzero, therefore the same argument shows that $Rb = R$, and so $cb = 1$ for some $c \in R$. However, since any left inverse is equal to any right inverse (Theorem 1.36), we conclude from $ba = 1$ and $cb = 1$ that $c = a$, and consequently that a is invertible. $\square$

Definition 4.71. A nonzero ring R is said to be **simple** if $\{0\}$ and R are its only two-sided ideals.

Theorem 4.70 implies that division rings are simple. There are other examples. For instance, Exercise 4.88 requires us to show that the ring of $n \times n$ matrices over a division ring is simple.

In the context of commutative rings, where there is no difference between one-sided and two-sided ideals, Theorem 4.70 gets the following form.

Corollary 4.72. *A commutative ring is a field if and only if it is a simple ring.*

4.3.2 Quotient Rings

We will introduce quotient rings in a similar fashion as quotient groups. The modifications that have to be made are rather obvious.

Lemma 4.73. *Let* $I \lhd R$. *If* $a, a', b, b' \in R$ *are such that* $a + I = a' + I$ *and* $b + I = b' + I$, *then* $ab + I = a'b' + I$.

Proof. By assumption, the elements $u_1 := a' - a$ and $u_2 := b' - b$ lie in I. We have to show that $a'b' - ab$ also lies in I. Since

$$a'b' - ab = (a + u_1)(b + u_2) - ab = u_1 b + a u_2 + u_1 u_2$$

and I is an ideal, this is indeed true. $\qquad\square$

The condition of Lemma 4.73 is actually characteristic for ideals; see Exercise 4.98. Armed with this lemma, the central theorem of this section is at hand.

Theorem 4.74. *Let* I *be an ideal of a ring* R. *The set* R/I *of all cosets of* I *in* R *is a ring under the addition and multiplication defined by*

$$(a + I) + (b + I) = (a + b) + I,$$
$$(a + I)(b + I) = ab + I.$$

The map

$$\pi : R \to R/I, \quad \pi(a) = a + I,$$

is a ring epimorphism with $\ker \pi = I$.

Proof. From the preceding section, we know that R/I is an additive group and that π is an additive group epimorphism with $\ker \pi = I$. Multiplication is well-defined by Lemma 4.73, and as in the proof of Theorem 4.33 we see that it is associative. The distributive laws in R/I follow from the distributive laws in R. The unity is the coset $1 + I$. Thus, R/I is a ring. Finally, π obviously satisfies $\pi(ab) = \pi(a)\pi(b)$ for all $a, b \in R$, so it is a ring epimorphism. $\qquad\square$

Definition 4.75. The ring R/I from Theorem 4.74 is called the **quotient ring** (or **factor ring**) of R by I, and the map π is called the **canonical epimorphism** (or **projection**) of R onto R/I.

As the basic example, consider the ring of integers $\mathbb{Z}$. All of its ideals are of the form $n\mathbb{Z}$ (Example 4.66) and multiplication in $\mathbb{Z}/n\mathbb{Z}$ coincides with multiplication in $\mathbb{Z}_n$ as defined in Section 2.2. Thus,

$$\mathbb{Z}/n\mathbb{Z} = \mathbb{Z}_n$$

is an equality of rings, not only of additive groups. Further examples will be provided in the next section and later on.

Theorem 4.35 states that normal subgroups are exactly the kernels of group homomorphisms. A similar theorem holds for rings.

Theorem 4.76. *A subset I of a ring R is an ideal if and only if I is the kernel of a homomorphism from R to some ring R′.*

Proof. Every ideal is the kernel of the canonical epimorphism. For the converse, assume that $I = \ker \varphi$ where $\varphi : R \to R'$ is a homomorphism. If $u, v \in I$, then

$$\varphi(u - v) = \varphi(u) - \varphi(v) = 0,$$

which shows that $u - v \in I$. Next, if $x \in R$ and $u \in I$, then

$$\varphi(xu) = \varphi(x)\varphi(u) = \varphi(x)0 = 0,$$

so $xu \in I$. Similarly, we see that $ux \in I$. Thus, I is an ideal. $\qquad\square$

4.3.3 Quotient Algebras

An **ideal of an algebra** is defined in the same way as an ideal of a ring. If I is an ideal of an algebra A, then, for every scalar λ and every element u of I, $\lambda u = \lambda(1u) = (\lambda 1)u$ lies in $AI \subseteq I$. This shows that an ideal of an algebra is a subspace.

The next theorem introduces the **quotient** (or **factor**) **algebra** and the corresponding canonical epimorphism.

Theorem 4.77. *Let I be an ideal of an algebra A. The set A/I of all cosets of I in A is an algebra under the addition, multiplication, and scalar multiplication defined by*

$$(a + I) + (b + I) = (a + b) + I,$$
$$(a + I)(b + I) = ab + I,$$
$$\lambda(a + I) = \lambda a + I.$$

The map

$$\pi : A \to A/I, \quad \pi(a) = a + I,$$

is an algebra epimorphism with $\ker \pi = I$.

Proof. In light of Theorems 4.36 and 4.74, all we need to show is that $\lambda(xy) = (\lambda x)y = x(\lambda y)$ holds for all scalars λ and cosets $x, y \in A/I$. Like everything else, this also follows immediately from the above definitions and the corresponding properties in A. $\qquad\square$

4.3.4 Maximal Ideals and Ideals of a Quotient Ring

Subrings and ideals of quotient rings can be described in the same way as (normal) subgroups of quotient groups. We will restrict ourselves to the description of ideals. The proofs of the next lemma and theorem are essentially the same as those of Lemma 4.37 and Theorem 4.38. Therefore we ask the reader to go through these proofs and make the necessary changes.

Lemma 4.78. *Let $\varphi : R \to R'$ be a ring homomorphism.*

(a) *If $I' \lhd R'$, then $\varphi^{-1}(I') \lhd R$.*
(b) *If $I \lhd R$ and φ is an epimorphism, then $\varphi(I) \lhd R'$.*

Theorem 4.79. *Let I be an ideal of a ring R. Every ideal of R/I is of the form J/I, where J is an ideal of R that contains I.*

Example 4.80. Either by applying Theorem 4.79 or by referring to Example 4.41, we see that every ideal of $\mathbb{Z}_n$ is of the form $k\mathbb{Z}_n$ where $k \mid n$. Thus, like $\mathbb{Z}$, the ring $\mathbb{Z}_n$ also has the property that all its additive subgroups are ideals. The reader should not be misled by these examples—most rings contain an abundance of additive subgroups that are not ideals.

Theorem 4.79 is especially useful when I has the following property.

Definition 4.81. Let $I \lhd R$. We say that I is a **maximal ideal** of R if $I \neq R$ and there *does not* exist an ideal J of R such that $I \subsetneq J \subsetneq R$.

For example, $2\mathbb{Z}$ is a maximal ideal of $\mathbb{Z}$. Indeed, if an ideal of $\mathbb{Z}$ contains at least one odd number along with all even numbers, then it contains all integers. More examples will be given in the next section.

Corollary 4.82. *An ideal I of a ring R is maximal if and only if the quotient ring R/I is simple.*

Proof. If I is maximal, then R/I is a nonzero ring (since $I \neq R$) and, according to Theorem 4.79, has no other ideals than I/I (that is, $\{0\}$) and R/I. This means that R/I is simple. Conversely, if I is not maximal, then there exists an ideal J such that $I \subsetneq J \subsetneq R$. Hence, J/I is an ideal of R/I which is readily seen to be different from both $\{0\}$ and R/I. Therefore, R/I is not simple. $\qquad\square$

We remark that a similar result holds for maximal normal subgroups—see Exercise 4.59.

Corollary 4.83. *An ideal I of a commutative ring R is maximal if and only if the quotient ring R/I is a field.*

Proof. It is obvious that R/I is also a commutative ring. This corollary therefore follows from Corollaries 4.72 and 4.82. □

In commutative rings, maximal ideals thus give rise to fields. This will turn out to be important in the study of field extensions.

Exercises

4.84. Explain why a nontrivial homomorphism from a simple ring to another ring is always injective.

Comment. In particular, homomorphisms between fields are thus automatically injective.

4.85. Let X be a nonempty subset of a ring R. Show that the set of all elements of the form

$$r_1 x_1 s_1 + r_2 x_2 s_2 + \cdots + r_n x_n s_n,$$

where $r_i, s_i \in R$ and $x_i \in X$, is the ideal generated by X (i.e., the ideal of R that contains X and is contained in any other ideal that contains X).

Comment. If R is commutative, we can simplify this description by omitting the elements r_i.

4.86. Prove that a ring R is simple if and only if for every $x \neq 0$ in R there exist $r_i, s_i \in R, i = 1, \ldots, n$, such that $\sum_{i=1}^{n} r_i x s_i = 1$.

4.87. Prove that the center of a simple ring is a field.

4.88. Let D be a division ring. Prove that the ring $M_n(D)$ is simple.

Hint. Make use of **matrix units**. These are matrices E_{ij} whose (i, j) entry is 1 and all other entries are 0 (see Example 1.153). The formula $E_{ij} A E_{kl} = a_{jk} E_{il}$, which holds for any matrix $A = (a_{ij})$, is particularly useful.

4.89. Prove that the characteristic of a simple ring is either 0 or a prime.

Comment. This statement is independent of Theorem 3.136 (c). That is, there are rings that have no zero-divisors and are not simple (e.g., $\mathbb{Z}$), and rings that have zero-divisors and are simple (e.g., $M_n(\mathbb{R})$).

4.90. Let V be a vector space over a field F. We say that $T \in \text{End}_F(V)$ has *finite rank* if im T is a finite-dimensional subspace of V. Prove that the set of all endomorphisms of finite rank forms an ideal of $\text{End}_F(V)$. With the help of Example 3.77 and Exercise 4.88, conclude from this that $\text{End}_F(V)$ is simple if and only if V is finite-dimensional.

4.91. Let a be an element of a ring R. Show that $L = \{x \in R \,|\, xa = 0\}$ is a left ideal of R, and $T = \{x \in R \,|\, ax = 0\}$ is a right ideal of R. Provide an example (e.g., in the ring $R = M_2(\mathbb{R})$) showing that $L \cap T$ is not necessarily a left or right ideal.

4.92. Let $L \neq \{0\}$ be a left ideal and $T \neq \{0\}$ be a right ideal of $M_2(\mathbb{R})$. Prove that $t\ell = 0$ cannot hold for all $\ell \in L$ and $t \in T$, and find an example where $\ell t = 0$ for all $\ell \in L, t \in T$.

4.93. Let $I = (X^2 + X) \triangleleft \mathbb{R}[X]$ and $J = (X^2 - X) \triangleleft \mathbb{R}[X]$. Determine IJ, $I \cap J$, and $I + J$. Here, $(q(X))$ denotes the principal ideal $\{q(X)f(X) \,|\, f(X) \in \mathbb{R}[X]\}$ (see Example 4.66).

4.94. Prove that

$$I = \{f \in C(\mathbb{R}) \,|\, f(x) = 0 \text{ for all } x < 0\}$$

and

$$J = \{f \in C(\mathbb{R}) \,|\, f(x) = 0 \text{ for all } x > 0\}$$

are ideals of the ring $C(\mathbb{R})$ satisfying $IJ = I \cap J = \{0\}$ and

$$I + J = \{f \in C(\mathbb{R}) \,|\, f(0) = 0\}.$$

4.95. Let I and J be ideals of a commutative ring R. Prove that $R = I + J$ implies $IJ = I \cap J$.

4.96. Let R_1 and R_2 be arbitrary rings. Prove that every ideal of the direct product $R_1 \times R_2$ is of the form $I_1 \times I_2$ where $I_1 \triangleleft R_1$ and $I_2 \triangleleft R_2$.

Comment. Normal subgroups of the direct product $G_1 \times G_2$ are not necessarily of the form $N_1 \times N_2$ with $N_1 \triangleleft G_1$ and $N_2 \triangleleft G_2$. For example, $\{(x, x) \,|\, x \in G\}$ is a subgroup of $G \times G$ for any group G. If G is Abelian, then it is of course a normal subgroup.

4.97. Determine all ideals of the ring $\mathbb{R} \times \mathbb{R} \times \mathbb{R}$. Which of them are maximal?

4.98. Let I be an additive subgroup of a ring R. Suppose that for all $a, a', b, b' \in R$, the conditions $a + I = a' + I$ and $b + I = b' + I$ imply that $ab + I = a'b' + I$. Prove that I is an ideal.

4.99. Is the ring $\mathbb{C}[X]/(X^2 + 1)$ an integral domain?

4.100. Is the ring $\mathbb{R}[X]/(X^2 + 1)$ an integral domain? Is it isomorphic to the ring $\mathbb{R}[X]/(X^2 - 1)$?

4.101. Let $I_1, \ldots, I_r$ be ideals of a ring R such that $I_i + I_j = R$ for all $i \neq j$. Prove that

$$\varphi : R \to R/I_1 \times \cdots \times R/I_r, \quad \varphi(a) = (a + I_1, \ldots, a + I_r),$$

is an epimorphism with $\ker \varphi = I_1 \cap \cdots \cap I_r$.

Sketch of proof. Checking that φ is a homomorphism with kernel $I_1 \cap \cdots \cap I_r$ is easy. It remains to prove that φ is surjective. Since $I_1 + I_i = R$, $i \geq 2$, there exist $t_i \in I_1$ and $u_i \in I_i$ such that $t_i + u_i = 1$. Hence,

$$1 = (t_2 + u_2) \cdots (t_r + u_r) = t + u_2 \cdots u_r,$$

where $t \in I_1$. This yields $\varphi(u_2 \cdots u_r) = (1, 0, \ldots, 0)$ (here, 1 denotes the unity $1 + I_1$ of the quotient ring R/I_1, and 0 denotes the zero element I_j of R/I_j). Similarly, we see that $\operatorname{im} \varphi$ contains all elements of the form $(0, \ldots, 0, 1, 0, \ldots, 0)$. Hence deduce that φ surjective.

4.102. Find $x \in \mathbb{Z}$ such that

$$x \equiv 2 \;(\mathrm{mod}\,3), \quad x \equiv 1 \;(\mathrm{mod}\,4), \quad x \equiv 3 \;(\mathrm{mod}\,7).$$

This should not be too difficult, one can always use the guessing and checking method. However, such a system of equations does not always have a solution. For example, Corollary 2.4 implies that an integer x satisfying

$$x \equiv 1 \;(\mathrm{mod}\,n_1) \text{ and } x \equiv 0 \;(\mathrm{mod}\,n_2)$$

exists only if n_1 and n_2 are relatively prime. What can be said in general? An answer to this question can be derived from the preceding exercise by taking $R = \mathbb{Z}$. Specifically, show that if $n_1, \ldots, n_r \in \mathbb{N}$ are pairwise relatively prime and $a_1, \ldots, a_r \in \mathbb{Z}$ are arbitrary, then there exists an integer x such that

$$x \equiv a_i \;(\mathrm{mod}\,n_i) \text{ for all } i = 1, \ldots, r.$$

If an integer y is another solution of this system of equations, then

$$x \equiv y \;(\mathrm{mod}\,n_1 \cdots n_r).$$

Comment. This result is called the **Chinese Remainder Theorem**. It was known, in some form, to Chinese mathematicians almost two millennia ago.

4.103. Although named and denoted in the same way, commutators in rings are defined differently as in groups. A **commutator** in a ring R is an element of the form $xy - yx$ with $x, y \in R$. We write this element as $[x, y]$. Let C_R denote the ideal of R generated by all commutators $[x, y]$ in R. Prove that R/C_R is commutative and that C_R is contained in every ideal I of R with the property that R/I is commutative. Describe C_R in the case where R is the ring of all upper triangular $n \times n$ real matrices.

4.104. Prove that every proper ideal I of a ring R is contained in a maximal ideal.

Hint. Use Zorn's lemma in a similar way as in Exercise 1.174. The fact that proper ideals do not contain 1 should be useful.

4.105. Let J_R denote the set of all elements in a nonzero commutative ring R that are not invertible. If J_R is closed under addition, then R is said to be a (commutative) **local ring**. Prove that in this case J_R is a maximal ideal of R (and hence R/J_R is a field). Find all $n \le 10$ such that $\mathbb{Z}_n$ is a local ring. Can you now make a general conjecture about when $\mathbb{Z}_n$ is local?

4.106. Prove that the set of all nilpotent elements of a commutative ring is an ideal. Show by an example that in general this does not hold for noncommutative rings.

4.107. A one-sided or two-sided ideal I of a ring R is said to be **nilpotent** if there exists an $n \in \mathbb{N}$ such that $u_1 \cdots u_n = 0$ for all $u_i \in I$. Show that the set of all *strictly* upper triangular $n \times n$ real matrices is a nilpotent (two-sided) ideal of the ring of all upper triangular $n \times n$ real matrices, which contains all other nilpotent ideals of this ring.

4.108. A one-sided or two-sided ideal I of a ring R is called **nil** if every element in I is nilpotent. Obviously, if I is nilpotent, then it is also nil. Examine the following example to show that the converse is not true. Let R be the (infinite) direct product of the rings $\mathbb{Z}_{2^n}$, i.e., $R = \mathbb{Z}_2 \times \mathbb{Z}_4 \times \mathbb{Z}_8 \times \cdots$ (the operations are defined componentwise, compare Example 1.175). Denote by S the set of all elements $(a_n) = (a_1, a_2, \dots)$ in R such that only finitely many a_n are different from 0. Show that $2S = \{2s \mid s \in S\}$ is a nil, but not nilpotent ideal of R.

4.109. Prove that the sum of two nilpotent two-sided ideals is again a nilpotent ideal, and similarly, the sum of two nil two-sided ideals is again a nil ideal.

4.110. One of the following two statements is definitely true and the proof is short and elementary. The other statement is called the **Köthe conjecture**. The question of whether or not it is true has been open since 1930. Find out which statement is which.

(a) If a ring R contains a nonzero nil one-sided ideal, then it also contains a nonzero nil two-sided ideal.
(b) If a ring R contains a nonzero nilpotent one-sided ideal, then it also contains a nonzero nilpotent two-sided ideal.

4.4 The Isomorphism Theorem and Examples of Quotient Structures

The meaning of isomorphisms and embeddings has already been discussed several times. Roughly speaking, we need isomorphisms to identify algebraic objects, and we need embeddings to view some objects inside some others. What about homomorphisms that are not injective? It does not immediately seem clear what they are good for. We have encountered them, however. Every quotient structure induces the canonical epimorphism which is injective only in the trivial case. In this section, we will prove the Isomorphism Theorem, which shows that in fact every homomorphism is intimately connected with some quotient structure, and then demonstrate through examples how this theorem can be used in describing concrete quotient groups and quotient rings.

4.4.1 The Isomorphism Theorem

The theorem which we are about to establish is of fundamental importance in algebra. It shows that every homomorphism gives rise to an isomorphism. In the literature, it is usually called the "First Isomorphism Theorem". However, we will omit "first" and simply call it the "Isomorphism Theorem". There are two more isomorphism theorems, which will also be briefly discussed. The three theorems are sometimes called "Noether Isomorphism Theorems," named after *Emmy Noether* (1882–1935), one of the most influential algebraists of modern times.

The homomorphism φ in the formulation of the theorem can be of any type, i.e., a group, a ring, a vector space, or an algebra homomorphism. Note that $\ker \varphi$ is a normal subgroup in the first case, an ideal in the second and the fourth case, and a subspace in the third case. In each case, we can thus form the quotient structure $A/\ker \varphi$. Recall also that $\operatorname{im} \varphi$ is an algebraic structure of the same type, i.e., a group, a ring, a vector space, and an algebra, respectively.

Theorem 4.111. (Isomorphism Theorem) *Let $\varphi : A \to A'$ be a homomorphism. Then*

$$A/\ker \varphi \cong \operatorname{im} \varphi.$$

Proof. We consider in detail the case where φ is a group homomorphism. As we know, $\operatorname{im} \varphi$ is then a subgroup of A' and $\ker \varphi$ is a normal subgroup of A, which enables us to form the quotient group $A/\ker \varphi$. Our goal is to find an isomorphism from $A/\ker \varphi$ to $\operatorname{im} \varphi$.

For any $a, a' \in A$, we have

$$
\begin{aligned}
a \ker \varphi = a' \ker \varphi \quad &\Longleftrightarrow \quad a^{-1}a' \in \ker \varphi \\
&\Longleftrightarrow \quad \varphi(a^{-1}a') = 1 \\
&\Longleftrightarrow \quad \varphi(a) = \varphi(a').
\end{aligned}
$$

This implies that the map

$$\overline{\varphi} : A/\ker \varphi \to \operatorname{im} \varphi, \quad \overline{\varphi}(a \ker \varphi) = \varphi(a), \tag{4.3}$$

is well-defined (since $a \ker \varphi = a' \ker \varphi$ implies $\varphi(a) = \varphi(a')$) and injective (since $\varphi(a) = \varphi(a')$ implies $a \ker \varphi = a' \ker \varphi$). Showing that $\overline{\varphi}$ is a homomorphism is straightforward:

$$
\begin{aligned}
\overline{\varphi}(a \ker \varphi \cdot b \ker \varphi) = \overline{\varphi}((ab) \ker \varphi) &= \varphi(ab) \\
&= \varphi(a)\varphi(b) = \overline{\varphi}(a \ker \varphi) \cdot \overline{\varphi}(b \ker \varphi).
\end{aligned}
$$

Since it is obviously surjective, $\overline{\varphi}$ is an isomorphism.

If φ is a ring, a vector space, or an algebra homomorphism, the above proof needs only obvious modifications. Note that $\ker \varphi$ is an ideal in the case when A and A' are rings or algebras, and a subspace in the case when A and A' are vector spaces. Also, cosets are written as $a + \ker \varphi$ instead of $a \ker \varphi$. The settings are thus different, but

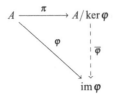

Fig. 4.1 Commutative diagram corresponding to the Isomorphim Theorem

working out the details of the proof for each of the three cases (ring, vector space, algebra) is just a simple exercise. Nevertheless, the reader is strongly advised not to skip it!   □

The Isomorphism Theorem provides a strong link between quotient structures and homomorphic images. Not that this connection is entirely new to us. We knew before that every quotient structure is the image of a homomorphism, namely the canonical epimorphism. What we have learned now is a sort of converse: every homomorphic image is isomorphic to some quotient structure. In some sense, homomorphisms and quotients structures are two sides of the same coin.

It is not only the Isomorphism Theorem itself that is illuminating, but also its proof. One can memorize it through the diagram shown in Figure 4.1. Indeed, by (4.3), the isomorphism $\overline{\varphi}$ is defined as the map such that the composition $\overline{\varphi} \circ \pi$, where π is the canonical epimorphism, is equal to the homomorphism φ—and this is exactly what the diagram indicates if we follow the arrows. Diagrams like this, i.e., diagrams representing maps by arrows in which all map compositions with the same start and endpoint are the same, are called **commutative diagrams**.

As an example of the applicability and efficiency of the Isomorphism Theorem, we now present an alternative proof of the classification theorem for cyclic groups (Theorem 3.9), which is considerably shorter and more conceptual than the one given in Section 3.1.

Example 4.112. Let G be a cyclic group and let $a \in G$ be such that $G = \langle a \rangle$. The map
$$\varphi : \mathbb{Z} \to G, \quad \varphi(n) = a^n,$$
is obviously a group epimorphism. If its kernel is trivial, then it is injective and hence an isomorphism. Otherwise, $\ker \varphi = n\mathbb{Z}$ for some $n \in \mathbb{N}$ (Theorem 2.2). The Isomorphism Theorem implies that
$$\mathbb{Z}_n = \mathbb{Z}/n\mathbb{Z} \cong \operatorname{im} \varphi = G.$$
Thus, either $G \cong \mathbb{Z}$ or $G \cong \mathbb{Z}_n$, which is the content of Theorem 3.9.

In the next two subsections, we will examine concrete examples of normal subgroups and ideals. We will see that the corresponding quotient structures can often be easily described by making use of the Isomorphism Theorem.

The other two isomorphism theorems are also important, but are more technical and not used so frequently as the first one. They cannot be formulated simultaneously for all algebraic structures. The group versions read as follows.

Theorem 4.113. (Second Isomorphism Theorem) *Let G be a group, let $H \leq G$, and let $N \triangleleft G$. Then $H \cap N \triangleleft H$, $N \triangleleft HN$, and*

$$H/(H \cap N) \cong HN/N.$$

Theorem 4.114. (Third Isomorphism Theorem) *Let G be a group and let $M, N \triangleleft G$ with $N \subseteq M$. Then*

$$G/M \cong (G/N)/(M/N).$$

The proofs are sketched in Exercises 4.147 and 4.150.

4.4.2 Examples of Normal Subgroups and Quotient Groups

We start with a trivial example.

Example 4.115. Let G be an arbitrary group. As we already mentioned, the subgroups $\{1\}$ and G are normal. It is immediate from the definition of a quotient group that $G/\{1\} \cong G$ and $G/G \cong \{1\}$. On the other hand, this also follows from the Isomorphism Theorem by taking φ to be the identity id_G and the trivial homomorphism, respectively.

Examples 4.3–4.5 can now be seen in a new light.

Example 4.116. Let G be the additive group $\mathbb{R}^2$ and let H be the x-axis, i.e., $H = \{(x_1, 0) \mid x_1 \in \mathbb{R}\}$. Since G is Abelian, we can form the quotient group G/H. Geometrically, its elements are horizontal lines. How are they added? The coset $(x_1, x_2) + H$ can be written as $(0, x_2) + H$ since

$$(x_1, x_2) - (0, x_2) = (x_1, 0) \in H.$$

The sum of the cosets $(0, x_2) + H$ and $(0, y_2) + H$ is the coset $(0, x_2 + y_2) + H$. The addition of (suitably written) cosets thus corresponds to the addition of real numbers. This leads to the conjecture that

$$G/H \cong \mathbb{R}, \tag{4.4}$$

and indeed, the map

$$(x_1, x_2) + H \mapsto x_2$$

is an isomorphism from G/H to $\mathbb{R}$, as can be immediately verified. One needs only to remember to check that this map is well-defined. It was thus easy to find an explicit isomorphism, but the Isomorphism Theorem provides a shortcut to establishing (4.4). All we have to find is an epimorphism from G to $\mathbb{R}$ with kernel H. The obvious choice is

$$\varphi : G \to \mathbb{R}, \quad \varphi(x_1, x_2) = x_2.$$

The main advantage in using the Isomorphism Theorem is that this way one avoids the question of well-definedness. Not that this question is difficult, but why bother when we got through with it once and for all in the proof of the Isomorphism Theorem.

The next example is conceptually similar, so we do not need to be so detailed.

Example 4.117. Let us describe the quotient group $\mathbb{C}^*/\mathbb{T}$. For any $a \in \mathbb{C}^*$, we can write the coset $a\mathbb{T}$ as $|a|\mathbb{T}$. The product of the cosets $|a|\mathbb{T}$ and $|b|\mathbb{T}$ is the coset $|a||b|\mathbb{T}$. The multiplication of cosets thus corresponds to the multiplication of positive real numbers. Hence, we guess that

$$\mathbb{C}^*/\mathbb{T} \cong \mathbb{R}^+.$$

To prove this, recall that the map $z \mapsto |z|$ is an epimorphism from $\mathbb{C}^*$ to $\mathbb{R}^+$ with kernel $\mathbb{T}$ (see Example 3.66). The desired conclusion thus follows from the Isomorphism Theorem.

Example 4.118. The alternating group A_n is a normal subgroup of the symmetric group S_n. This can be easily deduced directly from the definition of a normal subgroup. On the other hand, it also follows from the fact that A_n is the kernel of the epimorphism $\sigma \mapsto \text{sgn}(\sigma)$ from S_n to the group $(\{1, -1\}, \cdot)$ (see Example 3.68). Applying the Isomorphism Theorem to this situation, we obtain

$$S_n/A_n \cong \mathbb{Z}_2. \tag{4.5}$$

The Isomorphism Theorem actually implies that $S_n/A_n \cong \{1, -1\}$, but the group $\{1, -1\}$ is—as a cyclic group of order 2—isomorphic to $\mathbb{Z}_2$. In fact, it is easy to see that all groups of order 2 are isomorphic. Therefore, (4.5) also follows from the fact that $[S_n : A_n] = 2$ (see Example 4.5).

Example 4.119. The special linear group $\text{SL}_n(F)$ is the kernel of the epimorphism $A \mapsto \det(A)$ from the general linear group $\text{GL}_n(F)$ to the group $(F^*, \cdot)$ (Example 3.69). Consequently, $\text{SL}_n(F) \triangleleft \text{GL}_n(F)$ and

$$\text{GL}_n(F)/\text{SL}_n(F) \cong F^*.$$

Example 4.120. Let G_1 and G_2 be arbitrary groups. Consider their direct product $G_1 \times G_2$. Note that

$$\widetilde{G}_1 := \{(x_1, 1) \,|\, x_1 \in G_1\}$$

is a subgroup. From

$$(a_1, a_2)(x_1, 1)(a_1, a_2)^{-1} = (a_1, a_2)(x_1, 1)(a_1^{-1}, a_2^{-1}) = (a_1 x_1 a_1^{-1}, 1) \in \widetilde{G}_1$$

we see that it is normal, so we can form the quotient group. It should be of no surprise that

$$(G_1 \times G_2)/\widetilde{G}_1 \cong G_2.$$

This can be checked directly or by applying the Isomorphism Theorem to the epimorphism $\varphi : G_1 \times G_2 \to G_2$, $\varphi(x_1, x_2) = x_2$. Similarly,

$$(G_1 \times G_2)/\widetilde{G}_2 \cong G_1$$

where $\widetilde{G}_2$ is the subgroup consisting of elements $(1, x_2)$ with $x_2 \in G_2$.

Rewriting this example for the case where G_1 and G_2 are additive groups, we see that Example 4.116 is a special case of it.

Example 4.121. Let G be an arbitrary group. As already mentioned, its center $Z(G)$ is a normal subgroup. We can therefore form the quotient group $G/Z(G)$. Is it isomorphic to a group that we have already seen? In Example 3.72, we observed that $Z(G)$ is the kernel of the epimorphism $a \mapsto \varphi_a$ from G to the group of all inner automorphisms $\text{Inn}(G)$. Hence,

$$G/Z(G) \cong \text{Inn}(G)$$

by the Isomorphism Theorem.

4.4.3 Examples of Ideals and Quotient Rings

As we have seen, there are many similarities between basic group theory and basic ring theory. However, there are also some differences. We just wrote that the center $Z(G)$ of the group G is a normal subgroup. By analogy, one would expect that the center $Z(R)$ of the ring R is an ideal. However, this is not true, unless R is commutative. Indeed, $Z(R)$ contains the unity 1 of R, and is therefore an ideal only if it is equal to R (see Lemma 4.69).

After this non-example, we proceed to examples.

Example 4.122. The obvious examples of ideals of an arbitrary ring R are $\{0\}$ and R. It is clear that $R/\{0\} \cong R$ in $R/R \cong \{0\}$ (compare Example 4.115).

Example 4.123. As we know from Section 4.3, the prototype example of a quotient ring is $\mathbb{Z}/n\mathbb{Z} = \mathbb{Z}_n$. Recall that p is a prime if and only if $\mathbb{Z}_p$ is a field (Theorem 2.24). We can connect this with Corollary 4.83, which states that an ideal I of a commutative ring R is maximal if and only if R/I is a field. Hence, the following conditions are equivalent for every $p \in \mathbb{N}$:

(i) p is a prime.
(ii) $\mathbb{Z}_p$ $(= \mathbb{Z}/p\mathbb{Z})$ is a field.
(iii) $p\mathbb{Z}$ is a maximal ideal of $\mathbb{Z}$.

The next two examples are reinterpretations of Examples 3.75 and 3.76, obtained by applying the Isomorphism Theorem.

Example 4.124. Let R be an arbitrary ring. We denote by (X) the ideal of $R[X]$ consisting of all polynomials in $R[X]$ with zero constant term. By Example 3.75, (X) is the kernel of the epimorphism from $R[X]$ onto R which assigns to each polynomial its constant term. Therefore,

$$R[X]/(X) \cong R.$$

If $R = F$ is a field, then (X) is a maximal ideal of $F[X]$ by Corollary 4.83. Later, in Chapter 7, we will see that maximal ideals of $F[X]$ are of great significance in field theory. The ideal (X) is just the simplest example.

Example 4.125. Pick a real number x from the interval $[a, b]$. Example 3.76 shows that $I_x := \{f \in C[a, b] \mid f(x) = 0\}$ is the kernel of the epimorphism from $C[a, b]$ onto $\mathbb{R}$ which assigns to each function the function value at x. Hence,

$$C[a, b]/I_x \cong \mathbb{R}.$$

Corollary 4.83 tells us that I_x is a maximal ideal.

Our final example is analogous to Example 4.120.

Example 4.126. Let R_1 and R_2 be rings. Note that

$$\widetilde{R}_1 := \{(x_1, 0) \mid x_1 \in R_1\} \quad \text{and} \quad \widetilde{R}_2 := \{(0, x_2) \mid x_2 \in R_2\}$$

are ideals of the direct product $R_1 \times R_2$. Just as in Example 4.120, we see that

$$(R_1 \times R_2)/\widetilde{R}_1 \cong R_2 \quad \text{and} \quad (R_1 \times R_2)/\widetilde{R}_2 \cong R_1.$$

Corollary 4.82 shows that $\widetilde{R}_1$ is maximal if and only if R_2 is a simple ring.

4.4.4 A Word on Free Objects

This last subsection is a digression which the reader may choose to omit. We will give an informal introduction to some notions that are very important in algebra, but are usually not treated in an introductory course. Hopefully, the more interested readers will benefit from this short discussion.

Let A be a finitely generated commutative algebra over a field F. If $\{a_1, \ldots, a_n\}$ is a set of generators of A, then every element in A can be written as a linear combination of terms of the form $a_1^{k_1} \cdots a_n^{k_n}$, $k_i \geq 0$. That is to say, every element in A can be represented as $f(a_1, \ldots, a_n)$ for some polynomial $f(X_1, \ldots, X_n) \in F[X_1, \ldots, X_n]$; here, $f(a_1, \ldots, a_n)$ stands for the evaluation of this polynomial at $(a_1, \ldots, a_n)$. The map

$$\varepsilon : F[X_1, \ldots, X_n] \to A, \quad \varepsilon\big(f(X_1, \ldots, X_n)\big) = f(a_1, \ldots, a_n)$$

is therefore surjective. It is slightly tedious but nevertheless easy to check that ε is an algebra homomorphism. We call it the **evaluation homomorphism at** $(a_1, \ldots, a_n)$ (compare Example 3.75). Denoting its kernel by I, we have that I is an ideal of $F[X_1, \ldots, X_n]$ and

$$F[X_1, \ldots, X_n]/I \cong A$$

by the Isomorphism Theorem. This shows that every commutative algebra generated by n elements is isomorphic to the quotient algebra of the polynomial algebra in n variables by one of its ideals.

We have restricted ourselves to finitely generated algebras primarily for simplicity. One can talk about the algebra of polynomials in infinitely many variables. Each polynomial in this algebra involves only a finite number of variables, but different polynomials may involve different variables. Just as above, one can show that *every* commutative algebra is isomorphic to the quotient algebra of a polynomial algebra (in finitely or infinitely many variables) by an ideal.

Polynomial algebras thus play a special role in the theory of commutative algebras. In the theory of arbitrary, not necessarily commutative algebras, a similar role is played by what we call **free algebras**. Their elements are **noncommutative polynomials**. They are defined similarly to "ordinary" polynomials, except that the variables X_i do not commute (so $X_1 X_2$ and $X_2 X_1$ are different noncommutative polynomials). A precise definition would take us too far afield, but hopefully the reader has a good impression of what is involved. Every algebra is isomorphic to the quotient algebra of a free algebra by one of its ideals. The proof is just an adaptation of the proof in the commutative case. Incidentally, the usual polynomial algebras are also called **free commutative algebras**.

Similarly, every group is isomorphic to the quotient group of a **free group** by one of its normal subgroups. We will also avoid the precise definition of a free group. To give a basic idea, we only very loosely describe the free group in two generators, which we denote by F_2. Let us denote the generators by x and y. These two symbols play a similar formal role as variables in polynomials. The group F_2 consists of what we call **words**. These are formal expressions written as sequences of powers of x and y with nonzero integer exponents, i.e., expressions such as, for example,

$$x, \quad xy, \quad x^2 y^{-1} x^{-1}, \quad y^{-4} x^{-2} y^3 x y^{-1}, \quad \text{etc.}$$

We adjoin the symbol 1, also called the empty word, to F_2. The multiplication of words is defined in a self-evident way. For example,

$$x^3 y^{-2} \cdot x^{-4} y = x^3 y^{-2} x^{-4} y$$

and

$$y^{-2} x^5 y^{-2} x \cdot x^{-1} y^2 x^{-2} y = y^{-2} x^3 y.$$

The empty word 1 acts as an identity element. In this way, the set of all words F_2 becomes a group.

Finally we mention that every Abelian group is isomorphic to the quotient group of a **free Abelian group** by one of its normal subgroups. A free Abelian group is an Abelian group with basis—see Exercise 1.179.

Exercises

4.127. Let G be the group of all non-constant linear functions (see Exercise 1.58). Denote by N the subset of G consisting of all functions of the form $f(x) = x + b$, where $b \in \mathbb{R}$. Show that $N \lhd G$ and $G/N \cong \mathbb{R}^*$.

4.128. Prove that $\mathbb{R}/\mathbb{Z} \cong \mathbb{T}$.

Hint. $\varphi(x) = e^{2\pi i x}$.

4.129. To which familiar group is $\mathbb{R}^*/\{1, -1\}$ isomorphic?

4.130. To which familiar group is $\mathbb{R}^*/\mathbb{R}^+$ isomorphic?

4.131. To which familiar group is $\mathbb{C}^*/\mathbb{R}^+$ isomorphic?

4.132. To which familiar group is U_n/SU_n isomorphic?

Hint. What is the determinant of a unitary matrix?

4.133. Verify that the set of matrices $G = \left\{ \begin{bmatrix} 1 & m \\ 0 & 1 \end{bmatrix} \mid m \in \mathbb{Z} \right\}$ is an Abelian group under the usual matrix multiplication. Find a subgroup $N \leq G$ such that $G/N \cong \mathbb{Z}_3$.

4.134. Find a nontrivial subgroup N of the group $\mathbb{T}$ such that $\mathbb{T}/N \cong \mathbb{T}$.

Hint. If G is any group and φ is a surjective endomorphism of G, then $G/\ker \varphi \cong G$.

4.135. Find a non-Abelian group G such that the group $\text{Inn}(G)$ is Abelian.

4.136. Let F be a field of prime characteristic p. Give an alternative proof of Theorem 3.137 (b) by applying the Isomorphism Theorem to the homomorphism $\varphi : \mathbb{Z} \to F, \varphi(n) = n \cdot 1$.

4.137. Let R be a ring without unity and let $\mathbb{Z} \times R$ be the ring from Exercise 1.80. Prove that $I = \{0\} \times R$ is an ideal of $\mathbb{Z} \times R$ and that $(\mathbb{Z} \times R)/I \cong \mathbb{Z}$.

4.138. Prove that $I = \{f \in C[0, 1] \mid f(0) = f(1) = 0\}$ is an ideal of $C[0, 1]$ and that $C[0, 1]/I \cong \mathbb{R} \times \mathbb{R}$.

4.139. Let c be the ring of all convergent real sequences (see Section 2.3) and let c_0 be the set of all real sequences whose limit is 0. Show that c_0 is a maximal ideal of c and that $c/c_0 \cong \mathbb{R}$.

4.140. Prove that the ring $\mathbb{R}[X]/(X^2)$ is isomorphic to the ring of all 2×2 matrices of the form $\begin{bmatrix} a & b \\ 0 & a \end{bmatrix}$, $a, b \in \mathbb{R}$. (Recall that (X^2) denotes the ideal of $\mathbb{R}[X]$ consisting of all polynomials of the form $X^2 f(X)$, $f(X) \in \mathbb{R}[X]$.)

4.141. Prove that $M_2(2\mathbb{Z})$, the set of all 2×2 matrices with entries in $2\mathbb{Z}$, is an ideal of the ring $M_2(\mathbb{Z})$, and that $M_2(\mathbb{Z})/M_2(2\mathbb{Z}) \cong M_2(\mathbb{Z}_2)$. More generally, if I is an ideal of the ring R, then $M_n(I)$ is an ideal of the ring $M_n(R)$. To which ring is $M_n(R)/M_n(I)$ isomorphic?

4.142. Prove that $I = \left\{ \begin{bmatrix} 0 & k \\ 0 & 0 \end{bmatrix} \mid k \in \mathbb{Z} \right\}$ is an ideal of the ring $R = \left\{ \begin{bmatrix} n & k \\ 0 & n \end{bmatrix} \mid n, k \in \mathbb{Z} \right\}$. To which familiar ring is R/I isomorphic?

4.143. Find ideals I and J of the ring $R = \left\{ \begin{bmatrix} m & k \\ 0 & n \end{bmatrix} \mid m, n, k \in \mathbb{Z} \right\}$ such that $|R/I| = |R/J| = 4$ and $R/I \not\cong R/J$.

4.144. Find an ideal I of the ring $\mathbb{Z}[X]$ such that $\mathbb{Z}[X]/I \cong \mathbb{Z}_3[X]$.

4.145. Find an ideal J of the ring $\mathbb{Z}[X]$ such that $\mathbb{Z}[X]/J \cong \mathbb{Z}_3$.

Comment. Since $\mathbb{Z}_3$ is a field, J is a maximal ideal.

4.146. Let U be a subspace of a finite-dimensional vector space V. Prove that $\dim V/U = \dim V - \dim U$.

4.147. Prove Theorem 4.113 (Second Isomorphism Theorem) by applying the (First) Isomorphism Theorem to the homomorphism $\varphi : H \to HN/N, \varphi(h) = hN$.

4.148. Let U and W be subspaces of a vector space V. Show that

$$U/(U \cap W) \cong (U + W)/W.$$

Together with the result of Exercise 4.146, this implies the following well-known formula concerning subspaces of finite-dimensional spaces:

$$\dim(U + W) + \dim U \cap W = \dim U + \dim W.$$

Comment. This is the Second Isomorphism Theorem for vector spaces. The method of proof is the same as in the group case. This also applies to the ring version which reads as follows: If L is a subring and I is an ideal of a ring R, then $L/(L \cap I) \cong (L + I)/I$. A similar result holds for algebras.

4.149. Let $\varphi : G \to G'$ be a group homomorphism and let N be a normal subgroup of G such that $N \subseteq \ker \varphi$. Show that

$$\overline{\varphi} : G/N \to G', \quad \overline{\varphi}(aN) = \varphi(a),$$

is a (well-defined) homomorphism satisfying $\operatorname{im} \overline{\varphi} = \operatorname{im} \varphi$ and $\ker \overline{\varphi} = \ker \varphi/N$.

Comment. This is a generalization of the Isomorphism Theorem. The proof is only sligthy different.

4.150. Let G be a group, let $M, N \lhd G$, and let $N \subseteq M$. Deduce from the conclusion of the preceding exercise that the canonical epimorphism $\pi : G \to G/M$ induces an epimorphism $\overline{\pi} : G/N \to G/M$ such that $\ker \overline{\pi} = M/N$. Note that Theorem 4.114

(Third Isomorphism Theorem) follows by applying the (First) Isomorphism Theorem to $\overline{\pi}$.

Comment. The same method of proof yields the Third Isomorphism Theorem for other algebraic structures. The ring (algebra) version states that $R/I \cong (R/J)/(I/J)$ where $J \subseteq I$ are ideals of the ring (algebra) R, and the vector space version states that $V/U \cong (V/W)/(U/W)$ where $W \subseteq U$ are subspaces of the vector space V.

4.151. Let $\varphi : G \to G'$ and $\psi : G \to G''$ be group homomorphisms. Which of the following statements is true:

(a) If $\ker \varphi = \ker \psi$, then $\operatorname{im} \varphi \cong \operatorname{im} \psi$.
(b) If $\operatorname{im} \varphi \cong \operatorname{im} \psi$, then $\ker \varphi = \ker \psi$.

Provide a counterexample to the false statement.

4.5 Direct Products and Sums Revisited

Direct products and sums of different algebraic structures were introduced in Section 1.8, and afterward used in several examples and exercises. Being familiar with normal subgroups and ideals, we are now in a position to consider them from a different perspective.

4.5.1 External and Internal Direct Products of Groups

After the introduction of direct products of groups in Section 1.8, we mentioned in passing that they are more precisely termed *external* direct products (or external direct sums in the case of additive groups). We will now introduce *internal* direct products (and sums). The external direct products are used to build new, larger groups from old ones, while the internal direct products are used to break groups into smaller pieces. They both concern the same phenomenon, but are inverse to each other. The external direct product is about composing, while the internal direct product is about decomposing.

To motivate the definition of internal direct product, we will look at abstract properties of external direct products. Let $G_1, \ldots, G_s$ be groups and let

$$G := G_1 \times \cdots \times G_s$$

be their external direct product. For each i,

$$\widetilde{G}_i := \{(1, \ldots, 1, x_i, 1, \ldots, 1) \mid x_i \in G_i\} \tag{4.6}$$

is a normal subgroup of G (see Example 4.120). Informally, we may think of $\widetilde{G}_i$ as being equal to G_i. Strictly speaking, however, $\widetilde{G}_i$ is a subgroup of G isomorphic to

G_i (which, from the formal viewpoint, is not contained in G). It is intuitively clear that $\widetilde{G}_1, \ldots, \widetilde{G}_s$ carry complete information about G. Let us try to express this in precise terms. What are the key properties of $\widetilde{G}_i$? Firstly, since

$$(x_1, 1, \ldots, 1)(1, x_2, 1, \ldots, 1) \cdots (1, \ldots, 1, x_s) = (x_1, x_2, \ldots, x_s),$$

we see that G is equal to their product:

$$G = \widetilde{G}_1 \widetilde{G}_2 \cdots \widetilde{G}_s.$$

Secondly, each $\widetilde{G}_i$ has trivial intersection with the product of the other ones:

$$\widetilde{G}_i \cap \left(\widetilde{G}_1 \cdots \widetilde{G}_{i-1} \widetilde{G}_{i+1} \cdots \widetilde{G}_s \right) = \{1\}.$$

These are the two properties that constitute the definition of the internal direct product.

Definition 4.152. A group G is the **internal direct product** of normal subgroups $N_1, \ldots, N_s$ if the following conditions are satisfied:

(a) $G = N_1 N_2 \cdots N_s$.
(b) $N_i \cap (N_1 \cdots N_{i-1} N_{i+1} \cdots N_s) = \{1\}$ for all $i = 1, \ldots, s$.

The discussion before the definition shows that if G is the external direct product of groups $G_1, \ldots, G_s$, then G is the internal direct product of the normal subgroups $\widetilde{G}_i$ defined in (4.6). Since $\widetilde{G}_i$ is only formally different from G_i, in some sense this shows that external direct products are also internal. Theorem 4.154 will tell us that a kind of converse is also true. To prove it, we first have to get some insight into internal direct products.

From condition (b) it follows that $N_i \cap N_j = \{1\}$ for all $i \neq j$, and hence Theorem 4.30 (d) shows that

$$n_i n_j = n_j n_i \quad \text{for all } n_i \in N_i, n_j \in N_j, i \neq j. \tag{4.7}$$

We emphasize that the elements of N_i do not necessarily commute among themselves. They only commute with the elements of other subgroups N_j.

Condition (a) tells us that every element in G can be written as a product of elements of $N_1, \ldots, N_s$ in at least one way. We will now show that condition (b) implies that this can be done in exactly one way. Moreover, this uniqueness property is characteristic for internal direct products.

Lemma 4.153. *Let $N_1, \ldots, N_s$ be normal subgroups of a group G. The following conditions are equivalent:*

(i) *G is the internal direct product of $N_1, \ldots, N_s$.*
(ii) *Every element of G can be written, in a unique way except for the order of the factors, as a product $n_1 n_2 \cdots n_s$ where $n_i \in N_i$, $i = 1, \ldots, s$.*

Proof. (i) $\implies$ (ii). Assume that G is the internal direct product of $N_1, \ldots, N_s$. We have to prove that if $n_i, r_i \in N_i, i = 1, \ldots, s$, satisfy

$$n_1 n_2 \cdots n_s = r_1 r_2 \cdots r_s, \tag{4.8}$$

then $n_i = r_i$ for each i. If we multiply (4.8) from the left by r_1^{-1} and from the right by $(n_2 \cdots n_s)^{-1}$, we obtain

$$r_1^{-1} n_1 = (r_2 \cdots r_s)(n_2 \cdots n_s)^{-1}.$$

Since $N_2 \cdots N_s$ is a subgroup (see Remark 4.31), it follows that $r_1^{-1} n_1 \in N_2 \cdots N_s$. However, $r_1^{-1} n_1$ is an element of N_1, so condition (b) implies that $n_1 = r_1$. Therefore, (4.8) reduces to $n_2 \cdots n_s = r_2 \cdots r_s$. We can now repeat the argument and show that $n_2 = r_2$. Continuing in this fashion, we obtain that $n_i = r_i$ for each i.

(ii) $\implies$ (i). Assume now that every element in G can be written as $n_1 n_2 \cdots n_s$ for some uniquely determined $n_i \in N_i$. We must show that condition (b) holds. Take

$$a \in N_i \cap (N_1 \cdots N_{i-1} N_{i+1} \cdots N_s).$$

Then a can be written in two ways:

$$a = 1 \cdots 1 \cdot n_i \cdot 1 \cdots 1 \quad \text{and} \quad a = n_1 \cdots n_{i-1} \cdot 1 \cdot n_{i+1} \cdots n_s,$$

where $n_j \in N_j$. The uniqueness assumption implies that $n_j = 1$ for each j. Hence $a = 1$, proving (b). $\qquad\square$

We can now show that internal direct products are, up to isomorphism, equal to external direct products.

Theorem 4.154. *If a group G is the internal direct product of normal subgroups $N_1 \ldots, N_s$, then G is isomorphic to $N_1 \times \cdots \times N_s$, the external direct product of $N_1 \ldots, N_s$.*

Proof. Define $\varphi : N_1 \times \cdots \times N_s \to G$ by

$$\varphi\big((n_1, \ldots, n_s)\big) = n_1 \cdots n_s.$$

We want to show that φ is an isomorphism. The fact that φ is bijective is exactly the content of Lemma 4.153. Take $n_i, r_i \in N_i$ and set

$$u := \varphi\big((n_1, \ldots, n_s) \cdot (r_1, \ldots, r_s)\big).$$

Our goal is to show that u is equal to

$$\varphi\big((n_1, \overset{\bullet}{\ldots}, n_s)\big) \cdot \varphi\big((r_1, \ldots, r_s)\big) = n_1 \cdots n_s r_1 \cdots r_s.$$

From

$$(n_1, \ldots, n_s) \cdot (r_1, \ldots, r_s) = (n_1 r_1, \ldots, n_s r_s)$$

and the definiton of φ, we see that

$$u = n_1 r_1 n_2 r_2 n_3 r_3 \cdots n_s r_s.$$

According to (4.7), n_2 commutes with r_1, n_3 commutes with both r_2 and r_1, etc. This yields the desired conclusion that $u = n_1 \cdots n_s r_1 \cdots r_s$. $\square$

The notions of external and internal direct products are thus equivalent. It is therefore common to omit the adjectives "external" and "internal" and speak only about "direct products," as well as to use the notation $N_1 \times \cdots \times N_s$ also for the internal direct product. This does not mean that there is no difference between the two notions. Starting with given groups, it is trivial to form their external direct product, but finding normal subgroups inside a given group G whose internal direct product is G may be difficult.

We will now look at a few examples. For simplicity, we will restrict ourselves to the case where $s = 2$. The definition then reads as $G = N_1 N_2$ and $N_1 \cap N_2 = \{1\}$. By Lemma 4.153, this is equivalent to the condition that every element in G can be uniquely written as $n_1 n_2$ with $n_i \in N_i$. As a consequence, we have that the map $\varphi : G \to N_2$, $\varphi(n_1 n_2) = n_2$, is well-defined. With the help of (4.7), we see that φ is an epimorphism. Since $\ker \varphi = N_1$, the Isomorphism Theorem implies that

$$G/N_1 \cong N_2 \tag{4.9}$$

(compare this with Example 4.120). Similarly, of course, $G/N_2 \cong N_1$.

We begin with a trivial example.

Example 4.155. Every group G is the internal direct product of the normal subgroups G and $\{1\}$. Sometimes—for example, if G is simple—this trivial decomposition of G is also the only one that exists.

Example 4.156. The dihedral group D_4 consists of elements $1, r, s, rs$ that satisfy $r^2 = s^2 = 1$ and $rs = sr$. Observe that D_4 is the internal direct product of the (cyclic) subgroups $N_1 = \{1, r\}$ and $N_2 = \{1, s\}$.

Example 4.157. The group $\mathbb{C}^*$ is the internal direct subgroups of the subgroups $\mathbb{R}^+$ and $\mathbb{T}$. Indeed, every $z \in \mathbb{C}^*$ can be written as the product of $|z| \in \mathbb{R}^+$ and $\frac{z}{|z|} \in \mathbb{T}$, and it is clear that $\mathbb{R}^+ \cap \mathbb{T} = \{1\}$. From (4.9) we thus infer that $\mathbb{C}^*/\mathbb{T} \cong \mathbb{R}^+$, which was already observed in Example 4.117. Similarly, $\mathbb{C}^*/\mathbb{R}^+ \cong \mathbb{T}$.

Example 4.158. The special linear group $\mathrm{SL}_n(F)$ is a normal subgroup of the general linear group $\mathrm{GL}_n(F)$ (Example 4.119). One can show that the center Z of $\mathrm{GL}_n(F)$, which is of course a normal subgroup, consists of scalar matrices λI where $\lambda \in F^*$. Assume that $x \mapsto x^n$ is a bijective function from F to F (this holds, for instance, when $F = \mathbb{R}$ and n is odd). Let us show that $\mathrm{GL}_n(F)$ is then the internal direct product of Z and $\mathrm{SL}_n(F)$. Take $A \in \mathrm{GL}_n(F)$. By assumption, there exists a $\lambda \in F^*$ such that $\lambda^n = \det(A)$. The determinant of the matrix $\lambda^{-1} A$ then equals 1, so $\lambda^{-1} A \in \mathrm{SL}_n(F)$. Writing A as $\lambda I \cdot \lambda^{-1} A$, we thus see that $\mathrm{GL}_n(F)$ is the product of Z and $\mathrm{SL}_n(F)$.

The scalar matrix λI lies in $\mathrm{SL}_n(F)$ if and only if $\lambda^n = 1$. Since the function $x \mapsto x^n$ is injective, this holds only when $\lambda = 1$. Therefore, $Z \cap \mathrm{SL}_n(F) = \{1\}$.

When dealing with additive groups, we replace the term "direct product" with "**direct sum**," and instead of $N_1 \times \cdots \times N_s$ write $N_1 \oplus \cdots \oplus N_s$ (we use this notation for both the internal and external direct sum). Thus, if G is an additive group and $N_1, \ldots, N_s$ are subgroups of G, then

$$G = N_1 \oplus \cdots \oplus N_s$$

means that

$$G = N_1 + \cdots + N_s$$

and

$$N_i \cap (N_1 + \cdots + N_{i-1} + N_{i+1} + \cdots + N_s) = \{0\}$$

for all $i = 1, \ldots, s$. According to Lemma 4.153, in this case every element in G can be written as $n_1 + \cdots + n_s$ with $n_i \in N_i$, in a unique way.

Example 4.159. The group $\mathbb{Z}_6$ is the internal direct sum of the subgroups $\{0, 2, 4\}$ and $\{0, 3\}$. They are isomorphic to $\mathbb{Z}_3$ and $\mathbb{Z}_2$, respectively. Consequently, $\mathbb{Z}_6$ is isomorphic to the external direct sum $\mathbb{Z}_3 \oplus \mathbb{Z}_2$ (compare Exercise 3.16).

Example 4.160. The groups $\mathbb{Z}_2, \mathbb{Z}_3, \mathbb{Z}_4, \mathbb{Z}_5$, on the other hand, cannot be written as direct sums of their proper nontrivial subgroups. This is simply because the first two and the last one do not even have proper nontrivial subgroups, while $\mathbb{Z}_4$ has only one, namely $\{0, 2\}$.

Direct sums in vector spaces are considered in the same way as direct sums in additive groups, only the role of subgroups is replaced by subspaces. The additional operation of scalar multiplication does not cause any difficulties, and a version of Theorem 4.154 for vector spaces can be easily established. The details are left to the reader. It should be remarked, however, that although the definition of direct sums in vector spaces differs only slightly from that of direct sums in additive groups, there is a substantial difference concerning the existence; see Exercises 4.174 and 4.175.

4.5.2 Direct Products and Sums in Rings

The study of direct products and sums in rings is conceptually similar to that in groups, but different in details. Before starting, we remark that this subsection may be considered optional since its results will not play a significant role later in this book.

A ring R is, in particular, an additive group, so we can speak about expressing R as the direct sum of some of its additive subgroups. We will be interested in the situation where these subgroups are ideals, which we will denote by $I_1, \ldots, I_s$. By

Lemma 4.153, $R = I_1 \oplus \cdots \oplus I_s$ if and only if every element of R can be written in exactly one way as $u_1 + \cdots + u_s$, where $u_j \in I_j$.

We will see that direct sums of ideals are intimately connected with direct products of rings. Note, first of all, that if $R_1, \ldots, R_s$ are rings and $R = R_1 \times \cdots \times R_s$ is their direct product, then the sets

$$\widetilde{R}_j := \{(0, \ldots, 0, x_j, 0 \ldots, 0) \mid x_j \in R_j\}$$

are ideals of R (Example 4.126) and $R = \widetilde{R}_1 \oplus \cdots \oplus \widetilde{R}_s$. Theorem 4.161 below establishes a kind of converse to this observation.

Direct sums of ideals are best understood with the help of idempotents. Recall first that an element e of a ring R is called an idempotent if $e^2 = e$. We say that idempotents e and f are **orthogonal** if $ef = fe = 0$. For example, e and $1 - e$ are orthogonal idempotents for every idempotent e. By saying that idempotents $e_1, \ldots, e_s$ are *pairwise* orthogonal we mean, of course, that e_j and e_k are orthogonal whenever $j \neq k$. An idempotent lying in the center of R will be called a **central idempotent**. If e is a central idempotent, then $eR := \{ex \mid x \in R\}$ is an ideal of R (the above defined ideals $\widetilde{R}_j$ are of such form with e being $(0, \ldots, 0, 1, 0, \ldots, 0)$). Unless $e = 1$, eR is not a subring of R since it does not contain the unity 1. Nevertheless, it is a ring since it has its own unity, namely e. Indeed, $e(ex) = (e^2)x = ex$ for every $x \in R$, and hence, since e belongs to the center of R, also $(ex)e = ex$. This explains why the direct product of ideals I_j occurs in the next theorem. Namely, we may view them as rings.

Theorem 4.161. *Let $I_1, \ldots, I_s$ be ideals of a ring R. If $R = I_1 \oplus \cdots \oplus I_s$, then there exist pairwise orthogonal central idempotents $e_1, \ldots, e_s \in R$ such that $e_1 + \cdots + e_s = 1$ and $I_j = e_j R$ for all $j = 1, \ldots, s$. Moreover, R is isomorphic to the direct product $I_1 \times \cdots \times I_s$.*

Proof. By assumption, $I_j \cap I_k = \{0\}$ if $j \neq k$. Since the product of two ideals is contained in their intersection, it follows that

$$I_j I_k = \{0\} \quad \text{whenever } j \neq k. \tag{4.10}$$

Let $e_j \in I_j$ be such that $e_1 + \cdots + e_s = 1$. In light of (4.10), multiplying this equality from the left (resp. right) by $u_j \in I_j$, gives $u_j e_j = u_j$ (resp. $e_j u_j = u_j$). Thus,

$$u_j e_j = e_j u_j = u_j \quad \text{for all } u_j \in I_j. \tag{4.11}$$

Taking e_j for u_j we obtain $e_j^2 = e_j$. From (4.10), we see that $e_1, \ldots, e_s$ are pairwise orthogonal idempotents. Next, (4.11) implies that e_j commutes with all elements of I_j. Since, by (4.10), $e_j u_k = 0 = u_k e_j$ for all $u_k \in I_k$ with $k \neq j$, it follows that e_j lies in the center of R. By (4.11), I_j is contained $e_j R$. As I_j is an ideal containing e_j, the converse inclusion $e_j R \subseteq I_j$ also holds. Thus, $I_j = e_j R$.

We are left to prove that $I_1 \times \cdots \times I_s \cong R$. Let us show that

$$\varphi : I_1 \times \cdots \times I_s \to R, \quad \varphi\big((u_1, \ldots, u_s)\big) = u_1 + \cdots + u_s$$

is an isomorphism. Since every element in R can be uniquely written as $u_1 + \cdots + u_s$ with $u_j \in I_j$, φ is bijective. The proof that φ is additive is straightforward and we leave it to the reader. According to (4.11), e_j is the unity of the ring I_j, and so $(e_1, \ldots, e_s)$ is the unity of the ring $I_1 \times \cdots \times I_s$. As $e_1 + \cdots + e_s = 1$, we see that φ sends the unity to the unity. It remains to show that for any $u_j, v_j \in I_j$,

$$\varphi\big((u_1, \ldots, u_s) \cdot (v_1, \ldots, v_s)\big) = \varphi\big((u_1v_1, \ldots, u_sv_s)\big) = u_1v_1 + \cdots + u_sv_s$$

is equal to

$$\varphi\big((u_1, \ldots, u_s)\big) \cdot \varphi\big((v_1, \ldots, v_s)\big) = (u_1 + \cdots + u_s)(v_1 + \cdots + v_s).$$

Since, by (4.10), $u_j v_k = 0$ for all $j \neq k$, this is indeed true.   □

Example 4.162. If $e_1 = 1$ and $e_2 = 0$, we get the trivial case: $I_1 = R$ and $I_2 = \{0\}$. If R has no central idempotents different from 0 and 1 (in particular, if R has no zero-divisors), this is also the only possible case.

Example 4.163. In Example 4.159, we observed that the group $(\mathbb{Z}_6, +)$ is the direct sum of the subgroups $\{0, 2, 4\}$ and $\{0, 3\}$. These two subgroups are ideals of the ring $(\mathbb{Z}_6, +, \cdot)$. The corresponding idempotents are $e_1 = 4$ and $e_2 = 1 - e_1 = 3$.

Example 4.164. A direct verification shows that $e_1 = 6$, $e_2 = 10$, $e_3 = 15$ are pairwise orthogonal idempotents of the ring $\mathbb{Z}_{30}$, and that $e_1 + e_2 + e_3 = 1$. Therefore, $\mathbb{Z}_{30} = e_1\mathbb{Z}_{30} \oplus e_2\mathbb{Z}_{30} \oplus e_3\mathbb{Z}_{30}$. Note also that $e_1\mathbb{Z}_{30} \cong \mathbb{Z}_5$, $e_2\mathbb{Z}_{30} \cong \mathbb{Z}_3$, and $e_3\mathbb{Z}_{30} \cong \mathbb{Z}_2$. Consequently, $\mathbb{Z}_{30} \cong \mathbb{Z}_5 \times \mathbb{Z}_3 \times \mathbb{Z}_2$.

We remark that some authors call the direct product of (a finite number of) rings the *direct sum of rings*, and some use the term *direct product of ideals* for what we call the direct sum of ideals. It is therefore not uncommon to use the adjectives "external" and "internal" also in connection with direct products and sums in rings. We have chosen a terminology that made it possible to avoid them.

Finally, we mention that everything that was said about direct products and sums in rings carries over to direct products and sums in algebras.

Exercises

4.165. Suppose a group G is the direct product of normal subgroups $N_1 \ldots, N_s$ which are all Abelian. Prove that G is Abelian too.

Comment. In this and further exercises, we omit writing "internal".

4.166. Prove that $\mathbb{R}^*$ is the direct product of $\mathbb{R}^+$ and $\{1, -1\}$.

4.167. Prove that a non-Abelian group of order less than 12 is not equal to a direct product of proper nontrivial normal subgroups.

Hint. You can use the fact that all groups of order less than 6 are Abelian. We will show this in Section 6.1, when discussing consequences of Lagrange's Theorem. But you can try to prove this now!

4.168. Prove that D_{12} is the direct product of $\langle r^2, s \rangle$ and $\langle r^3 \rangle$.

4.169. Determine all $n \in \mathbb{N}$ such that the orthogonal group $O_n(\mathbb{R})$ is the direct product of the normal subgroups $SO_n(\mathbb{R})$ and $\{-I, I\}$.

4.170. Let $p_1, \ldots, p_s$ be distinct primes. Prove that the subgroup $\langle p_1, \ldots, p_s \rangle$ of the multiplicative group $\mathbb{R}^+$ is the direct product of the cyclic subgroups $\langle p_1 \rangle, \ldots, \langle p_s \rangle$.

4.171. Let G be a group. If $N \lhd G$ and $H \leq G$ are such that $G = NH$ and $N \cap H = \{1\}$, then we say that G is the **semidirect product** of N and H, and write $G = N \rtimes H$. Of course, if H is normal, then the semidirect product of N and H is the same as the direct product of N and H. Let $n \geq 3$. Prove that $S_n = A_n \rtimes \langle (1\,2) \rangle$, but there does not exist a a normal subgroup H of S_n such that S_n is the direct product of A_n and H.

4.172. Prove that the group $\mathbb{Z}$ is not a direct sum of proper nontrivial subgroups.

4.173. Let V_1, V_2, V_3 be distinct 1-dimensional subspaces of a 2-dimensional vector space V. Show that $V = V_1 + V_2 + V_3$ and $V_i \cap V_j = \{0\}$ for all $i \neq j$, but V is not the direct sum of $V_1, V_2,$ and V_3.

Comment. The aim of this exercise is to point out that condition (b) from the definition of the internal direct sum cannot be replaced by the condition that $N_i \cap N_j = \{1\}$ for all $i \neq j$.

4.174. Let V be a finite-dimensional vector space. Prove that, for every subspace U of V, there exists a subspace W such that $V = U \oplus W$.

Hint. A basis of U is a linearly independent subset of V and can be extended to a basis of V.

4.175. Let V be a finite-dimensional real vector space. Prove that, for every proper nonzero subspace U of V, there exist infinitely many subspaces W such that $V = U \oplus W$.

Comment. The assumption in this and the preceding exercise that V is finite-dimensional is actually redundant. Infinite-dimensional spaces also have the property that every linearly independent set can be extended to a basis.

4.176. Prove that the sum of pairwise orthogonal idempotents is an idempotent. Is the orthogonality assumption necessary?

4.177. Find s pairwise orthogonal nonzero idempotents in the ring $M_s(\mathbb{R})$ whose sum is the unity of $M_s(\mathbb{R})$. Can they be central idempotents?

4.178. Suppose that $e_1, \ldots, e_s$ are pairwise orthogonal idempotents of a ring R such that $e_1 + \cdots + e_s = 1$. Show that $R = e_1 R \oplus \cdots \oplus e_s R$.

Comment. We are not assuming that e_j are central. Therefore, $e_j R$ are (only) right ideals. In the special case where e_j are central, $e_j R$ are two-sided ideals and the statement of the exercise is the converse of the first statement of Theorem 4.161.

4.179. Let I be an ideal of a ring R. Prove that the following conditions are equivalent:

 (i) I is a ring (i.e., I has its own unity).
 (ii) There exists a central idempotent e in R such that $I = eR$.
 (iii) There exists an ideal J of R such that $R = I \oplus J$.

4.180. Let I be an ideal of a ring R. Prove that if there exists an ideal J of R such that $R = I \oplus J$, then J is unique.

4.181. Suppose a ring R is equal to the direct sum of s nonzero ideals $I_1, \ldots, I_s$. Prove that R has at least 2^s central idempotents.

4.182. Let p be a prime and k be a natural number. Prove that 0 in 1 are the only idempotents in $\mathbb{Z}_{p^k}$; moreover, every element in $\mathbb{Z}_{p^k}$ is either invertible or nilpotent.

4.183. Suppose that $n \in \mathbb{N}$ has at least two prime divisors. Prove that $\mathbb{Z}_n$ then contains an idempotent different from 0 and 1.

Hint. If n_1 and n_2 are relatively prime, then there exists an $x \in \mathbb{Z}$ such that $x \equiv 1 \pmod{n_1}$ and $x \equiv 0 \pmod{n_2}$ (see Exercise 4.102).

4.184. The set R of all real sequences is a ring under the usual addition and multiplication (see Example 1.175). Prove that R contains ideals I and J such that $R = I \oplus J$ and $R \cong I \cong J$ (so $R \cong R \times R$ by Theorem 4.161). Does the ring of all convergent sequences also contain such a pair of ideals?

4.185. Let $G = \{g_1, \ldots, g_n\}$ be any set. The set of all formal linear combinations $\sum_{i=1}^n \lambda_i g_i$, where λ_i are elements of a field F, becomes a vector space over F with basis G if we define addition and scalar multiplication in the obvious way (just as they would be defined if $\sum_{i=1}^n \lambda_i g_i$ was written as the n-tuple $(\lambda_1, \ldots, \lambda_n)$). Now assume that G is a group. Then this vector space becomes an algebra if we extend the multiplication on G to the whole vector space in the only possible way, that is, by taking into account the distributive laws and the equality $\lambda_i g_i \cdot \mu_j g_j = (\lambda_i \mu_j) g_i g_j$; here, $g_i g_j$ stands for the product of g_i in g_j in the group G, and $\lambda_i \mu_j$ stands for the product of λ_i and μ_j in the field F. We call it the **group algebra** of the group G over the field F, and denote it by $F[G]$.

 (a) Show that the element $u := \sum_{i=1}^n g_i$ satisfies $u g_j = g_j u = u$ for every $g_j \in G$.
 (b) Suppose that the characteristic of F is either 0 or does not divide $|G|$. Prove that $e := \frac{1}{|G|} u$ is a central idempotent of $F[G]$, the dimension of the ideal $e F[G]$ is 1, and the ideal $(1 - e) F[G]$ consists of all elements $\sum_{i=1}^n \lambda_i g_i$ such that $\sum_{i=1}^n \lambda_i = 0$.
 (c) Let $G = \{1, -1\}$, i.e., G is the (multiplicative) cyclic group with two elements. Prove that $F[G] \cong F \times F$, unless F has characteristic 2. In that case, $F[G]$ is isomorphic to the algebra of all matrices of the form $\begin{bmatrix} \alpha & \beta \\ 0 & \alpha \end{bmatrix}$ with $\alpha, \beta \in F$.

Part II
Algebra in Action

Chapter 5
Commutative Rings

We are at a new beginning. If the goal of Part I was gaining familiarity with algebraic structures, we now wish to understand them at a deeper level. This first chapter of Part II is primarily devoted to commutative rings that are in certain ways similar to the ring of integers $\mathbb{Z}$. The most prominent example is $F[X]$, the ring of polynomials over a field F. We will see that the theory of divisibility of integers, developed in Section 2.1, holds in essentially the same form not only for $F[X]$, but for more general rings called *principal ideal domains*. In the final part of the chapter, we will establish a theorem on modules over principal ideal domains, along with two surprising applications: one to group theory and the other to matrix theory.

5.1 Divisibility of Polynomials

We open the chapter with an elementary treatment of divisibility in the polynomial ring $F[X]$. Here, as usual, F stands for a field.

The notation and terminology from Section 2.4 will be used throughout. In particular, $\deg(f(X))$ denotes the degree of the polynomial $f(X)$. Recall (from Theorem 2.46) that for all $f(X), g(X) \in F[X]$,

$$\deg(f(X)g(X)) = \deg(f(X)) + \deg(g(X)). \tag{5.1}$$

This implies that $F[X]$ has no zero-divisors.

5.1.1 The Division Algorithm for $F[X]$

We said above that the polynomial ring $F[X]$ is similar to the ring of integers $\mathbb{Z}$. Why? This is not self-evident. After all, $F[X]$ seems much more complicated. Nevertheless, at an appropriate abstract level, it can be studied in much the same way

© Springer Nature Switzerland AG 2019
M. Brešar, *Undergraduate Algebra*, Springer Undergraduate Mathematics Series,
https://doi.org/10.1007/978-3-030-14053-3_5

as $\mathbb{Z}$. So, what are the common properties of the two rings? The first answer may be that they are both integral domains, i.e., they are commutative and have no zero-divisors. But so are fields, which are essentially different! However, not only are $\mathbb{Z}$ and $F[X]$ not fields, they are similar to each other in being "far" from fields in the sense that they have only few invertible elements. In $\mathbb{Z}$, these are only 1 and -1, and in $F[X]$, these are only the nonzero constant polynomials (as we see from (5.1)). The similarities mentioned so far are important, but not as far-reaching as the similarity of division algorithms. The algorithm for integers was given in Theorem 2.1. The next theorem describes the **division algorithm for polynomials**.

Theorem 5.1. *Let F be a field and let $f(X), g(X) \in F[X]$, with $g(X) \neq 0$. Then there exist unique $q(X), r(X) \in F[X]$ such that*

$$f(X) = q(X)g(X) + r(X),$$

and either $r(X) = 0$ or $\deg(r(X)) < \deg(g(X))$.

Proof. If $f(X)$ is either 0 or has lower degree than $g(X)$, then we take $q(X) = 0$ and $r(X) = f(X)$. We may therefore assume that $m := \deg(f(X))$ is greater than or equal to $n := \deg(g(X))$.

We proceed by induction on m. If $m = 0$ then $n = 0$, so we can take $q(X) = f(X)g(X)^{-1}$ and $r(X) = 0$. Thus, assume that $m > 0$. If a and b are the leading coefficients of $f(X)$ and $g(X)$, respectively, we can write

$$f(X) = aX^m + f_1(X), \quad g(X) = bX^n + g_1(X),$$

where $f_1(X)$ is either 0 or has degree less than m, and $g_1(X)$ is either 0 or has degree less than n. Hence, the polynomial

$$\begin{aligned} f(X) - ab^{-1}X^{m-n}g(X) &= aX^m + f_1(X) - ab^{-1}X^{m-n}(bX^n + g_1(X)) \\ &= f_1(X) - ab^{-1}X^{m-n}g_1(X) \end{aligned}$$

is either 0 or has degree less than m. By the induction assumption, there exist polynomials $q_1(X)$ and $r(X)$ such that

$$f(X) - ab^{-1}X^{m-n}g(X) = q_1(X)g(X) + r(X),$$

and either $r(X) = 0$ or $\deg(r(X)) < n$. Consequently, $f(X) = q(X)g(X) + r(X)$, where $q(X) = ab^{-1}X^{m-n} + q_1(X)$.

The proof of uniqueness is easy and essentially the same as in the case of integers. We therefore leave it to the reader. $\square$

We call $q(X)$ and $r(X)$ the **quotient** and **remainder** of the division, respectively. The case where the remainder $r(X)$ is 0 is of special interest. We say that $0 \neq g(X) \in F[X]$ **divides** $f(X) \in F[X]$ if there exists a $q(X) \in F[X]$ such that $f(X) = q(X)g(X)$. In this case, we write $g(X) \mid f(X)$. One of the basic facts concerning divisibility of polynomials is the following corollary, which the reader has probably already encountered in some form.

Corollary 5.2. *Let F be a field, let $f(X) \in F[X]$, and let $a \in F$. Then a is a root of $f(X)$ if and only if $X - a$ divides $f(X)$.*

Proof. By Theorem 5.1, there exist $q(X), r(X) \in F[X]$ such that

$$f(X) = q(X)(X - a) + r(X)$$

and $r(X)$ is a constant polynomial, i.e., $r(X) = c$ for some $c \in F$. Evaluating both sides at a gives $f(a) = c$. Thus, $f(a) = 0$ if and only if $r(X) = c = 0$, i.e., if and only if $X - a$ divides $f(X)$. $\qquad\square$

5.1.2 The Concept of an Irreducible Polynomial

We say that a polynomial in $F[X]$ is **reducible over** F if it can be written as a product of two polynomials in $F[X]$ of lower degree. A non-constant polynomial that is not reducible over F is said to be **irreducible over** F. Thus, $p(X) \in F[X]$ is irreducible over F if and only if the following two conditions are satisfied:

(a) $\deg(p(X)) \geq 1$.
(b) For every $g(X) \neq 0$ in $F[X]$, $g(X) \mid p(X)$ implies that $\deg(g(X)) = \deg(p(X))$ or $\deg(g(X)) = 0$.

The analogy with prime numbers is obvious and goes further. Every non-constant polynomial $f(X) \in F[X]$ can be written as

$$f(X) = p_1(X)p_2(X) \cdots p_n(X), \tag{5.2}$$

where $p_i(X)$ are irreducible over F. The reader can easily verify this by induction on $\deg(f(X))$. It is slightly more difficult to prove that $p_i(X)$ are unique up to the order in which they appear. We will not go into this now; the factorization into irreducibles will be considered in a more general setting in Section 5.3. Let us focus on the question of which polynomials are irreducible. We begin by pointing out three rather obvious facts.

Theorem 5.3. *Let F be a field and let $f(X) \in F[X]$.*

(a) *If $\deg(f(X)) = 1$, then $f(X)$ is irreducible over F.*
(b) *If $\deg(f(X)) \geq 2$ and $f(X)$ is irreducible over F, then $f(X)$ has no root in F.*
(c) *If $\deg(f(X)) = 2$ or $\deg(f(X)) = 3$, then $f(X)$ is irreducible over F if and only if $f(X)$ has no root in F.*

Proof. Of course, (a) is clear from the definition, and (b) follows from Corollary 5.2. Assume that $f(X)$ has degree 2 or 3, and has no root in F. The latter means that no polynomial of degree 1 divides $f(X)$. But then, in light of (5.1), $f(X)$ cannot be written as a product of two polynomials of lower degree. This proves (c). $\qquad\square$

The restriction on $\deg(f(X))$ in (c) is indeed necessary.

Example 5.4. The polynomial $f(X) = X^4 + 2X^2 + 1 \in \mathbb{R}[X]$ has no root in $\mathbb{R}$, but is reducible; namely, $f(X) = g(X)^2$ where $g(X) = X^2 + 1 \in \mathbb{R}[X]$.

5.1.3 The Irreducibility over $\mathbb{C}$ and $\mathbb{R}$

The question of irreducibility depends on the field F. For example, the polynomial $X^2 + 1$ is irreducible over $\mathbb{R}$ since it has no real root, but is reducible over $\mathbb{C}$ since it has a root i in $\mathbb{C}$; in fact, $X^2 + 1 = (X - i)(X + i)$. The *Fundamental Theorem of Algebra*, which will be proved in Section 7.10, states that every non-constant polynomial $f(X)$ in $\mathbb{C}[X]$ has a root in $\mathbb{C}$, and is hence, by Theorem 5.3 (b), reducible over $\mathbb{C}$ if $\deg(f(X)) \geq 2$. We can formulate this as follows.

Theorem 5.5. *The linear polynomials are the only irreducible polynomials over $\mathbb{C}$.*

The Fundamental Theorem of Algebra paves the way for describing irreducible polynomials over $\mathbb{R}$.

Theorem 5.6. *Besides linear polynomials, the only irreducible polynomials over $\mathbb{R}$ are quadratic polynomials without real roots, i.e., polynomials of the form $aX^2 + bX + c$ where $a \neq 0$ and $b^2 - 4ac < 0$.*

Proof. Let $p(X) \in \mathbb{R}[X]$ be irreducible over $\mathbb{R}$. We may assume that $\deg(p(X)) \geq 2$. By Theorem 5.3 (b), $p(X)$ has no real root. On the other hand, we can consider $p(X)$ as an element of $\mathbb{C}[X]$ and so it has a root $z \in \mathbb{C} \setminus \mathbb{R}$ by the Fundamental Theorem of Algebra. Corollary 5.2 tells us that there exists an $h(X) \in \mathbb{C}[X]$ such that

$$p(X) = (X - z)h(X).$$

As usual, we write $\overline{w}$ for the complex conjugate of $w \in \mathbb{C}$. Let $p(X) = \sum_{i=0}^{n} a_i X^i$. Since a_i is a real number, $a_i \overline{z}^i = \overline{(a_i z^i)}$ and hence $p(\overline{z}) = \overline{p(z)} = 0$. From $\overline{z} \neq z$ it follows that $h(\overline{z}) = 0$. Using Corollary 5.2 again, we obtain $h(X) = (X - \overline{z})k(X)$ for some $k(X) \in \mathbb{C}[X]$. Thus,

$$p(X) = (X - z)(X - \overline{z})k(X) = g(X)k(X), \tag{5.3}$$

where

$$g(X) := X^2 - (z + \overline{z})X + z\overline{z} \in \mathbb{R}[X].$$

The point here is that $g(X)$ has real coefficients. Since $g(X) \mid p(X)$, the quotient $k(X)$ must also have real coefficients. This can be checked directly, but follows also from Theorem 5.1. Indeed, taking $q(X), r(X) \in \mathbb{R}[X]$ such that $p(X) = q(X)g(X) + r(X)$ with $r(X) = 0$ or $\deg(r(X)) \leq 1$, we see by comparing with (5.3) that $g(X)\big(k(X) - q(X)\big)$ is equal to $r(X)$. Since $\deg(g(X)) = 2$, this is possible only when $k(X) = q(X)$ and $r(X) = 0$.

We have thus shown that $g(X)$ divides $p(X)$ in $\mathbb{R}[X]$. Since $p(X)$ is assumed to be irreducible over $\mathbb{R}$, its degree must be equal to $\deg(g(X)) = 2$. As we know from high school mathematics, a quadratic polynomial without real roots is of the form $aX^2 + bX + c$ with $b^2 - 4ac < 0$. $\qquad\qquad\qquad\qquad\qquad\qquad\qquad\qquad\qquad\qquad\qquad$ $\square$

5.1.4 The Irreducibility over $\mathbb{Q}$

In the remainder of this section, we consider the irreducibility of polynomials over $\mathbb{Q}$. Polynomials in $\mathbb{Q}[X]$ also have complex roots, but we cannot use this as effectively as we did for polynomials in $\mathbb{R}[X]$. Generally speaking, the question of irreducibility of polynomials with rational coefficients is not an easy one, and we cannot provide a definitive answer.

If we multiply a polynomial in $\mathbb{Q}[X]$ by a common multiple of denominators of its coefficients, we obtain a polynomial in $\mathbb{Z}[X]$. We may therefore restrict our discussion to polynomials with integer coefficients. Indeed, multiplying a polynomial by a nonzero element from the corresponding field does not affect its irreducibility.

A polynomial with integer coefficients is said to be **primitive** if the greatest common divisor of its coefficients is 1. In other words,

$$a_n X^n + a_n X^{n-1} + \cdots + a_1 X + a_0 \in \mathbb{Z}[X]$$

is primitive if $a_n, a_{n-1}, \ldots, a_0$ are relatively prime, meaning that there *does not* exist a prime p such that $p \mid a_i$ for every i. For example, $6X^3 - 3X^2 + 8$ is primitive, while $6X^3 - 4X^2 + 8$ is not. Note that if $f(X) \in \mathbb{Z}[X]$ is not primitive and d is the greatest common divisor of its coefficients, then we can write $f(X) = df_0(X)$ where $f_0(X)$ is primitive (e.g., $6X^3 - 4X^2 + 8 = 2(3X^3 - 2X^2 + 4)$).

The next lemma is due to *Carl Friedrich Gauss* (1777–1855).

Lemma 5.7. (Gauss' Lemma) *The product of two primitive polynomials is primitive.*

Proof. Suppose $f_0(X)$ and $g_0(X)$ are primitive, but $f_0(X)g_0(X)$ is not. Let p be a prime that divides all coefficients of $f_0(X)g_0(X)$. Then $f_0(X)g_0(X)$ lies in the ideal $p\mathbb{Z}[X]$ of the ring $\mathbb{Z}[X]$ (this is the principal ideal generated by the constant polynomial p). Since $f_0(X)$ and $g_0(X)$ are primitive, they do not belong to this ideal. The cosets

$$f_0(X) + p\mathbb{Z}[X], \; g_0(X) + p\mathbb{Z}[X] \in \mathbb{Z}[X]/p\mathbb{Z}[x]$$

are therefore nonzero, but their product

$$(f_0(X) + p\mathbb{Z}[X])(g_0(X) + p\mathbb{Z}[X]) = f_0(X)g_0(X) + p\mathbb{Z}[X]$$

equals 0. The ring $\mathbb{Z}[X]/p\mathbb{Z}[X]$ thus has zero-divisors. However, this cannot be true since, as we will show,

$$\mathbb{Z}[X]/p\mathbb{Z}[X] \cong \mathbb{Z}_p[X], \tag{5.4}$$

and $\mathbb{Z}_p[X]$ is, just as any ring of polynomials over a field, an integral domain.

It remains to prove (5.4). Let $\pi : \mathbb{Z} \to \mathbb{Z}/p\mathbb{Z} (= \mathbb{Z}_p)$ be the canonical epimorphism. It is straightforward to check that

$$\varphi\big(a_n X^n + a_{n-1} X^{n-1} + \cdots + a_0\big) = \pi(a_n) X^n + \pi(a_{n-1}) X^{n-1} + \cdots + \pi(a_0)$$

defines a ring epimorphism $\varphi : \mathbb{Z}[X] \to \mathbb{Z}_p[X]$ with $\ker \varphi = p\mathbb{Z}[X]$. Hence, (5.4) follows from the Isomorphism Theorem. $\qquad\square$

We could prove this lemma without using quotient rings, which, after all, were unknown in Gauss' time. The advantage of our abstract proof is that it is based on concepts and requires minimal computation. The downside, however, is that one might get an impression that this result is more profound than it actually is. We therefore encourage the reader to find a direct proof.

A polynomial $f(X) \in \mathbb{Z}[X]$ can be considered as an element of the ring $\mathbb{Q}[X]$. If it is irreducible over $\mathbb{Q}$, then, in particular, it cannot be written as a product of two polynomials in $\mathbb{Z}[X]$ of lower degree. The next theorem tells us that, perhaps surprisingly, the converse is also true for every non-constant polynomial. But first, a comment about the formulation of the theorem. The definition of irreducibility of polynomials in the ring $\mathbb{Z}[X]$ slightly differs from the above definition of irreducibility of polynomials in $F[X]$ where F is a field (see Example 5.40 below). That is why the assumption of the theorem is not stated as that $f(X)$ is irreducible over $\mathbb{Z}$.

Theorem 5.8. *If a non-constant polynomial $f(X) \in \mathbb{Z}[X]$ cannot be written as a product of two polynomials in $\mathbb{Z}[X]$ of lower degree, then $f(X)$ is irreducible over $\mathbb{Q}$.*

Proof. Suppose $f(X) = g(X)h(X)$ where $g(X), h(X) \in \mathbb{Q}[X]$. We must prove that one of the polynomials $g(X)$ and $h(X)$ has degree 0.

Pick $k, \ell \in \mathbb{N}$ such that $kg(X)$ and $\ell h(X)$ lie in $\mathbb{Z}[X]$. Let d, d_1, and d_2 be the greatest common divisor of the coefficients of $f(X)$, $kg(X)$, and $\ell h(X)$, respectively. Then we can write

$$f(X) = df_0(X), \quad kg(X) = d_1 g_0(X), \quad \text{and} \quad \ell h(X) = d_2 h_0(X),$$

where $f_0(X)$, $g_0(X)$, and $h_0(X)$ are primitive polynomials. Setting $a := k\ell d \in \mathbb{N}$ and $b := d_1 d_2 \in \mathbb{N}$, we can rewrite $k\ell f(X) = kg(X) \cdot \ell h(X)$ as

$$af_0(X) = bg_0(X)h_0(X). \tag{5.5}$$

By Gauss' Lemma, $g_0(X)h_0(X)$ is primitive. Therefore, the greatest common divisor of the coefficients of $bg_0(X)h_0(X)$ is b. Similarly, a is the greatest common divisor of the coefficients of $af_0(X)$. From (5.5), we thus infer that $a = b$ and $f_0(X) = g_0(X)h_0(X)$, which gives

$$f(X) = df_0(X) = dg_0(X)h_0(X). \tag{5.6}$$

By the assumption of the theorem, one of the polynomials $g_0(X)$ and $h_0(X)$ is constant. But then the same is true for one of $g(X)$ and $h(X)$. $\qquad\square$

The next theorem presents a useful test for irreducibility. It is named after *Gotthold Eisenstein* (1823–1852).

Theorem 5.9. (Eisenstein Criterion) *Let*

$$f(X) = a_n X^n + a_{n-1} X^{n-1} + \cdots + a_1 X + a_0 \in \mathbb{Z}[X]$$

be a polynomial of degree $n \geq 1$. If there exists a prime p such that

$$p \mid a_0, \ p \mid a_1, \ldots, p \mid a_{n-1}, \ p \nmid a_n, \ \text{and} \ p^2 \nmid a_0,$$

then $f(X)$ is irreducible over $\mathbb{Q}$.

Proof. Suppose $f(X)$ is reducible over $\mathbb{Q}$. Then, by Theorem 5.8, there exist polynomials with integer coefficients

$$g(X) = b_r X^r + b_{r-1} X^{r-1} + \cdots + b_1 X + b_0$$

and

$$h(X) = c_s X^s + c_{s-1} X^{s-1} + \cdots + c_1 X + c_0$$

such that $b_r \neq 0$, $c_s \neq 0$, $r < n$, $s < n$, and

$$f(X) = g(X)h(X). \tag{5.7}$$

In particular, $a_0 = b_0 c_0$. From $p \mid a_0$ and $p^2 \nmid a_0$ it thus follows that p divides exactly one of b_0 and c_0. We may assume that $p \mid b_0$ and $p \nmid c_0$. Since $a_n = b_r c_s$ by (5.7), and $p \nmid a_n$, we infer that $p \nmid b_r$. Hence, there exists $k \leq r$ such that $p \mid b_0, \ldots, p \mid b_{k-1}$, and $p \nmid b_k$. Using (5.7) again, we have

$$a_k = b_k c_0 + b_{k-1} c_1 + \cdots + b_0 c_k.$$

As p divides $b_0, \ldots, b_{k-1}$ as well as a_k (since $k < n$), it follows that p also divides $b_k c_0$. This, however, is in contradiction with $p \nmid b_k$ and $p \nmid c_0$. $\square$

Example 5.10. Let p be a prime and $n \geq 2$. The polynomial $X^n - p$, $n \geq 2$, satisfies the conditions of the theorem and is thus irreducible over $\mathbb{Q}$. In particular, it has no root in $\mathbb{Q}$ (by Theorem 5.3(b)). That is to say, $\sqrt[n]{p}$ is not a rational number. As the reader may know, this well-known fact can be easily proved directly.

Example 5.11. For any prime p, consider the polynomial

$$\Phi_p(X) = X^{p-1} + X^{p-2} + \cdots + X + 1.$$

Obviously, the Eisenstein Criterion is not directly applicable to it. There is, however, an indirect way. First notice that

$$\Phi_p(X)(X - 1) = X^p - 1. \tag{5.8}$$

Replacing X by $X + 1$ in this equality we get

$$\Phi_p(X+1)X = (X+1)^p - 1 = X^p + pX^{p-1} + \cdots + \binom{p}{p-2}X^2 + pX.$$

Canceling out X, we get an expression of $\Phi_p(X+1)$ for which the Eisenstein Criterion is applicable. This is because p divides $\binom{p}{k}$ whenever $1 \le k \le p-1$. Indeed, the binomial coefficient $\binom{p}{k} = \frac{p!}{k!(p-k)!}$ is a natural number and since p is a prime and $k!(p-k)!$ only involves factors smaller than p, $p \nmid k!(p-k)!$ and hence $p \mid \binom{p}{k}$. The polynomial $\Phi_p(X+1)$ is therefore irreducible over $\mathbb{Q}$. Hence so is $\Phi_p(X)$, since $\Phi_p(X) = g(X)h(X)$ implies $\Phi_p(X+1) = g(X+1)h(X+1)$.

By a repeated application of Corollary 5.2, we deduce from (5.8) that $\Phi_p(X)$ can be written as $\Pi_{\omega \in \Omega_p}(X-\omega)$, where $\Omega_p = \{\omega \in \mathbb{C} \mid \omega^p = 1\} \setminus \{1\}$. Now, take any $n \in \mathbb{N}$ and define

$$\Phi_n(X) := \Pi_{\omega \in \Omega_n}(X-\omega),$$

where Ω_n is the set of all **primitive nth roots of unity**. These are the complex numbers of the form $\omega = e^{2\pi i \frac{k}{n}}$, where $1 \le k < n$ and k and n relatively prime. It is not hard to see that they can also be described as the complex numbers with the property that $\omega^n = 1$ and $\omega^j \ne 1$ for all $1 \le j < n$. It turns out that $\Phi_n(X)$ is a polynomial with integer coefficients and that it is irreducible over $\mathbb{Q}$. The proof for a general n, however, is not as easy as for a prime. We call $\Phi_n(X)$ the nth **cyclotomic polynomial**.

Exercises

5.12. Let $f(X) = X^5 - X^4 + X - 1 \in F[X]$ and $g(X) = X^2 - X - 1 \in F[X]$. Determine the quotient $q(X)$ and the remainder $r(X)$ from the division algorithm in the case where:

(a) $F = \mathbb{Q}$.
(b) $F = \mathbb{Z}_2$.
(c) $F = \mathbb{Z}_3$.

5.13. Write $f(X) = X^5 - 81X$ as a product of irreducible polynomials over $\mathbb{R}$.

5.14. Write $f(X) = X^5 + 81X$ as a product of irreducible polynomials over $\mathbb{R}$.

5.15. Prove that $p(X) = 7X^6 + 30X^3 - 6X^2 + 60$ is irreducible over $\mathbb{Q}$.

5.16. Prove that $p(X) = \frac{3}{7}X^5 - \frac{7}{2}X^2 - X + 2$ is irreducible over $\mathbb{Q}$.

5.17. Prove that $f(X) = X^n + 1$ is irreducible over $\mathbb{Q}$ if and only if n is a power of 2.
Hint. The Eisenstein Criterion is sometimes applicable indirectly (see Example 5.11).

5.18. Write $f(X) = X^{16} - 6X^8 + 5$ as a product of irreducible polynomials over $\mathbb{Q}$.

5.19. Let $a_0, a_1, \ldots, a_n \in \mathbb{Z}$ and let p be a prime that does not divide a_n. Suppose $\sum_{i=0}^{n}(a_i + p\mathbb{Z})X^i$ is irreducible over $\mathbb{Z}_p$. Prove that $\sum_{i=0}^{n} a_i X^i$ is then irreducible over $\mathbb{Q}$.

Comment. The problem of irreducibility over $\mathbb{Q}$ can in this way be converted to the problem of irreducibility over $\mathbb{Z}_p$. The latter is sometimes easier. At least for small p, one can directly check whether a polynomial has a root in $\mathbb{Z}_p$. If it does not and is reducible, then it can be divided by a polynomial of degree at least 2 (which also has no roots in $\mathbb{Z}_p$). You can use this approach in the following three exercises.

5.20. Let a, b, c be odd integers. Prove that $p(X) = aX^4 + bX + c$ is irreducible over $\mathbb{Q}$.

5.21. Prove that $p(X) = 7X^5 + 3X^2 + 1$ is irreducible over $\mathbb{Q}$.

5.22. Prove that $p(X) = 36X^3 + 7X + 6$ is irreducible over $\mathbb{Q}$.

5.23. For each $n = 2, 3, 4, 5, 6$, write $f_n(X) = X^n + 1$ as a product of irreducible polynomials over $\mathbb{Z}_2$.

5.24. Write $f(X) = X^4 + 2$ as a product of irreducible polynomials over $\mathbb{Z}_3$.

5.25. Write $f(X) = X^4 - 2X^3 - 2X + 4$ as a product of irreducible polynomials over $\mathbb{Z}_7$.

5.26. Let F be a finite field. Prove that for every $n \in \mathbb{N}$, there exists an irreducible polynomial $p(X)$ over F of degree at least n.

Comment. As a matter of fact, we can find such a polynomial whose degree is exactly n. However, this is somewhat more difficult to prove (compare Exercise 7.118). The given exercise can be solved by adapting Euclid's proof of the infinitude of primes.

5.27. Let $f(X) \in \mathbb{Z}[X]$ and $g(X), h(X) \in \mathbb{Q}[X]$ be such that $f(X) = g(X)h(X)$. Inspecting the proof of Theorem 5.8, show that there exists a $q \in \mathbb{Q}$ such that $qg(X)$ and $q^{-1}h(X)$ have integer coefficients.

5.28. Let F be a field and let $a \in F$. Applying the binomial formula, show that every polynomial $f(X) \in F[X]$ can be written as a sum of terms of the form $b_k(X - a)^k$ with $b_k \in F$. Observe that this yields an alternative proof of Corollary 5.2.

5.29. Let F be any field. Suppose $a_1, \ldots, a_n \in F$ are distinct roots of a polynomial $f(X) \in F[X]$ of degree n. Prove that

$$f(X) = c(X - a_1) \cdots (X - a_n),$$

where c is the leading coefficient of $f(X)$.

5.30. Let F be any field and let $f(X) \in F[X]$ be a non-constant polynomial. Suppose there exist distinct $a_1, \ldots, a_n \in F$ such that $f(a_1) = \cdots = f(a_n)$. Prove that $\deg(f(X)) \geq n$.

5.31. Let F be an infinite field. Prove that $f(X), g(X) \in F[X]$ are equal if their associated polynomial functions are equal.

5.32. Theorem 5.6 was derived from the Fundamental Theorem of Algebra. Prove that, conversely, the Fundamental Theorem of Algebra follows from Theorem 5.6.

Hint. If you are unable to prove this yourself, look at the proof of the Fundamental Theorem of Algebra in Section 7.10.

5.33. Let $a_1, \ldots, a_n$ be distinct integers. Prove that

$$p(X) = (X - a_1) \cdots (X - a_n) - 1$$

is irreducible over $\mathbb{Q}$.

5.34. Let $a_1, \ldots, a_n$ be distinct integers. Prove that

$$p(X) = (X - a_1)^2 \cdots (X - a_n)^2 + 1$$

is irreducible over $\mathbb{Q}$.

5.2   Divisibility in Commutative Rings and Principal Ideals

The concept of divisibility, along with some associated notions, makes sense in every commutative ring.

Definition 5.35. Let R be a commutative ring.

1. A nonzero element $b \in R$ is said to **divide** an element $a \in R$ (or to be a **divisor** of a) if there exists a $q \in R$ such that $a = qb$. We write this as $b \mid a$.
2. Nonzero elements a and b are said to be **associates** (or **associated elements**) if $b \mid a$ and $a \mid b$.
3. An element $p \in R$ is said to be **irreducible** provided that:

   (a) $p \neq 0$ and p is non-invertible.
   (b) If $p = ab$ for some $a, b \in R$, then one of a and b is invertible.

4. Let a and b be elements, not both 0. An element $d \neq 0$ in R is said to be a **greatest common divisor** of a and b provided that:

   (a) $d \mid a$ and $d \mid b$.
   (b) If $c \neq 0$ in R is such that $c \mid a$ and $c \mid b$, then $c \mid d$.

   If a greatest common divisor of a and b is 1, then we say that a and b are **relatively prime**.

We said that these notions make sense, but not that they are very important in every commutative ring R. For instance, if $R = F$ is a field, then every nonzero element in F divides every element in F, which is hardly interesting. If the divisibility is too trivial to be studied in fields, it can be too complicated (and sometimes slightly unnatural) in rings with zero-divisors. Thus, we are essentially interested in the case where R is an integral domain, but not a field. Besides the two fundamental examples, the ring of integers $\mathbb{Z}$ and the ring of polynomials $F[X]$ over a field F, we also mention the ring of Gaussian integers $\mathbb{Z}[i]$. Also, the polynomial ring $R[X]$ is an integral domain whenever R is; this includes $\mathbb{Z}[X]$ and $F[X, Y] = (F[X])[Y]$.

We now seemingly change the subject matter. Recall that a **principal ideal** of a commutative ring R is an ideal generated by a single element (see Example 4.66). Specifically, the principal ideal generated by $a \in R$ is

$$(a) = \{ax \mid x \in R\}.$$

We can also write (a) as aR. For example, if $R = \mathbb{Z}$ then $(n) = n\mathbb{Z}$ is the set of all integer multiples of n. As another example, if $R = F[X]$, then (X) is the set of all polynomials with zero constant term. The ideals $\{0\}$ and R are also principal: $\{0\} = (0)$ and $R = (1)$. In the last equality, we may replace 1 by any invertible element. As a matter of fact, note that

$$a \text{ is invertible } \iff (a) = R. \tag{5.9}$$

What do principal ideals have to do with divisibility? The connection is actually straightforward, namely,

$$b \mid a \iff (a) \subseteq (b). \tag{5.10}$$

Indeed, $a = qb$ implies that $ax \in (b)$ for every $x \in R$, and $(a) \subseteq (b)$ implies that $a \in (b)$ and hence $a = qb$ for some $q \in R$. Although (5.10) is just a simple observation, it will be of key importance, enabling us to transform the study of divisibility into the study of ideals.

The purpose of this section is to provide some insight into the concepts introduced in Definition 5.35. We start with associated elements.

Lemma 5.36. *Let a and b be nonzero elements of an integral domain R. The following conditions are equivalent:*

(i) *a and b are associates.*
(ii) *$(a) = (b)$.*
(iii) *There exists an invertible element $u \in R$ such that $a = ub$.*

Proof. The equivalence of (i) and (ii) follows from (5.10) (and holds in all commutative rings, not only in integral domains). Assuming (i), we have $a = ub$ and $b = va$ for some $u, v \in R$. Hence, $a = u(va)$, and so $(1 - uv)a = 0$. Since R is an integral domain and $a \neq 0$, it follows that $uv = 1$. Thus, (iii) holds. Conversely, if $a = ub$ with u invertible, then $b = u^{-1}a$, and so (i) holds.   □

Example 5.37. The only invertible elements in $\mathbb{Z}$ are 1 and -1. Therefore, associated integers can differ only in sign.

Example 5.38. The only invertible elements in $F[X]$ are the nonzero constant polynomials. Thus, $f(X)$ and $g(X)$ are associates if and only $f(X) = ug(X)$ for some $u \in F \setminus \{0\}$.

Note that another way of stating that a and b are associates in R is that for all $0 \neq x, y \in R$, $x \mid a$ if and only if $x \mid b$, and $a \mid y$ if and only if $b \mid y$. From the point of view of divisibility, associated elements can therefore be identified. The presence of invertible elements different from 1 in R means that associated elements are not literally identical. This is a small technical nuisance, which will accompany us throughout the chapter.

Let us now turn to the notion of an irreducible element. First, a few examples.

Example 5.39. The irreducible elements in $\mathbb{Z}$ are the prime numbers and their negatives.

Example 5.40. The two definitions of irreducibility in $F[X]$, the one from this section and the one from the preceding one, coincide. One has to be more careful if F is not a field. For example, the constant polynomial p, where p is a prime, is irreducible in $\mathbb{Z}[X]$, but not in $\mathbb{Q}[X]$ where it is invertible.

Example 5.41. Since $2 = (1 + i)(1 - i)$, 2 is not an irreducible element of $\mathbb{Z}[i]$. On the other hand, 3 turns out to be irreducible (see Exercise 5.54).

The next lemma translates the notion of irreducibility into the language of principal ideals.

Lemma 5.42. *Let p be a nonzero element of an integral domain R. The following conditions are equivalent:*

(i) *p is irreducible.*
(ii) *The principal ideal (p) is maximal among principal ideals of R (i.e., $(p) \neq R$ and for every $a \in R$, $(p) \subseteq (a) \subsetneq R$ implies $(a) = (p)$).*

Proof. The condition $(p) \subseteq (a)$ is equivalent to the condition that $p = ab$ for some $b \in R$ (see (5.10)), and the condition $(a) \subsetneq R$ is equivalent to the condition that a is non-invertible (see (5.9)). In light of Lemma 5.36, we can thus reword (ii) as follows: if p is non-invertible and $p = ab$ with a non-invertible, then there exists an invertible $u \in R$ such that $p = au$. However, as $a \neq 0$ and R is an integral domain, $p = ab$ and $p = au$ can hold only when $b = u$. Thus, (ii) states the following: p is non-invertible and if $p = ab$ with a non-invertible, then b is invertible. But this is nothing but the definition of an irreducible element. $\qquad\square$

For concrete illustrations of this lemma, see Examples 4.123 and 4.124.

We proceed to the notion of a greatest common divisor. The first thing to point out is that it does not need to exist (see Exercise 5.59), and if it does, then it is not necessarily unique. The latter, however, is not a real problem. If d and d' are both greatest common divisors of a and b, then they are associates; conversely, if d is a greatest common divisor of a and b, then so is any of its associates. Since we do not distinguish between associates, greatest common divisors are nevertheless essentially unique. In some rings we slightly modify the definition to achieve true uniqueness. As we know, in $\mathbb{Z}$ we choose the (unique) *positive* integer satisfying the two conditions from Definition 5.35. Similarly, the **greatest common divisor of polynomials** $f(X)$ and $g(X)$ in $F[X]$ is defined as the (unique) *monic* polynomial satisfying these two conditions.

To connect greatest common divisors with principal ideals, we first say a few words about the more general **finitely generated ideals**, i.e., ideals generated by finitely many elements. Let R be an arbitrary commutative ring. We let $(a_1, \ldots, a_n)$ denote the ideal generated by the elements $a_1, \ldots, a_n \in R$. It contains all the principal ideals (a_i), and hence also their sum $(a_1) + \cdots + (a_n)$. On the other hand, this sum is an ideal that contains all the elements a_i, so we actually have

$$(a_1, \ldots, a_n) = (a_1) + \cdots + (a_n).$$

Thus, $(a_1, \ldots, a_n)$ consists of all elements of the form $a_1 x_1 + \cdots + a_n x_n$ where $x_i \in R$. In short, a finitely generated ideal is a finite sum of principal ideals.

Example 5.43. The ideal $(6, 15)$ of $\mathbb{Z}$ consists of integers of the form $6x + 15y$, $x, y \in \mathbb{Z}$. Since it contains $3 = 6 \cdot (-2) + 15 \cdot 1$, it also contains all integer multiples of 3. On the other hand, all its elements are divisible by 3. Hence, $(6, 15)$ is actually the principal ideal (3). The two generators 6 and 15 can thus be replaced by the single generator 3.

Example 5.44. Note that the ideal (X, Y) of the polynomial ring $F[X, Y]$ consists of all polynomials in X and Y with zero constant term. Is it a principal ideal? Suppose it was. Let $f(X, Y) \in F[X, Y]$ be such that $(X, Y) = (f(X, Y))$. In particular, $X \in (f(X, Y))$ and hence $X = f(X, Y)g(X, Y)$ for some polynomial $g(X, Y)$. Writing the two polynomials on the right-hand side in the form $\sum_i h_i(X)Y^i$, we see that $f(X, Y)$ does not contain any monomial involving Y. Similarly, $Y \in (f(X, Y))$ implies that $f(X, Y)$ does not contain any monomial involving X. Therefore, $f(X, Y)$ is a constant polynomial, implying that $(X, Y) = (f(X, Y))$ is either $\{0\}$ or $F[X, Y]$. This contradiction shows that (X, Y) is not a principal ideal.

The existence of the greatest common divisor of two integers was established in Theorem 2.3. The same proof, only dressed up in a different language, shows the following.

Theorem 5.45. *Let a and b be elements of a commutative ring R, not both 0. If (a, b) is a principal ideal, then a greatest common divisor d of a and b exists and can be written in the form $d = ax + by$ for some $x, y \in R$.*

Proof. By assumption, there exists a $d \in R$ such that

$$(a, b) = (d).$$

From $a, b \in (d)$ it follows that $d \mid a$ in $d \mid b$. Since $d \in (a, b)$, we can write $d = ax + by$ for some $x, y \in R$. If $c \mid a$ and $c \mid b$, i.e., $a = cz$ and $b = cw$ for some $z, w \in R$, then $d = c(zx + wy)$, so $c \mid d$. Thus, d is a greatest common divisor of a and b. $\square$

The existence of $\gcd(m, n)$ for a pair of integers m and n thus follows from the fact that every ideal of $\mathbb{Z}$ is principal (see Example 4.66). In the next section, we will see that there are other rings in which all ideals are principal. It should be remarked, however, that the condition that the ideal (a, b) is principal is only sufficient, but not necessary for the existence of a greatest common divisor of a and b. This is evident from Example 5.44: although the ideal (X, Y) is not principal, X and Y obviously have a greatest common divisor, namely 1.   •

Finally, we remark that the proof of Theorem 5.45 shows that any element d satisfying $(a, b) = (d)$ is a greatest common divisor of a and b. For instance, $\gcd(6, 15) = 3$ by Example 5.43.

Exercises

5.46. Prove that $1, -1, i$, and $-i$ are the only invertible elements of $\mathbb{Z}[i]$. What are the associates of $m + ni$?

5.47. What are the associates of a polynomial $f(X)$ in $\mathbb{Z}[X]$?

5.48. What are the associates of a polynomial $f(X, Y)$ in $F[X, Y]$?

5.49. Let d be an integer that is not a perfect square. Set

$$\mathbb{Z}[\sqrt{d}] := \{m + n\sqrt{d} \mid m, n \in \mathbb{Z}\}.$$

(If $d < 0$, then $\sqrt{d} = i\sqrt{-d}$; $\mathbb{Z}[\sqrt{-1}]$ is thus the ring of Gaussian integers $\mathbb{Z}[i]$). Define $N : \mathbb{Z}[\sqrt{d}] \to \mathbb{Z}$ by

$$N(m + n\sqrt{d}) := (m + n\sqrt{d})(m - n\sqrt{d}) = m^2 - dn^2.$$

We call N the **norm map**. Prove:

(a) $\mathbb{Z}[\sqrt{d}]$ is a subring of $\mathbb{C}$ (if $d > 0$, then of course $\mathbb{Z}[\sqrt{d}] \subseteq \mathbb{R}$).
(b) The subfield of $\mathbb{C}$ generated by $\mathbb{Z}[\sqrt{d}]$ is equal to

$$\mathbb{Q}(\sqrt{d}) := \{r + s\sqrt{d} \mid r, s \in \mathbb{Q}\}.$$

(c) $N(xy) = N(x)N(y)$ for all $x, y \in \mathbb{Z}[\sqrt{d}]$.
(d) $x \in \mathbb{Z}[\sqrt{d}]$ is invertible if and only if $N(x) = \pm 1$.
(e) If $N(x) = \pm p$, where p is a prime, then x is irreducible.

Comment. The rings $\mathbb{Z}[\sqrt{d}]$ are important examples of integral domains, particularly because of their connection to number theory. In the following exercises, they will serve to illustrate the concepts considered in this section.

5.50. Let $d < -1$. Prove that 1 and -1 are the only invertible elements of $\mathbb{Z}[\sqrt{d}]$.

5.51. Prove that $(1 + \sqrt{2})^n$, $n \in \mathbb{Z}$, are distinct invertible elements of $\mathbb{Z}[\sqrt{2}]$.

5.52. Let $x \in \mathbb{Z}[i]$. Prove that $N(x) \in 2\mathbb{Z}$ if and only if $1 + i$ divides x.

5.53. Find all divisors of 2 in $\mathbb{Z}[i]$.

5.54. With the help of Exercise 5.49 (e), we easily find examples of irreducible elements in $\mathbb{Z}[i]$, e.g., $1 + i, 4 - i, -3 + 2i, 7 + 8i$, etc. Prove that 3 is also irreducible, although $N(3) = 9$ is not a prime.

Hint. If $xy = 3$ then $N(x)N(y) = 9$.

5.55. Prove that 5 is an irreducible element of $\mathbb{Z}[\sqrt{-2}]$. What about 2, 3, and 7?

5.56. Prove that elements a and b of a nonzero commutative ring are relatively prime if $ax + by = 1$ for some $x, y \in R$.

5.57. Prove that $2\sqrt{2}$ and 9 are relatively prime in $\mathbb{Z}[\sqrt{2}]$ by:

(a) Using the preceding exercise.
(b) Using the norm map.

5.58. Find a greatest common divisor of $a = 3 - 4i$ and $b = 1 - 3i$ in $\mathbb{Z}[i]$.

5.59. Prove that $a = 6$ and $b = 2 + 2\sqrt{-5}$ do not have a greatest common divisor in $\mathbb{Z}[\sqrt{-5}]$.

Hint. If d was a greatest common divisor of a and b, d would be divisible by $c = 2$ and $c' = 1 + \sqrt{-5}$. Hence, $N(c)$ and $N(c')$ would divide $N(d)$, and on the other hand, $N(d)$ would divide $N(a)$ and $N(b)$.

5.60. Determine which of the following polynomials in $\mathbb{R}[X, Y]$ are irreducible:

(a) $X^4 - Y^2$.
(b) $X^3 - Y^2$.
(c) $X^2Y^2 + X^2 + Y^2 + 1$.
(d) $X^2Y^2 + X^2 + Y^2 - 1$.

5.61. Find an irreducible polynomial of degree 2 in $\mathbb{C}[X, Y]$.

5.62. Prove that $\{m + ni \in \mathbb{Z}[i] \mid m + n \in 2\mathbb{Z}\}$ is a principal ideal of $\mathbb{Z}[i]$ and find its generator.

5.63. Let R be a commutative ring. Denote by I the set of all polynomials in $R[X]$ such that the sum of their coefficients is 0. Prove that I is a principal ideal of $R[X]$ and find its generator.

Comment. The assumption that R is commutative is actually redundant here. We define a principal ideal of an arbitrary, not necessarily commutative, ring R as an ideal generated by a single element. The principal ideal generated by $a \in R$ consists of elements of the form $\sum_i x_i a y_i, x_i, y_i \in R$ (compare Exercise 4.85). If a lies in the center of R, then we can write every such element simply as $ax, x \in R$.

The reader might now wonder why we have restricted ourselves to commutative rings from the very beginning. Not that divisibility cannot be studied in noncommutative rings, but it is incomparably more complicated. In particular, $a = qb$ and $a = bq$ are then two different conditions, so we have to distinguish between right divisors and left divisors.

5.64. Prove that $(2, X)$ is not a principal ideal of $\mathbb{Z}[X]$.

Comment. The rings $\mathbb{Z}[X]$ and $F[X, Y]$ (see Example 5.44) thus contain ideals that are not principal. The next exercise states that much more is true.

5.65. Let R be an integral domain that is not a field. Prove that $R[X]$ contains an ideal that is not principal.

5.66. Let I be the set of all real sequences that have only finitely many nonzero terms. Prove that I is an ideal of the ring of all real sequences (from Example 1.175) which is not finitely generated.

5.67. The principal ideal (X) of $\mathbb{R}[X]$ can be described as the set of all polynomials $f(X)$ satisfying $f(0) = 0$. Now let $\mathrm{Map}(\mathbb{R})$ be the ring of all functions from $\mathbb{R}$ to $\mathbb{R}$. Prove that the set $J = \{f \in \mathrm{Map}(\mathbb{R}) \mid f(0) = 0\}$ is a principal ideal of $\mathrm{Map}(\mathbb{R})$ and find its generator. Prove also that $K = \{f \in C(\mathbb{R}) \mid f(0) = 0\}$ is not a principal ideal of the ring of continuous functions $C(\mathbb{R})$. Moreover, K is not even finitely generated. *Hint.* If $f_1, \ldots, f_n \in K$, then $\sqrt{|f_1| + \cdots + |f_n|} \in K$.

5.3 Principal Ideal Domains and Related Classes of Rings

The last results of the preceding section indicate that having no other ideals than principal ideals is a useful property of a ring. Let us give integral domains with this property a name.

Definition 5.68. An integral domain in which every ideal is principal is called a **principal ideal domain** (or PID).

The obvious example is the ring of integers $\mathbb{Z}$. In the first part of the section, we will provide further examples, and after that study properties of PIDs.

5.3.1 *Examples (Euclidean Domains)*

How did we prove that all ideals of the ring $\mathbb{Z}$ are principal? Additive subgroups of $\mathbb{Z}$ are obviously automatically ideals, and we actually proved that every subgroup H of $(\mathbb{Z}, +)$ is of the form $H = n\mathbb{Z}, n \geq 0$ (Theorem 2.2). The proof was based on the division algorithm for integers. Assuming that $H \neq \{0\}$, we took the smallest positive n in H, and then concluded from $m = qn + r$ that each $m \in H$ is a multiple of n, for otherwise $r = m - qn$ would have been a positive integer in H smaller than n. We also have a division algorithm for the polynomials. It should not be a big surprise that essentially the same proof shows that all ideals of the polynomial ring $F[X]$ are principal. Indeed, taking a nonzero ideal I of $F[X]$, we choose a polynomial $g(X)$ in I of lowest degree, and then apply the division algorithm to show that $I = (g(X))$.

The easy details are left to the reader. We can now ask ourselves whether there are other integral domains having a sort of division algorithm such that the same method of proof shows that all their ideals are principal. The following definition makes precise what we are looking for.

Definition 5.69. An integral domain R is called a **Euclidean domain** if there exists a function $\delta : R \setminus \{0\} \to \mathbb{N} \cup \{0\}$ satisfying the following two conditions:

(a) For each pair of elements $a, b \in R$, $b \neq 0$, there exist $q, r \in R$ such that $a = qb + r$ with $r = 0$ or $\delta(r) < \delta(b)$.
(b) $\delta(a) \leq \delta(ab)$ for all $a, b \in R \setminus \{0\}$.

Before providing examples, we first show that such a ring has only principal ideals.

Theorem 5.70. *Every Euclidean domain is a principal ideal domain.*

Proof. The zero ideal $\{0\}$ is generated by the zero element 0. Take a nonzero ideal I of a Euclidean domain R. Choose $b \in I$ such that $\delta(b) \leq \delta(x)$ for every $x \neq 0$ in I. We claim that $I = (b)$. Since $b \in I$, (b) is contained in I. It remains to show that every a from I lies in (b). Let $q, r \in R$ be such that $a = qb + r$ with $r = 0$ or $\delta(r) < \delta(b)$. Since a and qb lie in I, so does $r = a - qb$. Therefore, the possibility $\delta(r) < \delta(b)$ must be excluded. Hence, $r = 0$ and $a = qb \in (b)$. $\square$

The following theorem lists some of the most important examples of Euclidean domains.

Theorem 5.71. *The following rings are Euclidean domains, and hence also principal ideal domains:*

1. *The ring of integers $\mathbb{Z}$.*
2. *The ring of polynomials $F[X]$ over a field F.*
3. *The ring of Gaussian integers $\mathbb{Z}[i]$.*
4. *The ring of formal power series $F[[X]]$ over a field F.*

Proof. 1. Define δ by $\delta(a) = |a|$. The division algorithm for integers shows that δ satisfies condition (a). This is clear if $b > 0$, while for $b < 0$ we have $a = q(-b) + r = (-q)b + r$ with $r = 0$ or $1 \leq r < -b$. It is obvious that δ satisfies condition (b).
  2. As the division algorithm for polynomials suggests, we define δ by

$$\delta(f(X)) = \deg(f(X)).$$

Both conditions (a) and (b) are then fulfilled.
  3. Define δ by

$$\delta(m + ni) = m^2 + n^2.$$

That is, $\delta(a)$ is equal to $|a|^2$ where $|a|$ is the absolute value of the complex number a (incidentally, in the setting of Exercise 5.49, $\delta(a) = N(a)$). Since $|ab|^2 = |a|^2|b|^2$ holds for all complex numbers a and b, we in particular have $\delta(ab) = \delta(a)\delta(b)$ for all $a, b \in \mathbb{Z}[i]$. Hence, δ satisfies condition (b). Let us turn to condition (a). Take $a, b \in \mathbb{Z}[i], b \neq 0$. Considering a and b as elements of the field of complex numbers, we have $a = cb$ where $c := ab^{-1} \in \mathbb{C}$. Note that c can be written as $u + vi$ with u, v lying in $\mathbb{Q}$, but not necessarily in $\mathbb{Z}$. Choose $k, \ell \in \mathbb{Z}$ such that $|u - k| \leq \frac{1}{2}$ and $|v - \ell| \leq \frac{1}{2}$. Set $q := k + \ell i \in \mathbb{Z}[i]$. It remains to show that $r := a - qb \in \mathbb{Z}[i]$ satisfies $|r|^2 < |b|^2$. This follows from

$$|r|^2 = |a - qb|^2 = |cb - qb|^2 = |c - q|^2 |b|^2$$
$$= \left((u - k)^2 + (v - \ell)^2\right)|b|^2 \leq (\frac{1}{4} + \frac{1}{4})|b|^2 = \frac{1}{2}|b|^2.$$

4. A nonzero element $b \in F[[X]]$ can be written as $b = \sum_{k \geq n} b_k X^k$, where $n \geq 0$ and $b_n \neq 0$. We then define $\delta(b) = n$. Verifying that δ satisfies condition (b) is straightforward. To verify condition (a), we first note that

$$\delta(b) \leq \delta(a) \implies b \mid a. \tag{5.11}$$

Indeed, if b is as above and $a = \sum_{k \geq m} a_k X^k$ with $m \geq n$, then $a = qb$ where $q = \sum_{k \geq m-n} q_k X^k$ and q_k are given by

$$q_{m-n} = b_n^{-1} a_m, \quad q_{m-n+1} = b_n^{-1}(a_{m+1} - q_{m-n}b_{n+1}), \text{ etc.}$$

From (5.11), (a) easily follows: if $a = 0$ or $\delta(b) > \delta(a)$, then we take $q = 0$ and $r = a$, and if $\delta(b) \leq \delta(a)$, then $b \mid a$ (so $a = qb$ and $r = 0$). $\square$

There are other examples of Euclidean domains. The obvious ones are fields (we can take any constant function for δ), but they should be considered trivial examples. It turns out that some of the rings $\mathbb{Z}[\sqrt{d}]$ (from Exercise 5.49) are Euclidean domains, and some are not. For example, $\mathbb{Z}[\sqrt{-2}]$ and $\mathbb{Z}[\sqrt{2}]$ are (Exercises 5.90 and 5.91), but $\mathbb{Z}[\sqrt{-3}]$ and $\mathbb{Z}[\sqrt{-5}]$ are not (Exercise 5.96). As another non-example, we point out that $F[X, Y]$ is not a PID (see Example 5.44), and hence also not a Euclidean domain. More generally, $R[X]$ is not a PID if R is not a field (Exercise 5.65).

As one would expect, not every PID is a Euclidean domain. However, providing examples is not so easy. The most standard one is the ring of all complex numbers of the form $m + n\frac{1+\sqrt{19}i}{2}$ where $m, n \in \mathbb{Z}$. The proof that this ring is a PID but not a Euclidean domain is not very hard, but takes some space. We shall therefore omit it. Some of the readers may find this a little bit disturbing. Examples matter! If we have not given an actual proof that there exist PIDs that are not Euclidean domains, why do not we just study Euclidean domains? It is difficult to give a clear answer, but there are some good reasons. First of all, studying PIDs is not much harder, most of the arguments are identical. It is also more interesting (and instructive) that a theory based on a simple and clear condition—all ideals are principal—can be developed. The definition of a Euclidean domain is, on the other hand, rather technical and less appealing. Examples matter in mathematics, but elegance matters too.

5.3.2   Properties (Unique Factorization Domains)

Some of the nice properties of PIDs follow immediately from the results that were obtained in the preceding section and earlier. The following theorem connects two approaches to maximal ideals that were discussed in Sections 4.3 and 5.2.

Theorem 5.72. *Let p be a nonzero element of a principal ideal domain R. The following statements are equivalent:*

(i) *p is irreducible.*
(ii) *(p) is a maximal ideal of R.*
(iii) *$R/(p)$ is a field.*

Proof. Since all ideals of R are principal, Lemma 5.42 shows that conditions (i) and (ii) are equivalent. The equivalence of (ii) and (iii) follows from Corollary 4.83. □

The special case where $R = \mathbb{Z}$ was examined already in Example 4.123. Now we know that not only does every prime number p give rise to a field (namely, the field $\mathbb{Z}/(p) = \mathbb{Z}_p$), but fields arise from any irreducible element of a PID. In particular, we have the following corollary, which will turn out to be important in the theory of field extensions.

Corollary 5.73. *Let F be a field. Then $p(X) \in F[X]$ is irreducible over F if and only if $F[X]/(p(X))$ is a field.*

In a Euclidean domain R, a greatest common divisor of two elements in R can be computed by the **Euclidean algorithm**, which works in the same way as in the classical situation where $R = \mathbb{Z}$ (the only difference being that the condition $r_{j+1} < r_j$ is replaced by the condition $\delta(r_{j+1}) < \delta(r_j)$). We do not have such a concrete procedure for finding greatest common divisors in PIDs, but we know that they always exist. Namely, the following theorem holds.

Theorem 5.74. *Let a and b be elements of a principal ideal domain R, not both 0. Then a greatest common divisor d of a and b exists and can be written in the form $d = ax + by$ for some $x, y \in R$.*

Proof. This is immediate from Theorem 5.45. □

The Fundamental Theorem of Arithmetic states that natural numbers different from 1 can be uniquely factored into products of primes. Our goal is to establish a theorem of this kind for every PID. To this end, we first introduce the following notion.

Definition 5.75. Let R be a commutative ring. We say that $p \in R$ is a **prime element** provided that:

(a) $p \neq 0$ and p is non-invertible.
(b) If $a, b \in R$ are such that $p \mid ab$, then $p \mid a$ or $p \mid b$.

Euclid's Lemma (Corollary 2.6) tells us that prime numbers are prime elements of the integral domain $\mathbb{Z}$. On the other hand, prime numbers are irreducible elements of $\mathbb{Z}$. What is the relationship between prime and irreducible elements in general?

Lemma 5.76. *In any integral domain, every prime element is irreducible. In a principal ideal domain, an element is prime if and only if it is irreducible.*

Proof. Let p be prime, and suppose $p = xy$. Since, in particular, p divides xy, either $p \mid x$ or $p \mid y$. Assuming the former, we have $x = pu$ and hence $p = puy$. By cancellation, $uy = 1$, showing that y is invertible. This proves that p is irreducible.

Assume now that p is an irreducible element of a PID R. A simple adaptation of the proof of Euclid's Lemma will show that p is prime. Indeed, suppose that p divides ab, but does not divide a. Since p is irreducible, p and a are relatively prime. By Theorem 5.74, there exist $x, y \in R$ such that $px + ay = 1$. Multiplying by b, we obtain $pxb + (ab)y = b$. Since p divides ab, it follows that it divides b as well. This proves that p is prime. $\square$

Exercises 5.94 and 5.95 provide two examples showing that the second statement does not hold for all integral domains.

We continue with a lemma treating the undesirable situation where an element of an integral domain cannot be factored into irreducibles (for a concrete example, see Exercise 5.93).

Lemma 5.77. *Let a be a nonzero and non-invertible element of an integral domain R. If a cannot be written as a product of irreducible elements, then R contains an infinite sequence of elements $a = a_1, a_2, a_3, \ldots$ such that*

$$(a_1) \subsetneq (a_2) \subsetneq (a_3) \subsetneq \ldots . \tag{5.12}$$

Proof. The phrase "an element is a product of irreducible elements" of course includes the case where there is a single factor. So, $a = a_1$ itself is not irreducible. Accordingly, there exist $a_2, b_2 \in R$, none of which is invertible, such that $a_1 = a_2 b_2$. Thus, a_2 and b_2 divide a_1, but none of them is its associate. In view of our assumption, at least one of a_2 and b_2 cannot be written as a product of irreducible elements. Let this be a_2. Hence, $a_2 = a_3 b_3$ where a_3 and b_3 are non-invertible and at least one of them, say a_3, is not equal to a product of irreducible elements. Continuing in this way, we obtain a sequence $a_1, a_2, a_3, \ldots$ such that $a_{n+1} \mid a_n$ but a_{n+1} and a_n are not associates. Note that this can be equivalently expressed as (5.12). $\square$

Our next goal is to show that the situation described in this lemma cannot occur if R is a PID. We could do this directly; however, we now have a good opportunity to mention another type of ring: a commutative ring R is said to be **Noetherian**, or is said to satisfy the **ascending chain condition** on ideals, if any chain

$$I_1 \subseteq I_2 \subseteq I_3 \subseteq \ldots$$

of ideals of R terminates, i.e., there exists an $n \in \mathbb{N}$ such that $I_n = I_{n+1} = \ldots$ (in particular, (5.12) then cannot occur). Noetherian rings are considered the most important class of commutative rings. In an introductory abstract algebra course, however, it is difficult to go far enough to experience their significance. For our purposes, all we need to know about them is the following lemma.

Lemma 5.78. *Every principal ideal domain is a Noetherian ring.*

Proof. Let $I_1 \subseteq I_2 \subseteq I_3 \subseteq \ldots$ be a chain of ideals of R. We claim that $I := \bigcup_{j=1}^{\infty} I_j$ is an ideal. It is clear that $RI_j \subseteq I_j, j \in \mathbb{N}$, implying that $RI \subseteq I$. We have to show that I is a subgroup under addition. Given $x, y \in I$, there exist $k, \ell \in \mathbb{N}$ such $x \in I_k$ and $y \in I_\ell$. We may assume that $k \leq \ell$. But then $I_k \subseteq I_\ell$, and so $x - y \in I_\ell \subseteq I$.

Since R is a principal ideal domain, there exists an $a \in R$ such that $I = (a)$. Being an element of $I = \bigcup_{j=1}^{\infty} I_j$, a belongs to some I_n. Hence, $(a) \subseteq I_n$. On the other hand, $I_n \subseteq I = (a)$. Thus, $I = I_n$, so the chain becomes stationary at I_n. $\square$

Note that this proof can be easily adapted to show that a commutative ring R is Noetherian provided that every ideal of R is *finitely generated*. It is not hard to see— let this be a challenge for the reader—that the converse implication is also true. Thus, commutative Noetherian rings are exactly the commutative rings in which all ideals are finitely generated. PIDs, in which all ideals are singly generated, are their very special examples. It turns out that the polynomial rings $F[X_1, \ldots, X_n]$ are always Noetherian, but are PIDs only when $n = 1$ (see Example 5.44). Moreover, **Hilbert's Basis Theorem** states that $R[X]$ is Noetherian whenever R is Noetherian.

Now we have enough prerequisites to prove a version of the Fundamental Theorem of Arithmetic for PIDs. As the reader may have guessed, this theorem states that every nonzero and non-invertible element of a PID can be factored into irreducible elements in an essentially unique way. What exactly do we mean by "essentially"? For instance, in $\mathbb{Z}$ we do not want to distinguish between factorizations such as $2 \cdot 2 \cdot 7$ and $(-2) \cdot (-7) \cdot 2$. How to express this in abstract terms? We answer this question through the following definition.

Definition 5.79. An integral domain R is said to be a **unique factorization domain** (or UFD) if, for every nonzero and non-invertible $a \in R$, the following holds:

(a) a can be written as $a = p_1 p_2 \cdots p_s$ where p_i are irreducible elements of R.
(b) The factorization in (a) is *unique up to associates and the order of the factors*. By this we mean that if $a = q_1 q_2 \cdots q_t$ is another factorization of a into irreducible elements, then $s = t$ and there is a permutation σ of the set $\{1, \ldots, s\}$ such that p_i and $q_{\sigma(i)}$ are associates for every $i = 1, \ldots, s$.

Of course, p_i and p_j with $i \neq j$ may be associates or even equal. Rearranging the factors, we can write $a = u p_1^{k_1} \ldots p_r^{k_r}$ where u is invertible and p_i and p_j are not associates whenever $i \neq j$.

The Fundamental Theorem of Arithmetic shows that the ring $\mathbb{Z}$ is a unique factorization domain. We now generalize this as follows.

Theorem 5.80. *Every principal ideal domain is a unique factorization domain.*

Proof. Let a be a nonzero and non-invertible element of a PID R. Lemmas 5.77 and 5.78 show that (a) holds. To prove (b), we argue as in the proof of the Fundamental Theorem of Arithmetic. Suppose

$$a = p_1 p_2 \cdots p_s = q_1 q_2 \cdots q_t$$

where p_i, q_i are irreducible. In particular, p_1 divides $q_1 \cdots q_t$. Since, by Lemma 5.76, p_1 is prime, it divides one of the q_i. The order of the factors is irrelevant for us, so we may assume that $p_1 \mid q_1$. As q_1 is irreducible and p_1 is non-invertible, it follows that p_1 and q_1 are associates. Thus, $q_1 = u p_1$ for some invertible $u \in R$. Canceling p_1 in

$$p_1 p_2 \cdots p_s = (u p_1) q_2 \cdots q_t$$

we obtain

$$p_2 p_3 \cdots p_s = (u q_2) q_3 \cdots q_t.$$

Note that $u q_2$ is also irreducible. We can therefore repeat the above argument. In a finite number of steps (or, more formally, by induction on s), we reach the desired conclusion.   □

The class of UFDs turns out to be much larger than the class of PIDs. In particular, if R is a UFD, then so is $R[X]$. Note that this implies that the polynomial ring $F[X_1, \ldots, X_n]$ is a UFD. However, not every integral domain is a UFD (see Exercise 5.96).

The following sequence of implications summarizes this section's findings:

$$\mathbb{Z}, F[X], \mathbb{Z}[i] \implies \text{Euclidean domain} \implies \text{PID} \implies \text{UFD}.$$

Exercises

5.81. Solve Exercise 5.58 by using the Euclidean algorithm.

5.82. Find the greatest common divisor of $f(X) = X^5 + X^3 + 2X^2 + 2$ and $g(X) = X^6 - X^3 - 6$ in $\mathbb{Q}[X]$.

5.83. Find the greatest common divisor of $f(X) = X^5 + X^4 + X + 1$ and $g(X) = X^4 + X^2$ in $\mathbb{Z}_2[X]$.

5.84. Find all $a \in \mathbb{R}$ such that $f(X) = X^5 + 1$ and $g(X) = X^3 + a$ are relatively prime in $\mathbb{R}[X]$. Consider the same question with $\mathbb{R}$ replaced by $\mathbb{C}$.

5.85. Let a be a nonzero element of a Euclidean domain R. Prove:

(a) If $\delta(a) = 0$, then a is invertible.
(b) If $\delta(ab) = \delta(b)$ for some $b \neq 0$, then a is invertible.
(c) If $\delta(a) = 1$, then a is invertible or irreducible.

5.86. Prove that nonzero elements a and b of a Euclidean domain are associates if and only if $\delta(a) = \delta(b)$ and $a \mid b$.

5.87. As a special case of Theorem 5.80, we have that every nonzero and non-invertible element of a Euclidean domain is a product of irreducible elements. Give an alternative (simpler) proof of this fact, independent of Lemmas 5.77 and 5.78.

Hint. The statement (b) of Exercise 5.85 may be useful.

5.88. Show that the elements q and r from the definition of a Euclidean domain are not always uniquely determined.

Hint. They certainly are uniquely determined in $\mathbb{Z}$ and $F[X]$, so an example must be found elsewhere.

5.89. Let R be an integral domain. Suppose a function $\delta : R \setminus \{0\} \to \mathbb{N} \cup \{0\}$ satisfies condition (a) of Definition 5.69. Prove that the function $\overline{\delta} : R \setminus \{0\} \to \mathbb{N} \cup \{0\}$,

$$\overline{\delta}(a) = \min\{\delta(ax) \mid x \in R \setminus \{0\}\},$$

then satisfies both conditions (a) and (b).

Comment. This shows that, strictly speaking, condition (b) is redundant in Definition 5.69. However, it is useful to have it, so it is simpler to require it than struggling with replacing δ by $\overline{\delta}$.

5.90. Prove that $\mathbb{Z}[\sqrt{-2}]$ is a Euclidean domain.

Sketch of proof. The proof is similar to that for $\mathbb{Z}[i]$. Define δ by $\delta(m + n\sqrt{-2}) = m^2 + 2n^2$. Given $a, b \in \mathbb{Z}[\sqrt{-2}]$, $b \neq 0$, write ab^{-1} as $u + v\sqrt{-2}$ with $u, v \in \mathbb{Q}$, then pick integers k and ℓ that are closest to u and v, respectively, and check that $q = k + \ell\sqrt{-2}$ yields the solution of the problem.

5.91. Prove that $\mathbb{Z}[\sqrt{2}]$ is a Euclidean domain.

Hint. $\delta(m + n\sqrt{2}) = |m^2 - 2n^2|$.

5.92. Is $R[X]$ a Euclidean domain if R is a Euclidean domain? Is $R[X]$ a PID if R is a PID?

5.93. Let R be the subring of $\mathbb{Q}[X]$ consisting of all polynomials with integer constant term. Prove that the polynomial $X \in R$ cannot be written as a product of irreducible elements of R. Moreover, find a sequence of elements in R satisfying the conclusion of Lemma 5.77.

5.94. Show that $1 + \sqrt{-3}$ is an irreducible, but not a prime element of $\mathbb{Z}[\sqrt{-3}]$.

Hint. Use $N(xy) = N(x)N(y)$ to show that it is irreducible, and use

$$(1 + \sqrt{-3}) \cdot (1 - \sqrt{-3}) = 2 \cdot 2$$

to show that it is not prime. (Incidentally, the latter equality also shows that 4 has two essentially different factorizations in $\mathbb{Z}[\sqrt{-3}]$).

5.95. Show that $2 + \sqrt{-5}$ is an irreducible, but not a prime element of $\mathbb{Z}[\sqrt{-5}]$.

5.96. Prove that in a UFD, an element is prime if and only if it is irreducible.

Comment. Together with the previous two exercises, this implies that $\mathbb{Z}[\sqrt{-3}]$ and $\mathbb{Z}[\sqrt{-5}]$ are not UFDs (and hence neither PIDs nor Euclidean domains).

5.97. Prove that in a UFD, every pair of elements a and b (not both 0) has a greatest common divisor.

Comment. In view of Exercise 5.59, this gives another proof that $\mathbb{Z}[\sqrt{-5}]$ is not a UFD.

5.98. Let a, b, c be elements of a UFD. Suppose that a and b are relatively prime and both divide c. Prove that ab divides c as well.

5.99. An ideal P of a commutative ring R is called a **prime ideal** if $P \neq R$ and for all $a, b \in R$, $ab \in P$ implies $a \in P$ or $b \in P$. Prove:

(a) P is a prime ideal if and only if R/P is an integral domain.
(b) Every maximal ideal is a prime ideal.
(c) If R is an integral domain but not a field, then $\{0\}$ is a prime ideal of R that is not maximal.
(d) If p is a nonzero element of an integral domain R, then p is a prime element if and only if (p) is a prime ideal.
(e) A nonzero ideal of a PID is prime if and only if it is maximal.

5.100. Prove that (X) is a prime ideal of $F[X, Y]$. Is it maximal?

5.101. Consider $n \in \mathbb{Z}$ as a constant polynomial in $\mathbb{Z}[X]$. When is (n) a prime ideal of $\mathbb{Z}[X]$? Is it maximal?

Hint. Answers can be obtained by a direct approach. However, you may find a shortcut by inspecting the proof of Gauss' Lemma.

5.102. Prove that $p(X) = X^2 + X + 1$ is irreducible over $\mathbb{Z}_2$. Hence, $\mathbb{Z}_2[X]/(p(X))$ is a field by Corollary 5.73. How many elements does it have?

Comment. In Section 7.6, we will provide a deeper insight into questions like this.

5.103. Find a polynomial $p(X) \in \mathbb{R}[X]$ such that $\mathbb{R}[X]/(p(X)) \cong \mathbb{C}$.

5.104. Let p and q be irreducible elements of a PID R. Prove that $R/(pq) \cong R/(p) \times R/(q)$ if and only if p and q are not associates.

5.4 Basics on Modules

The main theme of the rest of the chapter is the structure theorem for modules over a PID and its applications. The purpose of this preliminary section is merely to introduce some of the basic notions concerning modules. All of them are more or less straightforward generalizations of the notions with which we are already familiar. We will therefore omit some details and be more condensed than usual.

We defined modules in Section 3.5. The discussion there, however, was rather brief and sparse, based on the viewpoint that the notion of a module over a ring is equivalent to the notion of a homomorphism from a ring to the endomorphism ring of an additive group. This will not be so important for us in what follows. Let us start our consideration of modules again from scratch.

5.4.1   The Concept of a Module

A **module over a ring** R, or an **R-module**, is an additive group M together with an external binary operation $(a, v) \mapsto av$ from $R \times M$ to M, called **module multiplication**, which satisfies

$$a(u + v) = au + av, \quad (a + b)v = av + bv, \quad (ab)v = a(bv), \quad 1v = v$$

for all $a, b \in R$ and $u, v \in M$. We point out the same three examples that were succinctly mentioned in Section 3.5.

Example 5.105. (a) It is clear from the definitions that a module over a field R is nothing but a vector space over R.

(b) Every additive group M becomes a $\mathbb{Z}$-module if we define the operation $(n, v) \mapsto nv$ in the usual (and only possible!) way, i.e.,

$$nv := \underbrace{v + v + \cdots + v}_{n\text{ times}}, \quad (-n)v := n(-v), \quad \text{and} \quad 0v := 0$$

for all $n \in \mathbb{N}$ and $v \in M$ (compare Remark 1.94). Since, on the other hand, every $\mathbb{Z}$-module is an additive group by definition, we may regard the concepts "additive group" and "$\mathbb{Z}$-module" as equivalent. Actually, we can say, seemingly more generally, that the concepts "Abelian group" and "$\mathbb{Z}$-module" are equivalent. Indeed, the difference between "Abelian group" and "additive group" is just notational. Every Abelian group becomes an additive group if we start writing its operation as $+$.

(c) Every ring R can be regarded as an R-module if we consider the multiplication in R as the module multiplication. This different viewpoint of multiplication is not just a cosmetic change. We will see that the submodules of the module R differ from the subrings of the ring R, the module homomorphisms of the module R differ from the ring homomorphisms of the ring R, etc.

We remark that vector spaces are not very typical examples of modules (just as fields are not typical examples of rings). However, considering vector spaces as modules may help the reader to grasp the notions we are about to introduce. Examples (b) and (c) indicate better where we are heading.

5.4.2   Submodules

A subset N of an R-module M is said to be a **submodule** of M if N is itself an R-module under the operations of M. Equivalently, N is an additive subgroup of M and $rx \in N$ for all $r \in R$ and $x \in N$.

Example 5.106. (a) If R is a field, a submodule is the same as a subspace.
(b) If $R = \mathbb{Z}$, a submodule is the same as a subgroup.
(c) Submodules of the R-module R are precisely the left ideals of R. Of course, if R is commutative, then we can omit the adjective "left". However, although this chapter is devoted to commutative rings, in this section it would be unnatural to exclude noncommutative rings from discussion.

Obvious examples of submodules of M are $\{0\}$ and M. A submodule different from $\{0\}$ is said to be a *nonzero* or *nontrivial* submodule, and a submodule different from M is said to be a *proper* submodule. If N_1 and N_2 are submodules of M, then clearly so are their intersection $N_1 \cap N_2$ and their sum

$$N_1 + N_2 = \{x_1 + x_2 \mid x_i \in N_i, i = 1, 2\}.$$

5.4.3   Module Homomorphisms

Let M and M' be R-modules. A map $\varphi : M \to M'$ is a **module homomorphism** if

$$\varphi(u + v) = \varphi(u) + \varphi(v) \quad \text{and} \quad \varphi(au) = a\varphi(u)$$

for all $u, v \in M$ and $a \in R$. Needless to say, we can also talk about module isomorphisms, endomorphisms, etc., and we write $M \cong M'$ to denote that the modules M and M' are isomorphic. Also, we define the **kernel** and the **image** of φ in the usual way,

$$\ker \varphi = \{u \in M \mid \varphi(u) = 0\} \quad \text{and} \quad \operatorname{im} \varphi = \{\varphi(u) \mid u \in M\}.$$

One immediately checks that $\ker \varphi$ is a submodule of M and $\operatorname{im} \varphi$ is a submodule of M'.

Example 5.107. (a) Module homomorphisms between vector spaces are linear maps.
(b) Module homomorphisms between $\mathbb{Z}$-modules are additive maps.
(c) Let I be a left ideal of a ring R. For any $c \in I$, the map

$$\varphi : I \to I, \quad \varphi(u) = uc,$$

is a module endomorphism of the R-module I. Indeed, $uc \in I$ since $u, c \in I$, φ is clearly additive and satisfies $\varphi(au) = auc = a\varphi(u)$ for all $a \in R$ and $u \in I$.

5.4.4 Quotient Modules

Quotient vector spaces were introduced in Theorem 4.36. A brief inspection of the proof shows that replacing vector spaces by modules (i.e, fields by rings) causes no problem. Thus, if N is a submodule of an R-module M, then the set

$$M/N = \{u + N \mid u \in M\}$$

of all cosets of N in M is an R-module under the addition and module multiplication defined by

$$(u + N) + (v + N) = (u + v) + N,$$
$$a(u + N) = au + N.$$

Also, the map

$$\pi : M \to M/N, \quad \pi(u) = u + N,$$

is a module epimorphism with $\ker \pi = N$. We call M/N the **quotient** (or **factor**) **module** of M by N, and we call π the **canonical epimorphism**.

Example 5.108. (a) If R is a field, M/N is the quotient vector space.
(b) If $R = \mathbb{Z}$, M/N is the quotient additive group.
(c) Let I be a left ideal of a ring R. The module multiplication on the quotient additive group R/I is defined by

$$a(b + I) = ab + I.$$

This is new! The quotient ring R/I, which we can form if I is a two-sided ideal, is something else.

The **Isomorphism Theorem** (Theorem 4.111) holds for modules too. Thus, if $\varphi : M \to M'$ is a module homomorphism, then

$$A/\ker \varphi \cong \operatorname{im} \varphi. \tag{5.13}$$

The proof is the same as for vector space homomorphisms.

5.4.5 Direct Sums of Modules

The **(external) direct sum** of R-modules $M_1, \ldots, M_s$, denoted $M_1 \oplus \cdots \oplus M_s$, is the Cartesian product $M_1 \times \cdots \times M_s$ with addition and module multiplication defined by

$$(u_1, \ldots, u_s) + (v_1, \ldots, v_s) = (u_1 + v_1, \ldots, u_s + v_s),$$
$$a(u_1, \ldots, u_s) = (au_1, \ldots, au_s)$$

for all $u_i, v_i \in M_i$, $i = 1, \ldots, s$, and $a \in R$. It is immediate that $M_1 \oplus \cdots \oplus M_s$ is an R-module.

Example 5.109. (a) If R is a field, $M_1 \oplus \cdots \oplus M_s$ is the usual direct sum of vector spaces.

(b) If $R = \mathbb{Z}$, $M_1 \oplus \cdots \oplus M_s$ is the usual direct sum of additive groups.

(c) The direct sum $R \oplus \cdots \oplus R$ of s copies of the R-module R is usually denoted by R^s. It should not be confused with the direct product $R \times \cdots \times R$ of s copies of the ring R, which is sometimes also denoted by R^s. Indeed the module R^s and the ring R^s are identical as additive groups, but are different algebraic structures.

We say that an R-module M is the **internal direct sum** of submodules $N_1, \ldots, N_s$ if M is the internal direct sum of the additive subgroups $N_1, \ldots, N_s$. Recall from Section 4.5 that this means that

$$M = N_1 + \cdots + N_s$$

and

$$N_i \cap (N_1 + \cdots + N_{i-1} + N_{i+1} + \cdots + N_s) = \{0\} \qquad (5.14)$$

for all $i = 1, \ldots, s$, or, equivalently, that every element in M can be written as $x_1 + \cdots + x_s$, with $x_i \in N_i$, in a unique way. We also remark that an equivalent version of (5.14) is that for all $x_i \in N_i$, $x_1 + \cdots + x_s = 0$ implies that each $x_i = 0$.

In the same fashion as for groups, we see that if M is equal to the internal direct sum of submodules $N_1, \ldots, N_s$, then M is isomorphic to the external direct sum of modules $N_1, \ldots, N_s$. The difference between internal and external direct sums is therefore inessential, and we use the same notation $\oplus$ for both. Thus,

$$M = N_1 \oplus \cdots \oplus N_s$$

can either mean that M is the internal direct sum of submodules $N_1, \ldots, N_s$, or that M is the external direct sum of modules $N_1, \ldots, N_s$. It should always be clear from the context which one is meant.

We say that a submodule N_1 of an R-module M is a **direct summand** if there exists a submodule N_2 of M such that $M = N_1 \oplus N_2$ (i.e., $M = N_1 + N_2$ and $N_1 \cap N_2 = \{0\}$). Trivial examples of direct summands are M and $\{0\}$ (as $M = M \oplus \{0\}$).

Example 5.110. (a) If R is a field, every submodule is a direct summand (see Exercise 4.174 and the comment to Exercise 4.175).

(b) The vector space case should not mislead us. A subgroup of an additive group is only exceptionally a direct summand. For instance, since $m\mathbb{Z} \cap n\mathbb{Z} \neq \{0\}$ whenever $m \neq 0$ and $n \neq 0$, none of the proper nontrivial subgroups of $(\mathbb{Z}, +)$ is a direct summand. In Examples 4.159 and 4.160, we saw that the group $(\mathbb{Z}_6, +)$ has proper nontrivial subgroups that are direct summands, whereas $(\mathbb{Z}_4, +)$ does not.

(c) Let R be commutative. Theorem 4.161 shows that a submodule I of the R-module R (= ideal of the ring R) is a direct summand if and only if there exists an idempotent $e \in I$ such that $I = eR$ (then $J = (1 - e)R$ satisfies $R = I \oplus J$). If R is an integral domain, the only idempotents in R are 0 and 1 (since $e^2 = e$ implies $e(1 - e) = 0$), and so proper nonzero submodules cannot be direct summands.

5.4.6 Module Generators

Let u be an element of an R-module M. Note that

$$Ru := \{au \mid a \in R\}$$

is a submodule of M; moreover, it is the smallest submodule containing u. We call it the **cyclic submodule generated by** u. If $M = Ru$ for some $u \in M$, then we say that M is a **cyclic module**.

Example 5.111. (a) If R is a field, a nonzero cyclic submodule is the same as a 1-dimensional subspace.

(b) If $R = \mathbb{Z}$, a cyclic submodule is the same as a cyclic subgroup, and a cyclic module is the same as a cyclic group.

(c) If R is commutative, cyclic submodules of the R-module R are exactly the principal ideals of R. Note that R itself is cyclic, generated by 1.

We continue to assume that u is an element of an R-module M. The set

$$\mathrm{ann}_R(u) := \{a \in R \mid au = 0\}$$

is clearly a left ideal of R. We call it the **annihilator of** u. With its help, we will now show that, up to isomorphism, the cyclic modules are exactly the quotient modules from Example 5.108 (c). Indeed, take a cyclic module Ru, and define $\varphi : R \to Ru$ by $\varphi(a) = au$. Note that φ is a module epimorphism with $\ker \varphi = \mathrm{ann}_R(u)$. From the Isomorphism Theorem (5.13) it follows that

$$R/\mathrm{ann}_R(u) \cong Ru. \tag{5.15}$$

Thus, every cyclic R-module is isomorphic to a module of the form R/I where I is a left ideal of R. Conversely, R/I is cyclic as it is generated by $1 + I$.

Note that in the special case where $R = \mathbb{Z}$, the discussion in the preceding paragraph recovers the usual classification of cyclic groups (compare Example 4.112).

Now let X be a nonempty subset of an R-module M. The **submodule generated by** X, i.e., the smallest submodule of M containing X, is readily seen to consist of all finite sums of elements of the form au where $a \in R$ and $u \in X$. If it is equal to M, then we say that X **generates** M, or that X is a **generating set** of M. A cyclic module is generated by a set containing only one element. If M is generated by some finite set $\{u_1, \ldots, u_s\}$, then we say that it is **finitely generated**. Note that, in this case, M is equal to a finite sum of cyclic submodules: $M = Ru_1 + \cdots + Ru_s$.

A basis of a module is defined in the same way as a basis of a vector space. Thus, a subset B of an R-module M is a **basis** of M if B generates M and for any distinct $u_1, \ldots, u_s \in B$ and any $a_i \in R$,

$$a_1 u_1 + \cdots + a_s u_s = 0 \implies a_i = 0 \text{ for every } i. \tag{5.16}$$

Not every module has a basis. Those that do are called **free modules**. We will be more interested in a different kind of module in which (5.16) is never fulfilled, not even when $s = 1$. Under the assumption that R is an integral domain, we say that M is a **torsion module** if, for every $u \in M$, there exists an $a \neq 0$ in R such that $au = 0$. In other words, $\mathrm{ann}_R(u) \neq \{0\}$ for every $u \in M$.

Example 5.112. (a) As we know, every vector space has a basis. Thus, vector spaces are free modules. We mention in passing that the term "vector space" is also used for modules over (possibly noncommutative) division rings. The same proof that works for vector spaces over fields shows that vector spaces over division rings are also always free.

(b) The simplest example of a free additive group (free $\mathbb{Z}$-module) is $\mathbb{Z}$. Its basis is $\{1\}$. More generally, the direct sum of s copies of $\mathbb{Z}$, $\mathbb{Z}^s = \mathbb{Z} \oplus \cdots \oplus \mathbb{Z}$, is free (compare Exercise 1.179). Its standard basis consists of s elements

$$u_1 = (1, 0, \ldots, 0), \ u_2 = (0, 1, 0, \ldots, 0), \ \ldots, \ u_s = (0, \ldots, 0, 1). \tag{5.17}$$

On the other hand, any finite additive group G is a torsion $\mathbb{Z}$-module. This is because every element $u \in G$ has finite order, so there exists an $n \in \mathbb{N}$ such that $nu = 0$.

(c) There is nothing special about $R = \mathbb{Z}$—for every ring R, R^s is a free R-module, having a basis with s elements defined in (5.17). Conversely, if M is a free R-module with a basis consisting of s elements, then M is isomorphic to R^s. The proof is the same as for vector spaces, so we leave it as an exercise for the reader. However, not everything about free modules is so straightforward. Generally speaking, they are considerably more complicated than vector spaces. Some basic information about this is given in the comment to Exercise 5.126. Let us point out here that if M is a free module and N is a submodule of M, then not only can the quotient module M/N not be free, it can even be a torsion module. Indeed, take an integral domain R and a proper nonzero ideal I of R. The R-module R is, of course, free, but the quotient module R/I is torsion since $\mathrm{ann}_R(u) \supseteq I$ for every $u \in R/I$.

Exercises

5.113. Recall from Section 3.5 that "our" modules are actually *left* modules, and that there are analogously defined *right* modules. Just as left ideals can be regarded as left modules, right ideals can be regarded as right modules. Distinguishing between left and right modules is thus necessary. However, the distinction is inessential for modules over commutative rings. Indeed, every left module M over a commutative ring R can be made into a right module over R by defining $ua := au$. Prove it!

5.114. Let M be an additive group, viewed as a $\mathbb{Z}$-module. Show that for every $u \in M$, $\text{ann}_{\mathbb{Z}}(u) = m\mathbb{Z}$ if u has finite order m, and $\text{ann}_{\mathbb{Z}}(u) = \{0\}$ if u has infinite order.

5.115. Let R be the polynomial ring $F[X]$ and let I be the principal ideal (X^3). Determine $\text{ann}_R(u)$ for every $u \in R/I$.

5.116. Let M be an R-module. We call

$$\text{ann}_R(M) := \bigcap_{u \in M} \text{ann}_R(u)$$

the **annihilator of** M. Prove that $\text{ann}_R(M)$ is a two-sided ideal of R and that $\text{ann}_R(M) \neq \{0\}$ if M is a finitely generated torsion module and R is an integral domain.

5.117. Prove that the direct sum of two finitely generated modules is finitely generated.

5.118. Prove that the direct sum of two free modules is free.

5.119. Prove that the direct sum of two torsion modules is torsion. (Recall that torsion modules are defined over integral domains.)

5.120. Verify that Theorems 3.47 and 3.48 hold for module homomorphisms too.

5.121. Prove the **Second Isomorphism Theorem** for modules: if N_1 and N_2 are submodules of a module M, then

$$N_1/(N_1 \cap N_2) \cong (N_1 + N_2)/N_2.$$

Hint. Exercises 4.147 and 4.148.

5.122. Prove the **Third Isomorphism Theorem** for modules: if N_1 and N_2 are submodules of a module M and $N_2 \subseteq N_1$, then

$$M/N_1 \cong (M/N_2)/(N_1/N_2).$$

Hint. Exercise 4.150.

5.123. Let M be a module over an integral domain R. Show that

$$\text{tor}(M) := \{u \in M \mid \text{ann}_R(u) \neq \{0\}\}$$

is a submodule of M. It is called the **torsion submodule**. We say that M is **torsion-free** if $\text{tor}(M) = \{0\}$. Prove that every free module is torsion-free, and that $\mathbb{Q}$ is a torsion-free $\mathbb{Z}$-module which is not free.

5.124. Give an example of a $\mathbb{Z}$-module which is neither torsion nor torsion-free.

5.125. Let M be a torsion-free R-module. Show that $\{u_1, \ldots, u_s\}$ is a basis of M if and only if $M = Ru_1 \oplus \cdots \oplus Ru_s$. Explain by an example why the assumption that M is torsion-free is not superfluous.

5.126. Let R be commutative. Prove that every nonzero submodule of the R-module R is free if and only if R is a principal ideal domain.

Comment. The R-module R is, of course, free. The "only if" part thus shows that free modules may contain submodules that are not free. There are more surprises. For example, for some rings R, a free R-module may have bases of different cardinalities. For rings R for which this cannot happen—it turns out that this includes all commutative rings—we define the **rank** of a free R-module to be the cardinality of any of its bases. Of course, if R is a field, this is the same as the dimension.

5.127. An R-module $M \neq \{0\}$ is said to be **simple** if $\{0\}$ and M are its only submodules. Prove:

(a) M is simple if and only if $Rv = M$ for every $v \neq 0$ in M. Hence, simple modules are cyclic.
(b) If R is a division ring, then every nonzero cyclic R-module is simple.
(c) If R is not a division ring, then the R-module R is cyclic but not simple.

5.128. Which additive groups are simple viewed as $\mathbb{Z}$-modules?

5.129. Verify that the following operations define modules and find out whether or not they are simple:

(a) Let R be the ring $\mathbb{Z}_8$, let M be the additive group $\{0, 2, 4, 6\} \leq \mathbb{Z}_8$, and let the operation from $R \times M$ to M be the usual multiplication in $\mathbb{Z}_8$.
(b) Let R be the ring $M_n(\mathbb{R})$, let M be the additive group $\mathbb{R}^n$, and let the operation from $R \times M$ to M be the usual multiplication of matrices and vectors.
(c) Let R be the ring of continuous functions $C[0, 1]$, let M be the additive group $\{f \in C[0, 1] \mid f(0) = 0\}$, and let the operation from $R \times M$ to M be the usual multiplication of functions.

5.130. Explain why a nonzero homomorphism from a simple module to an arbitrary module is injective, and a nonzero homomorphism from an arbitrary module to a simple module is surjective.

5.131. Show that the set $\mathrm{End}_R(M)$ of all module endomorphisms of an R-module M is a ring under the usual operations, i.e., $(f + g)(u) := f(u) + g(u)$ and $(fg)(u) := f(g(u))$. Prove that $\mathrm{End}_R(M)$ is a division ring if M is simple.

Comment. This statement is known as **Schur's Lemma**. Although elementary and easy to prove, it is of fundamental importance in the branch of algebra called representation theory.

5.132. Let R be any ring. Prove that $\varphi : R \to R$ is a module endomorphism if and only if there exists an $c \in R$ such that $\varphi(a) = ac$ for all $a \in R$. Deduce from this that if R is commutative, then R and $\mathrm{End}_R(R)$ are isomorphic rings.

5.133. Let M and M' be simple R-modules. Prove that $M \oplus \{0\}$ and $\{0\} \oplus M'$ are the only nonzero proper submodules of $M \oplus M'$ if and only if M and M' are not isomorphic.

5.134. The following is an important generalization of the R-module R. Let R be a subring of a ring S. Then S can be viewed as an R-module if we consider the given ring multiplication rs, where $r \in R$ and $s \in S$, as the module multiplication. Determine whether or not S is a finitely generated R-module in the case where:

(a) $S = \mathbb{Q}$ and $R = \mathbb{Z}$.
(b) $S = \mathbb{R}[X, Y]$ and $R = \mathbb{R}[X]$.
(c) S is the ring of all functions from $\mathbb{R}$ to $\mathbb{R}$ and R is the subring of all even functions, i.e., functions f satisfying $f(-x) = f(x)$ for all $x \in \mathbb{R}$.
(d) $S = \mathbb{R}[X]$ and R is the subring of S generated by all constant polynomials and some non-constant polynomial $g(X)$.

5.5 Modules over Principal Ideal Domains

In this section, we will determine the structure of finitely generated torsion modules over PIDs. A particular emphasis will be given to an important special case, the structure of finite Abelian groups.

The section is mathematically demanding, for two reasons. The first reason is that the proof of the main theorem is complex and much longer than any of the proofs we have encountered so far. The second reason is that this theorem deals with modules, with which we have far less experience than with other algebraic structures. To ease the exposition, we will discuss the corollary concerning finite Abelian groups first, preceding the proof. Hopefully, this will help the reader to build intuition before tackling the more abstract subject matter.

5.5.1 Finite Abelian Groups

For convenience, we assume that all Abelian groups considered in this section are additive groups. This is, of course, just a matter of notation and does not affect the generality of our discussion.

What examples of finite Abelian groups do we know? The basic ones are finite *cyclic* groups. Since we do not want to distinguish between isomorphic groups, we can think of them as the groups $(\mathbb{Z}_n, +)$ (see Theorem 3.9). Further examples can be obtained by taking (finite) *direct sums* of finite cyclic groups. These are also finite

Abelian groups, and not all of them are cyclic (for example, $\mathbb{Z}_2 \oplus \mathbb{Z}_2$ is not). Thus, groups of the form

$$\mathbb{Z}_{n_1} \oplus \mathbb{Z}_{n_2} \oplus \cdots \oplus \mathbb{Z}_{n_s} \tag{5.18}$$

provide a large family of examples of finite Abelian groups. What about other, maybe not-so-obvious examples? It turns out that, surprisingly, there are none. This is the main message of this subsection. A little bit more can be said, however. If we write $n \in \mathbb{N}$ as $n = k\ell$ with k, ℓ relatively prime, then $\mathbb{Z}_n \cong \mathbb{Z}_k \oplus \mathbb{Z}_\ell$. This is the content of Exercise 3.20, and can be also deduced from Claim 2 in the proof of Theorem 5.143 below. Hence, by induction, if $n = p_1^{k_1} \cdots p_r^{k_r}$ where p_i are distinct primes, then $\mathbb{Z}_n \cong \mathbb{Z}_{p_1^{k_1}} \oplus \cdots \oplus \mathbb{Z}_{p_r^{k_r}}$. Therefore, the group in (5.18) is isomorphic to the group

$$\mathbb{Z}_{p_1^{k_1}} \oplus \mathbb{Z}_{p_2^{k_2}} \oplus \cdots \oplus \mathbb{Z}_{p_t^{k_t}}, \tag{5.19}$$

where $k_i \in \mathbb{N}$ and p_i are *not necessarily distinct* primes (the same primes can occur in factorizations of different n_i).

Every nontrivial finite Abelian group is thus isomorphic to a group of the form (5.19). We state this result in an equivalent way that involves internal direct sums.

Theorem 5.135. **(Fundamental Theorem of Finite Abelian Groups)** *A nontrivial finite Abelian group is a direct sum of cyclic subgroups of prime power order.*

(When using multiplicative notation, we of course replace "sum" by "product".)

Theorem 5.135 is a special case of Theorem 5.143, which describes finitely generated torsion modules over PIDs. Some of the readers may be interested only in Theorem 5.135, and perhaps do not even wish to learn about modules which one may find too advanced for an undergraduate course. Therefore, we will now provide a little "module-group dictionary," which should make it possible to understand— without studying the preceding section and in fact without any knowledge of modules—the proof of Theorem 5.135 by reading the proof of Theorem 5.143. The first thing one has to know is that a module over the ring of integers $\mathbb{Z}$ is just an additive group. In this context, a submodule is a subgroup, a quotient module is a quotient group, and a cyclic submodule is a cyclic subgroup. Referring to the notation of Theorem 5.143 and its proof, one thus takes $\mathbb{Z}$ for R and a finite Abelian (additive) group for M, can write $\langle w_i \rangle$ instead of Rw_i, and can simply take the whole group M for the generating set $\{u_1, \ldots, u_s\}$. Concerning the terminology, "order" is then (up to sign) the usual order of a group element, "prime element" is (up to sign) a prime number, "relatively prime elements" are relatively prime integers, and "torsion" only means that every element has finite order. The results on modules and PIDs that are referred to in the proof are to be replaced by some basic facts about Abelian groups and the integers. Making these adjustments, the proof then does not require anything not already contained in Part I. Indeed there are some small changes one would normally make if considering only finite Abelian groups, but they are small and unimportant. There are other proofs of the Fundamental Theorem of Finite Abelian Groups, but none of them is essentially simpler than the proof of Theorem 5.143.

If G is a finite Abelian group and H_i are subgroups of G, then

$$G = H_1 \oplus \cdots \oplus H_t \implies |G| = |H_1| \cdots |H_t|.$$

This simple observation is important in order to fully understand the meaning of Theorem 5.135.

Example 5.136. Let G be an Abelian group with $|G| = 42$. We know that G is isomorphic to a group of the form (5.19). Determining p_i and k_i is easy; namely, from $p_1^{k_1} p_2^{k_2} \cdots p_t^{k_t} = 42 = 7 \cdot 3 \cdot 2$ we see that the only possibility is that $\{p_1, p_2, p_3\} = \{7, 3, 2\}$ and $k_1 = k_2 = k_3 = 1$. The order of summands in a direct sum is irrelevant, since $G_1 \oplus G_2 \cong G_2 \oplus G_1$ holds for any additive groups G_1 and G_2. Therefore, G is isomorphic to the group

$$\mathbb{Z}_7 \oplus \mathbb{Z}_3 \oplus \mathbb{Z}_2.$$

Thus, up to isomorphism, there is only one Abelian group of order 42.

Example 5.137. Now let G be an Abelian group with $|G| = 16 = 2^4$. Since the primes p_i in (5.19) are not necessarily distinct, this case is more complicated. Observe that G is isomorphic to one of the following groups:

$$\mathbb{Z}_{16}, \quad \mathbb{Z}_8 \oplus \mathbb{Z}_2, \quad \mathbb{Z}_4 \oplus \mathbb{Z}_4, \quad \mathbb{Z}_4 \oplus \mathbb{Z}_2 \oplus \mathbb{Z}_2, \quad \mathbb{Z}_2 \oplus \mathbb{Z}_2 \oplus \mathbb{Z}_2 \oplus \mathbb{Z}_2.$$

Thus, there are five possibilities for G. As a matter of fact, at this point we should not say five but at most five. That is, if two of these five groups were isomorphic, there would be less than five possibilities. However, the next theorem shows that this is not the case.

Theorem 5.138. *Let p be a prime. If $k_1 \geq \cdots \geq k_u \geq 1$ and $\ell_1 \geq \cdots \geq \ell_v \geq 1$ are such that*

$$\mathbb{Z}_{p^{k_1}} \oplus \cdots \oplus \mathbb{Z}_{p^{k_u}} \cong \mathbb{Z}_{p^{\ell_1}} \oplus \cdots \oplus \mathbb{Z}_{p^{\ell_v}},$$

then $u = v$ and $k_i = \ell_i$ for all i.

Proof. Denote the group on the left-hand side by G and the group on the right-hand side by G'. They both have order p^n where

$$n = k_1 + \cdots + k_u = \ell_1 + \cdots + \ell_v. \tag{5.20}$$

We proceed by induction on n. The $n = 1$ case is trivial, so we assume that $n > 1$ and that the theorem holds for all groups of order p^t with $t < n$. For any additive group K, write $pK := \{px \mid x \in K\}$. It is clear that pK is a subgroup of K, and easy to see that

$$p\mathbb{Z}_{p^m} \cong \mathbb{Z}_{p^{m-1}}$$

for every $m \in \mathbb{N}$ (if $m = 1$, this is the trivial group $\{0\}$). From $G \cong G'$ it follows that $pG \cong pG'$; indeed, if $x \mapsto \varphi(x)$ is an isomorphism from G to G', then $px \mapsto \varphi(px) = p\varphi(x)$ is an isomorphism from pG to pG'. Consequently,

$$\mathbb{Z}_{p^{k_1-1}} \oplus \cdots \oplus \mathbb{Z}_{p^{k_w-1}} \cong \mathbb{Z}_{p^{\ell_1-1}} \oplus \cdots \oplus \mathbb{Z}_{p^{\ell_z-1}},$$

where w is the largest integer such that $k_w > 1$, and z is the largest integer such that $\ell_z > 1$. By the induction hypothesis it follows that $w = z$ and $k_i - 1 = \ell_i - 1$ for $i = 1, \ldots, w$. Thus, $k_i = \ell_i$ whenever k_i and ℓ_i are greater than 1. It remains to prove that $u = v$, i.e., that G and G' contain the same number of copies of the group $\mathbb{Z}_p$. Taking into account that $k_i = \ell_i$ for $i = 1, \ldots, w = z$, this follows from (5.20).   □

Theorems 5.135 and 5.138 together give a complete classification of finite Abelian groups. Provided that we are able to factorize $n \geq 2$ into a product of primes, we can now list all different Abelian groups of order n. By this we mean that every Abelian group of order n is isomorphic to one of the groups from the list, and no two groups from the list are isomorphic.

Example 5.139. Let $n=200=5^2 \cdot 2^3$. An Abelian group of order $5^2 = 25$ is isomorphic either to $\mathbb{Z}_{25}$ or $\mathbb{Z}_5 \oplus \mathbb{Z}_5$, and an Abelian group of order $2^3 = 8$ is isomorphic either to $\mathbb{Z}_8$, $\mathbb{Z}_4 \oplus \mathbb{Z}_2$, or $\mathbb{Z}_2 \oplus \mathbb{Z}_2 \oplus \mathbb{Z}_2$. Therefore, an Abelian group of order $n=200$ is isomorphic to one of the following six groups:

$$\mathbb{Z}_{25} \oplus \mathbb{Z}_8, \quad \mathbb{Z}_{25} \oplus \mathbb{Z}_4 \oplus \mathbb{Z}_2, \quad \mathbb{Z}_{25} \oplus \mathbb{Z}_2 \oplus \mathbb{Z}_2 \oplus \mathbb{Z}_2,$$
$$\mathbb{Z}_5 \oplus \mathbb{Z}_5 \oplus \mathbb{Z}_8, \quad \mathbb{Z}_5 \oplus \mathbb{Z}_5 \oplus \mathbb{Z}_4 \oplus \mathbb{Z}_2, \quad \mathbb{Z}_5 \oplus \mathbb{Z}_5 \oplus \mathbb{Z}_2 \oplus \mathbb{Z}_2 \oplus \mathbb{Z}_2.$$

Another way of summarizing Theorems 5.135 and 5.138 is to say that, for every nontrivial finite Abelian group G, there exist cyclic subgroups $H_1, \ldots, H_t$ such that $G = H_1 \oplus \cdots \oplus H_t$, $|H_i| = p_i^{k_i}$ for some prime p_i and $k_i \in \mathbb{N}$, and, moreover, the numbers $p_i^{k_i}$ are uniquely determined. This does not mean that H_i are uniquely determined—their number t and their orders $p_i^{k_i}$ are, but not H_i themselves, as the following simple example shows.

Example 5.140. The group $G = \mathbb{Z}_2 \oplus \mathbb{Z}_2$ has three subgroups of order 2:

$$H_1 = \{(0,0), (1,0)\}, \quad H_2 = \{(0,0), (0,1)\}, \quad \text{and} \quad H_3 = \{(0,0), (1,1)\}.$$

Note that $G = H_i \oplus H_j$ whenever $i \neq j$.

5.5.2 Finitely Generated Torsion Modules over PIDs

In what follows, we consider a module M over a PID R. For every $u \in M$, $\mathrm{ann}_R(u)$ is then a principal ideal. If $\mathrm{ann}_R(u) = (a)$, we say that u has **order** a. In other words, u has order a if $au = 0$ and, for all $b \in R$, $bu = 0$ implies $a \mid b$. Recall from Lemma 5.36 that a is unique up to multiplication by an invertible element.

Example 5.141. If $R = \mathbb{Z}$ and u has order $a \neq 0$, then the order of u in the group-theoretic sense is $|a|$ (compare Exercise 5.114).

We will be particularly interested in elements whose order is a power of a prime element (we remark that, by Lemma 5.76, "prime element" is the same as "irreducible element," but we will use the former term in this section). The next lemma considers various conditions that will be encountered in the proof of the main theorem.

Lemma 5.142. *Let M be a module over a principal ideal domain R, let u be nonzero element in M, and let $p \in R$ be prime.*

(a) *If $p^k u = 0$, then u has order p^ℓ for some $1 \leq \ell \leq k$.*
(b) *u has order p^ℓ if and only if $p^\ell u = 0$ and $p^{\ell-1} u \neq 0$.*
(c) *If u has order p^ℓ and N is a submodule of M, then $u + N \in M/N$ is either 0 or has order p^j for some $1 \leq j \leq \ell$.*
(d) *If u has order p^ℓ and $b \in R$ is such that b and p are relatively prime, then bu also has order p^ℓ.*
(e) *If u has order p^ℓ and $v \in M$ has order p^k with $k \geq \ell$, then, for all $j, m \in \mathbb{N}$ with $j \leq k$, $p^j u = p^m v$ implies $m \geq j$.*

Proof. (a) If u has order a, then $p^k \in (a)$ by our assumption. Since a has a unique factorization in R (Theorem 5.80), this is possible only when $a = zp^\ell$ with z invertible and $1 \leq \ell \leq k$.

(b) Since p is not invertible, $p^{\ell-1} \notin (p^\ell)$. Therefore, $p^{\ell-1} u \neq 0$ if u has order p^ℓ. This proves the "only if" part. The "if" part follows from (a).

(c) Since $p^\ell (u + N) = p^\ell u + N = 0$, this also follows from (a).

(d) It is clear that $p^\ell (bu) = b(p^\ell u) = 0$. In view of (b), it is therefore enough to show that $p^{\ell-1}(bu) \neq 0$. Since b and p are relatively prime, so are b and p^ℓ. Applying Theorem 5.74, we see that there exist $x, y \in R$ such that $bx + p^\ell y = 1$. From $p^\ell u = 0$ it follows that $bxu = u$, and hence $p^{\ell-1} bxu = p^{\ell-1} u \neq 0$. Consequently, $p^{\ell-1}(bu) \neq 0$.

(e) Suppose, on the contrary, that $m < j$. Since $j \leq k$, we then have $k - m - 1 \geq 0$. Multiplying $p^j u = p^m v$ by p^{k-m-1} we obtain

$$p^{j+k-m-1} u = p^{k-1} v.$$

As $j + k - m - 1 \geq k \geq \ell$, the element on the left-hand side is 0, and so $p^{k-1}v$ is 0 too. However, this contradicts our assumption that v has order p^k. □

We are now in a position to prove the main theorem. The proof is long but instructive. In particular, it nicely illustrates how effective the usage of quotient structures can be.

Theorem 5.143. (Fundamental Theorem of Finitely Generated Torsion Modules over a PID) *Let M be a nonzero finitely generated torsion module over a principal ideal domain R. Then M is a direct sum of cyclic submodules:*

$$M = Rw_1 \oplus \cdots \oplus Rw_t$$

for some $w_i \in M$. Moreover, the elements w_i can be chosen so that their orders are powers of prime elements.

Proof. To make the proof easier to read, we split it into a series of claims. The first one is essentially Exercise 5.116. In order to state it, we introduce, for every $a \in R$,

$$M(a) := \{u \in M \mid au = 0\}.$$

Note that $M(a)$ is a submodule of M.

Claim 1. *There exists an* $r \neq 0$ *in* R *such that* $M = M(r)$.

Proof of Claim 1. Let $\{u_1, \ldots, u_s\}$ be a generating set of M. By the torsion assumption, for every i there exists an $r_i \neq 0$ in R such that $r_i u_i = 0$. Hence, $r := r_1 \cdots r_s$ is not 0 and satisfies $ru_i = 0$ for every i. Since $M = Ru_1 + \cdots + Ru_s$, this implies that $ru = 0$ for every $u \in M$; that is, $M = M(r)$. □

Claim 2. *If* $r = qq'$ *with* q, q' *relatively prime, then* $M = M(q) \oplus M(q')$.

Proof of Claim 2. By Theorem 5.74, there exist $x, y \in R$ such that $q'x + qy = 1$. Hence, for every $u \in M$,

$$u = 1u = q'(xu) + q(yu).$$

Since $q'v \in M(q)$ and $qv \in M(q')$ for any $v \in M$, this shows that $M = M(q) + M(q')$. Moreover, it also shows that $qu = q'u = 0$ implies $u = 0$, meaning that $M(q)$ and $M(q')$ have trivial intersection. □

Claim 3. *There exist prime elements* $p_1, \ldots, p_m \in R$ *and* $k_1, \ldots, k_m \in \mathbb{N}$ *such that*

$$M = M(p_1^{k_1}) \oplus \cdots \oplus M(p_m^{k_m}).$$

Proof of Claim 3. Note first that r is not invertible as $M \neq \{0\}$. Since R is a UFD (Theorem 5.80), we have $r = zp_1^{k_1}p_2^{k_2} \cdots p_m^{k_m}$ for some invertible $z \in R$, $k_i \in \mathbb{N}$, and prime p_i such that p_i and p_j are not associates if $i \neq j$. Note that $zp_1^{k_1}$ and $p_2^{k_2} \cdots p_m^{k_m}$ are then relatively prime, so we can employ Claim 2 to obtain

$$M = M(p_1^{k_1}) \oplus M(p_2^{k_2} \cdots p_m^{k_m}).$$

A simple induction completes the proof. □

The proofs of the first three claims were quite easy, but they took us far. We now see that it suffices to consider the case where $M = M(p^k)$ for some prime p and $k \in \mathbb{N}$. Indeed, if the conclusion of the theorem holds for each of the summands $M(p_i^{k_i})$, then it also holds for M. Thus, from now on we assume that there exist a prime p and $k \in \mathbb{N}$ such that $p^k u = 0$ for every $u \in M$. We also assume that k is the smallest natural number with this property. According to the statements (a) and (b) of Lemma 5.142, we can reword these assumptions as that every nonzero element in M has order p^ℓ with $1 \leq \ell \leq k$, and there exists at least one element whose order is exactly p^k.

The proof that M is a direct sum of cyclic submodules will be by induction on the number s of generators of M. If $s = 1$, then M itself is cyclic. Hence, we may assume that $s > 1$ and that the theorem holds for all modules with less than s generators. As above, we let $\{u_1, \ldots, u_s\}$ denote a generating set of M. Note that at least one of the generators u_i must have order p^k. Without loss of generality, let this be u_1. We now consider the quotient module M/Ru_1. For every $u \in M$, we write

$$\bar{u} := u + Ru_1 \in M/Ru_1.$$

Claim 4. *There exist* $\tau_2, \ldots, \tau_t \in M/Ru_1$ *such that* $M/Ru_1 = R\tau_2 \oplus \cdots \oplus R\tau_t$.

Proof of Claim 4. Since $\{u_1, u_2, \ldots, u_s\}$ is a generating set of M, $\{\bar{u}_1, \bar{u}_2, \ldots, \bar{u}_s\}$ is a generating set of M/Ru_1. However, $\bar{u}_1 = 0$. Therefore, M/Ru_1 is generated by less than s elements, and the claim follows from the induction hypothesis (we have denoted the first index by 2 rather than by 1 only for convenience). □

Our next goal is to represent the cosets τ_i in a special way. By Lemma 5.142 (c), each nonzero element τ in M/Ru_1 has order p^j for some $1 \leq j \leq k$, and every $w \in M$ such that $\bar{w} = \tau$ has order p^ℓ with $\ell \geq j$. We will now show that w can be chosen so that $\ell = j$ (equivalently, $p^j w = 0$).

Claim 5. *Suppose* $\tau \in M/Ru_1$ *has order* p^j *for some* $j \leq k$. *Then there exists a* $w \in M$ *such that* $\bar{w} = \tau$ *and* w *has also order* p^j.

Proof of Claim 5. Pick any $u \in M$ such that $\bar{u} = \tau$. The claim will be proved by finding an $x \in Ru_1$ such that $p^j(u - x) = 0$. Indeed, $w := u - x$ then satisfies $\bar{w} = \tau$ and has order p^j (by the remark preceding the claim).

From $p^j \bar{u} = 0$ we infer that $p^j u = au_1$ for some $a \in R$. Since R is a UFD, we can write $a = p^m b$ with $m \geq 0$ and b, p relatively prime. Thus,

$$p^j u = p^m b u_1. \tag{5.21}$$

Lemma 5.142 (d) shows that bu_1 has order p^k, and so Lemma 5.142 (e) tells us that $m \geq j$. Hence it follows from (5.21) that $x := p^{m-j} bu_1$ has the desired properties. □

Returning to Claim 4, we now see that, for every $i = 2, \ldots, t$, we can choose $w_i \in R$ such that $\bar{w}_i = \tau_i$ and w_i has the same order as τ_i. This means that for every $a \in R$,

$$a\bar{w}_i = 0 \iff aw_i = 0, \quad i = 2, \ldots, t. \tag{5.22}$$

For w_1, we simply take u_1.

Claim 6. *If* $a_1, \ldots, a_t \in R$ *are such that* $a_1 w_1 + \cdots + a_t w_t = 0$, *then each* $a_i w_i = 0$.

Proof of Claim 6. Since $\bar{w}_1 = 0$, our assumption implies that $a_2 \bar{w}_2 + \cdots + a_t \bar{w}_t = 0$. By Claim 4, M/Ru_1 is a direct sum of $R\bar{w}_2, \ldots, R\bar{w}_t$, and so $a_i \bar{w}_i = 0$, $i = 2, \ldots, t$. From (5.22) it follows that $a_i w_i = 0$, $i = 2, \ldots, t$, and hence also $a_1 w_1 = 0$. □

Claim 7. $M = Rw_1 + \cdots + Rw_t$.

Proof of Claim 7. Take $u \in M$. Then $\bar{u} \in M/Ru_1$, so there exist $a_2, \ldots, a_t \in R$ such that $\bar{u} = a_2 \bar{w}_2 + \cdots + a_t \bar{w}_t$. Hence, $u - a_2 w_2 - \cdots - a_t w_t \in Ru_1$. Since $u_1 = w_1$, this proves our claim. □

Claims 6 and 7 show that $M = Rw_1 \oplus \cdots \oplus Rw_t$. The proof is thus complete. □

Remark 5.144. If $w \in M$ has order a, then we see from (5.15) that the cyclic module Rw is isomorphic to $R/(a)$. Theorem 5.143 can be therefore stated as follows: if M is a nonzero finitely generated torsion module over a PID R, then there exist prime elements $p_1, \ldots, p_t$ in R and natural numbers $k_1, \ldots, k_t$ such that

$$M \cong R/(p_1^{k_1}) \oplus \cdots \oplus R/(p_t^{k_t}). \qquad (5.23)$$

Just as for finite Abelian groups, we can show that the elements $p_i^{k_i}$ are essentially unique. The proof is not much harder than the proof of Theorem 5.138. The main reason for restricting ourselves to the group case was to make the exposition simpler.

Remark 5.145. With some additional effort, one can prove that every finitely generated (not necessarily torsion) module M over a PID R is a direct sum of cyclic submodules. Some of these submodules may be torsion and some not. As in (5.23), the torsion submodules can be chosen to be isomorphic to $R/(p^k)$ where p is prime and $k \in \mathbb{N}$. Those that are not torsion are free, and hence isomorphic to the R-module R (see Example 5.112). Thus, M is isomorphic to a module of the form

$$R^s \oplus R/(p_1^{k_1}) \oplus \cdots \oplus R/(p_t^{k_t}).$$

As a special case we have that every finitely generated Abelian group is isomorphic to a group of the form

$$\mathbb{Z}^s \oplus \mathbb{Z}_{p_1^{k_1}} \oplus \cdots \oplus \mathbb{Z}_{p_t^{k_t}}$$

for some $s, k_i \in \mathbb{N}$ and prime p_i.

Exercises

5.146. Is $\mathbb{Z}_{12} \cong \mathbb{Z}_4 \oplus \mathbb{Z}_3$? Is $\mathbb{Z}_{12} \cong \mathbb{Z}_6 \oplus \mathbb{Z}_2$?

5.147. Show that $\mathbb{Z}_{36} \oplus \mathbb{Z}_{24} \cong \mathbb{Z}_{72} \oplus \mathbb{Z}_{12}$.

5.148. Show that $\mathbb{Z}_{78} \oplus \mathbb{Z}_{18} \cong \mathbb{Z}_{26} \oplus \mathbb{Z}_9 \oplus \mathbb{Z}_6$.

5.149. Show that

$$\mathbb{Z}_{11} \oplus \mathbb{Z}_{25} \oplus \mathbb{Z}_5 \oplus \mathbb{Z}_9 \oplus \mathbb{Z}_9 \oplus \mathbb{Z}_3 \oplus \mathbb{Z}_4 \oplus \mathbb{Z}_2 \cong \mathbb{Z}_{9900} \oplus \mathbb{Z}_{90} \oplus \mathbb{Z}_3.$$

5.150. Prove that every finite Abelian group is isomorphic to a group of the form

$$\mathbb{Z}_{n_1} \oplus \mathbb{Z}_{n_2} \oplus \cdots \oplus \mathbb{Z}_{n_r},$$

where $n_{i+1} \mid n_i$ for all $i = 1, \ldots, r - 1$.

Hint. The preceding exercise considers a special case.

5.151. Determine, up to isomorphism, all Abelian groups of order 324.

5.152. Determine the number of non-isomorphic Abelian groups of order:

(a) $32 = 2^5$.
(b) $42336 = 7^2 \cdot 3^3 \cdot 2^5$.
(c) $211680 = 7^2 \cdot 5 \cdot 3^3 \cdot 2^5$.

5.153. Determine the number of elements of orders 2, 4, and 8 in each of the Abelian groups of order 8, i.e., in $\mathbb{Z}_8$, $\mathbb{Z}_4 \oplus \mathbb{Z}_2$, and $\mathbb{Z}_2 \oplus \mathbb{Z}_2 \oplus \mathbb{Z}_2$.

5.154. Determine, up to isomorphism, all Abelian groups of order 800 that contain elements of order 16 but no element of order 25.

5.155. As usual, let $\mathbb{Z}_n^*$ denote the group of all invertible elements of the ring $(\mathbb{Z}_n, +, \cdot)$. Determine the group of the form (5.19) to which $\mathbb{Z}_n^*$ is isomorphic, in the case where:

(a) $n = 5$.
(b) $n = 8$.
(c) $n = 10$.
(d) $n = 16$.

Comment. It turns out that the groups $\mathbb{Z}_p^*$, where p is a prime, are cyclic. See Exercise 7.117.

5.156. Let p and q be distinct primes. Suppose pq divides the order of a finite Abelian group G. Prove that G contains a cyclic subgroup of order pq. Is this still true if $p = q$?

5.157. Let G and G' be Abelian groups such that $|G| = p^k$ and $|G'| = q^\ell$ where p and q are distinct primes and $k, \ell \in \mathbb{N}$. Prove that the group $G \oplus G'$ is cyclic if and only if G and G' are cyclic.

5.158. Let G be a finite Abelian group. Prove that the following conditions are equivalent:

 (i) G is cyclic.
 (ii) If a prime p divides $|G|$, then G contains $p - 1$ elements of order p.
(iii) G contains less than $p^2 - 1$ elements of order p for every prime p.
(iv) G does not contain a subgroup isomorphic to $\mathbb{Z}_p \oplus \mathbb{Z}_p$ for some prime p.

5.159. Prove that every Abelian group of order 200 contains a subgroup of order 20.

5.160. Let now G be an arbitrary finite Abelian group and let m be any natural number that divides $|G|$. Prove that G contains a subgroup of order m.

Comment. This is not always true if G is non-Abelian. See Exercise 6.23.

5.161. Let G, H, and H' be finite Abelian groups. Prove that $G \oplus H \cong G \oplus H'$ implies $H \cong H'$.

5.162. Give an example showing that the conclusion of the preceding exercise in general does not hold for infinite Abelian groups. More specifically, find an infinite Abelian group G such that $G \oplus G \cong G$ (and hence $G \oplus G \cong G \oplus \{0\}$).

5.163. Let R be a ring. Prove:

(a) If $|R| = p$ for some prime p, then R is isomorphic to the ring $\mathbb{Z}_p$.
(b) If the additive group $(R, +)$ is cyclic, then R is commutative.
(c) If $|R| = p^2$ for some prime p, then R is commutative.

Conclude from these statements that all rings having less than 8 elements are commutative. Find a noncommutative ring with 8 elements!

Hint. The centralizer of any ring element a is a subgroup under addition which contains a and 1. This may help you in proving (c). When searching for a noncommutative ring with a certain property, one usually first looks at matrix rings. However, since $|M_2(S)| = |S|^4$ holds for every finite ring S, the ring of all matrices of size 2×2 cannot have 8 elements. Search among its subrings!

5.164. Show that a finitely generated torsion $\mathbb{Z}$-module is a finite set (and hence a finite Abelian group).

5.165. Let u be a nonzero element of a torsion module M over a PID R. Prove that the cyclic R-module Ru is simple if and only if u has order p for some prime p.

5.166. Let M be a nonzero module over a PID R such that $\mathrm{ann}_R(M) \neq \{0\}$ (see Exercise 5.116). Since $\mathrm{ann}_R(M)$ is an ideal, there exists an $a \neq 0$ in R such that $\mathrm{ann}_R(M) = (a)$. Prove that for every $b \in R$, the following conditions are equivalent:

(i) a and b are relatively prime.
(ii) For every $u \in M$, $bu = 0$ implies $u = 0$.

5.167. Let M be a nonzero finitely generated torsion module over a PID R. Prove that the following conditions are equivalent:

(i) $N \cap N' \neq \{0\}$ for all nonzero submodules N and N' of M.
(ii) $M \cong R/(p^k)$ for some prime element $p \in R$ and $k \in \mathbb{N}$.

5.6 The Jordan Normal Form of a Linear Map

There are two basic examples of principal ideal domains: the ring of integers $\mathbb{Z}$ and the polynomial ring $F[X]$ where F is a field. In the preceding section, we discussed the meaning of the Fundamental Theorem of Finitely Generated Torsion Modules over a PID R in the case where $R = \mathbb{Z}$. We will now show that this theorem also has an important application in the case where $R = F[X]$. Specifically, we will use it to derive the existence of the Jordan normal form of a vector space endomorphism.

Students usually first encounter the Jordan normal form in a linear algebra course. By many, it is considered one of the most difficult and torturous chapters, conceptually involved and technically demanding. The big advantage of the approach to the Jordan normal form that we are about to demonstrate is that it is technically quite simple— once we have mastered the content of the preceding section. However, due to its abstract nature, it is not so easy to grasp the whole idea of this approach. One way or another, there is no royal road to the Jordan normal form, to paraphrase Euclid (the reader has probably heard this story: when asked by King Ptolemy whether there was some easier way to learn geometry than reading the *Elements*, Euclid replied that there was no royal road to geometry).

We start with a brief introduction to the problem that will be addressed. The readers with a good linear algebra background may skip this part and move directly to Theorem 5.174.

Let V be an n-dimensional vector space over a field F, and let $T \in \mathrm{End}_F(V)$ (i.e., T is a linear map from V to V). Choose a basis $B = \{b_1, \ldots, b_n\}$ of V. For each j there exist unique $t_{ij} \in F$ such that

$$T(b_j) = \sum_{i=1}^{n} t_{ij} b_i.$$

We then call

$$T_B := \begin{bmatrix} t_{11} & t_{12} & \ldots & t_{1n} \\ t_{21} & t_{22} & \ldots & t_{2n} \\ \vdots & \vdots & \ddots & \vdots \\ t_{n1} & t_{n2} & \ldots & t_{nn} \end{bmatrix}$$

the **matrix of T with respect to the basis** B. Recall from Example 3.77 that $\mathrm{End}_F(V)$ is an algebra over F isomorphic to the matrix algebra $M_n(F)$, via the isomorphism $T \mapsto T_B$. This will not be essentially needed in what follows, but is important for understanding and appreciating Theorem 5.174 below.

Example 5.168. A map $T : F^2 \to F^2$ is linear if and only if there exist $t_{ij} \in F$ such that

$$T\big((x_1, x_2)\big) = (t_{11}x_1 + t_{12}x_2, t_{21}x_1 + t_{22}x_2).$$

The matrix of T with respect to the standard basis $B = \{(1, 0), (0, 1)\}$ is

$$T_B = \begin{bmatrix} t_{11} & t_{12} \\ t_{21} & t_{22} \end{bmatrix}.$$

Needless to say, if we choose a different basis B', the matrix $T_{B'}$ will also be different.

We say that $\lambda \in F$ is an **eigenvalue** of T if there exists a $v \neq 0$ in V such that $T(v) = \lambda v$. Every such vector v is called an **eigenvector** of T belonging to λ. If there exists a basis $B = \{b_1, \ldots, b_n\}$ of V consisting of eigenvectors of T, then T is said to be **diagonalizable**. The matrix T_B is then diagonal with eigenvalues on the diagonal.

Example 5.169. Let $T : F^2 \to F^2$ be given by

$$T\big((x_1, x_2)\big) = (x_2, x_2).$$

Notice that $b_1 := (1, 1)$ is an eigenvector of T corresponding to the eigenvalue 1 and $b_2 := (1, 0)$ is an eigenvector of T corresponding to the eigenvalue 0. Since $B := \{b_1, b_2\}$ is a basis of F^2, T is diagonalizable and

$$T_B = \begin{bmatrix} 1 & 0 \\ 0 & 0 \end{bmatrix}.$$

Example 5.170. Is $T : F^2 \to F^2$, defined by

$$T\big((x_1, x_2)\big) = (x_2, -x_1),$$

diagonalizable? This question is more subtle than one might first think. By a simple calculation, we see that every eigenvalue λ of T must satisfy $\lambda^2 = -1$. Therefore, if we take, for example, $F = \mathbb{R}$, then T is not diagonalizable simply because it has no eigenvalues in $\mathbb{R}$. If, however, $F = \mathbb{C}$, then i and $-i$ are eigenvalues of T, with corresponding eigenvectors $b_1 := (1, i)$ and $b_2 := (1, -i)$. Since they are linearly independent, $B := \{b_1, b_2\}$ is a basis of $\mathbb{C}^2$. Thus, T is diagonalizable and

$$T_B = \begin{bmatrix} i & 0 \\ 0 & -i \end{bmatrix}.$$

When discussing eigenvalues and related notions, we usually restrict ourselves to vector spaces over fields that are similar to $\mathbb{C}$ in the following sense.

Definition 5.171. A field F is said to be **algebraically closed** if every non-constant polynomial $f(X) \in F[X]$ has a root in F.

The field $\mathbb{C}$ is, of course, algebraically closed by the Fundamental Theorem of Algebra. We will not give further examples at this point. Algebraically closed fields will be discussed in Section 7.5.

No matter whether F is algebraically closed or not, we can easily find linear maps that are not diagonalizable.

Example 5.172. Let $N : V \to V$ be a nonzero *nilpotent* linear map. This means that there is a $k \geq 2$ such that $N^k = 0$ and $N^{k-1} \neq 0$. Note that $N(v) = \lambda v$ implies $N^j(v) = \lambda^j v$ for every $j \in \mathbb{N}$. Taking k for j we thus see N has no eigenvalues different from 0. Since $N \neq 0$, this implies that N is *not* diagonalizable. It is worth mentioning that, slightly more generally, the sum of a nonzero nilpotent linear map and a scalar multiple of the identity also is not diagonalizable. The easy proof is left to the reader.

To give a simple concrete example, let V be the space of all polynomials in $F[X]$ of degree at most 2, and let $N : V \to V$ be defined by $N(f(X)) = f'(X)$ where $f'(X)$ is the derivative of $F[X]$ (i.e., if $f(X) = a_0 + a_1 X + a_2 X^2$, then $f'(X) = a_1 + 2a_2 X$). Since $N^3 = 0$, N is not diagonalizable. However, assuming that the characteristic of F is not 2, the matrix of N with respect to the basis $B := \{1, X, \frac{1}{2}X^2\}$,

$$N_B = \begin{bmatrix} 0 & 1 & 0 \\ 0 & 0 & 1 \\ 0 & 0 & 0 \end{bmatrix},$$

also has a nice simple form and is not far from being a diagonal matrix. (If the characteristic of F is 2, then $N^2 = 0$ and the matrix of N with respect to the standard basis $\{1, X, X^2\}$ is $\begin{bmatrix} 0 & 1 & 0 \\ 0 & 0 & 0 \\ 0 & 0 & 0 \end{bmatrix}$.)

Let U be a subspace of V. We say that U is **invariant** under $T \in \mathrm{End}(V)$ if $T(u) \in U$ for every $u \in U$. The restriction of T to U is then an endomorphism of U.

Suppose there exist subspaces V_1 and V_2 such that $V = V_1 \oplus V_2$ and both V_1 and V_2 are invariant under T. Take a basis B_1 of V_1 and a basis B_2 of V_2. Then $B := B_1 \cup B_2$ is a basis of V. If A_1 is the matrix of the restriction of T to V_1 with respect to B_1 and A_2 is the matrix of the restriction of T to V_2 with respect to B_2, then the block matrix

$$\begin{bmatrix} A_1 & 0 \\ 0 & A_2 \end{bmatrix}$$

is the matrix of T with respect to B. We may thus think of T as being composed of two independent parts. One also says that T is a *direct sum* of two linear maps, one acting on V_1 and the other acting on V_2. A generalization to direct sums of more than two maps is straightforward.

If our goal is to represent a linear map T by a matrix which is as close as possible to a diagonal matrix, then, in view of what has been noticed so far, the result we will prove is the best we could hope for. We need a few more preliminaries before we can state it.

Let

$$f(X) = a_0 + a_1 X + \cdots + a_m X^m \in F[X].$$

The definition of the evaluation of $f(X)$ at an element (see Section 2.4) makes sense for elements from any algebra over F. In particular, for $T \in \mathrm{End}_F(V)$, we define

$$f(T) := a_0 I + a_1 T + \cdots + a_m T^m.$$

The next lemma follows from the Cayley–Hamilton Theorem (see Exercise 5.181), with which the reader is probably familiar from a linear algebra course. However, unlike the lemma, this theorem is quite a deep result. We will therefore give a simple direct proof, rather than referring to the theorem.

Lemma 5.173. *If V is a finite-dimensional vector space over F, then, for every $T \in \mathrm{End}_F(V)$, there exists a nonzero polynomial $f(X) \in F[X]$ such that $f(T) = 0$.*

Proof. If $\dim_F V = n$, then

$$\dim_F \mathrm{End}_F(V) = \dim_F M_n(F) = n^2.$$

Hence, $I, T, \ldots, T^{n^2}$ are linearly dependent since their number exceeds the dimension of $\mathrm{End}(V)$. Let $a_i \in F$ be such that

$$a_0 I + a_1 T + \cdots + a_{n^2} T^{n^2} = 0$$

and at least one $a_i \neq 0$. Then

$$f(X) = a_0 + a_1 X + \cdots + a_{n^2} X^{n^2}$$

is a nonzero polynomial satisfying $f(T) = 0$.   □

We are now in a position to state and prove the theorem to which this section is devoted.

Theorem 5.174. *Let V be a finite-dimensional vector space over an algebraically closed field F, and let $T : V \to V$ be a linear map. Then there exists a basis B of V with respect to which the matrix T_B of T has block form*

$$T_B = \begin{bmatrix} J_1 & 0 & \ldots & 0 \\ 0 & J_2 & \ldots & 0 \\ \vdots & \vdots & \ddots & \vdots \\ 0 & 0 & \ldots & J_n \end{bmatrix},$$

where each J_i is a square matrix of the form

$$J_i = \begin{bmatrix} \lambda_i & 1 & 0 & \ldots & 0 & 0 \\ 0 & \lambda_i & 1 & \ldots & 0 & 0 \\ 0 & 0 & \lambda_i & \ldots & 0 & 0 \\ \vdots & \vdots & \vdots & \ddots & \vdots & \vdots \\ 0 & 0 & 0 & \ldots & \lambda_i & 1 \\ 0 & 0 & 0 & \ldots & 0 & \lambda_i \end{bmatrix}$$

for some $\lambda_i \in F$.

Proof. We begin by making an $F[X]$-module out of V. Since V is a vector space, the additive group structure on V is already given. We define the module multiplication as follows: for all $f(X) \in F[X]$ and $v \in V$, let

$$f(X)v := f(T)(v).$$

We first point out that $f(T)$ is also a linear map from V to V, so $f(T)(v)$ belongs to V. Checking the module axioms is straightforward, but we urge the reader to go through the details. We remark that in order to prove that $f_1(X)(f_2(X)v) = (f_1(X)f_2(X))v$ is fulfilled, one has to use the easily verified fact that $f(X) = f_1(X)f_2(X)$ yields $f(T) = f_1(T)f_2(T)$.

Observe that every basis of the vector space V is also a generating set of the module V. Therefore, V is a finitely generated module. Lemma 5.173 tells us that it is also a torsion module. Since $F[X]$ is a PID (Theorem 5.71), we are in a position to apply the Fundamental Theorem of Finitely Generated Torsion Modules over a PID

(Theorem 5.143). Thus, there exist $w_1, \ldots, w_t \in V$ such that the cyclic submodules $V_i := F[X]w_i$ satisfy

$$V = V_1 \oplus \cdots \oplus V_t \tag{5.24}$$

and orders of w_i are powers of prime (=irreducible) elements in $F[X]$. We will, step by step, translate these module-theoretic notions into the language of linear algebra. First, what does is it mean that V_i is a submodule? By the very definition, V_i is a subgroup under addition, from $av_i \in V_i$ for every constant polynomial a and every $v_i \in V_i$ it follows that V_i is a subspace, and from $T(v_i) = Xv_i \in V_i$ for every $v_i \in V_i$ it follows that V_i is invariant under T. We now see (5.24) in a new light. The vector space V is a direct sum of the subspaces V_i that are invariant under T. Therefore, if we choose, for every i, *any* basis B_i of V_i, the matrix of T with respect to $B := B_1 \cup \cdots \cup B_t$ has block form given in the statement of the theorem.

It remains to show that B_i can be chosen so that the matrix of the restriction of T to V_i is equal to J_i (for some $\lambda_i \in F$). This will follow from the fact that w_i has order $p_i(X)^{k_i}$ where $k_i \in \mathbb{N}$ and $p_i(X)$ is an irreducible polynomial in $F[X]$. Since F is algebraically closed, Theorem 5.3 shows that only linear polynomials are irreducible in $F[X]$. Without loss of generality, we may assume that $p_i(X)$ is monic (recall that orders of elements are unique only up to multiplication by invertible elements). Thus, $p_i(X) = X - \lambda_i$ for some $\lambda_i \in F$. Let us write N_i for $T - \lambda_i I$. Note that the condition that w_i has order $(X - \lambda_i)^{k_i}$ means that $N_i^{k_i}(w_i) = 0$ and $N_i^{k_i-1}(w_i) \neq 0$ (see Lemma 5.142 (b)). We claim that

$$B_i := \{N_i^{k_i-1}(w_i), \ldots, N_i(w_i), w_i\}$$

is a basis of V_i. Assume there are $\alpha_i \in F$, not all 0, such that

$$\alpha_{k_i-1}N_i^{k_i-1}(w_i) + \cdots + \alpha_1 N_i(w_i) + \alpha_0 w_i = 0. \tag{5.25}$$

Let r be such that $\alpha_r \neq 0$ and $\alpha_i = 0$ if $i < r$ ($r = 0$ if $\alpha_0 \neq 0$). Applying $N_i^{k_i-1-r}$ to (5.25) we obtain $\alpha_r N_i^{k_i-1}(w_i) = 0$, which is a contradiction. This proves that B_i is linearly independent. To prove that B_i spans V_i, we first remark that the condition that V_i is the cyclic module generated by w_i means that V_i is the linear span of the vectors of the form $T^j(w_i)$, $j \geq 0$. Now, writing $T^j = (N_i + \lambda_i I)^j$ and using the binomial theorem (along with $N_i^{k_i}(w_i) = 0$) we see that each $T^j(w_i)$ lies in the linear span of B_i. We have thereby proved that B_i is a basis of V_i. Finally, from

$$T(N_i^{\ell}(w_i)) = (N_i + \lambda_i I)(N_i^{\ell}(w_i)) = N_i^{\ell+1}(w_i) + \lambda_i N_i^{\ell}(w_i)$$

we see that B_i is a basis with respect to which the matrix of the restriction of T to V_i is of the desired form. $\qquad\square$

The matrix T_B from the theorem is called the **Jordan normal form** of T, and the matrix J_i is called the **Jordan block** corresponding to the eigenvalue λ_i. Of course, if all Jordan blocks J_i are of size 1×1, then T_B is a diagonal matrix.

As we learn in linear algebra, matrices can be considered as linear maps, and two square matrices are similar if and only if they represent the same linear map with respect to (possibly) different bases. Therefore, an alternative (and perhaps more standard) way to state Theorem 5.174 is that every matrix $T \in M_n(F)$ is similar to the matrix T_B described in the statement of the theorem.

We have now reached the end of the chapter. Admittedly, the last two sections were not an easy read. However, hopefully the reader will agree that it was fascinating to see that a mathematical theorem can have two applications that seem so different and unrelated. The classification of finite Abelian groups and the existence of the Jordan normal form were essentially obtained by the same proof. We have truly experienced the power of abstraction.

Exercises

5.175. Let F be any field and let $T : F^2 \to F^2$ be a diagonalizable linear map. True or False:

(a) If $(1, 0)$ is an eigenvector of T, then so is $(0, 1)$.
(b) If $\ker T \neq \{0\}$ and $T \neq 0$, then T has two distinct eigenvalues.

5.176. Determine a linear map $T : \mathbb{C}^2 \to \mathbb{C}^2$ which is not invertible, has an eigenvalue 1, and satisfies $T\big((1, 0)\big) = (1, 1)$. What are the eigenvectors of T? What is the Jordan normal form of T?

5.177. Determine a linear map $T : \mathbb{C}^2 \to \mathbb{C}^2$ whose only eigenvalue is 1 and satisfies $T\big((1, 0)\big) = (1, 1)$. What are the eigenvectors of T? What is the Jordan normal form of T?

5.178. Prove that if λ is an eigenvalue of a linear map T, then $f(\lambda)$ is an eigenvalue of $f(T)$ for every polynomial $f(X) \in F[X]$.

5.179. Let V be an arbitrary vector space (finite-dimensional or not). Suppose that $T : V \to V$ is a linear map such that every nonzero vector in V is its eigenvector. Prove that T is a scalar multiple of the identity I.

Hint. Assume there exist nonzero $v_1, v_2 \in V$ such that $T(v_1) = \lambda_1 v_1$, $T(v_2) = \lambda_2 v_2$, and $\lambda_1 \neq \lambda_2$, and consider $T(v_1 + v_2)$.

5.180. Theorem 5.174 in particular shows that, under the assumption that F is algebraically closed and V is a nonzero finite-dimensional vector space over F, every linear map $T : V \to V$ has an eigenvalue. Show that this can be also deduced from Lemma 5.173.

5.181. Let V be a finite-dimensional vector space, and let $T : V \to V$ be a linear map. We assume the reader is familiar with the **characteristic polynomial** $f_T(X) = \det(XI - T_B)$ of T and its basic properties. Show that the **Cayley–Hamilton Theorem**, which states that $f_T(T) = 0$, can be derived from Theorem 5.174 and the fact that every field can be embedded into an algebraically closed field (see Subsection 7.5.4).

5.182. The **minimal polynomial** of a linear map $T : V \to V$ is the monic polynomial $m(X) \in F[X]$ of lowest degree such that $m(T) = 0$. Prove that $m(X)$ divides every polynomial $f(X) \in F[X]$ such that $f(T) = 0$.

Comment. The Cayley–Hamilton Theorem implies that the degree of $m(X)$ does not exceed $n = \dim_F V$ (while the proof of Lemma 5.173 only shows that it does not exceed n^2).

5.183. Let $N : V \to V$ be a linear map such that $N^k = 0$ and $N^{k-1} \neq 0$. Prove that X^k is the minimal polynomial of N.

5.184. Let $T : V \to V$ be a linear map such that the $F[X]$-module considered in the proof of Theorem 5.174 is cyclic. That is to say, there exists a $w \in V$ such that V is equal to the linear span of the vectors $T^j(w), j \geq 0$ (in this case we say that w is a **cyclic vector** for T). Prove that a linear map $S : V \to V$ commutes with T if and only if $S = f(T)$ for some $f(X) \in F[X]$.

Hint. By assumption, there is an $f(X) \in F[X]$ such that $S(w) = f(T)(w)$.

5.185. Prove that eigenvectors belonging to distinct eigenvalues of a linear map $T : V \to V$ are linearly independent. Hence deduce that the number of distinct eigenvalues cannot exceed the dimension n of V, and that T is diagonalizable if it has n distinct eigenvalues.

5.186. Is the sum of two diagonalizable linear maps diagonalizable?

5.187. Is the product of two diagonalizable linear maps diagonalizable?

5.188. Let V be a finite-dimensional vector space, and let $E : V \to V$ be an idempotent linear map (i.e., $E^2 = E$). Prove that $V = \operatorname{im} E \oplus \ker E$ and hence derive that E is diagonalizable.

Comment. An idempotent linear map is called a *projection*. You may remember this term from linear algebra.

5.189. Let V be a finite-dimensional vector space. Prove that a linear map $T : V \to V$ is diagonalizable if and only if T is a linear combination of pairwise orthogonal idempotent linear maps whose sum is I.

5.190. Let V be a finite-dimensional vector space over an algebraically closed field F, and let $T : V \to V$ be a linear map. Show that there exist $\lambda_i \in F$ and linear maps $E_i, N_i : V \to V, i = 1, \ldots, t$, such that:

(a) $T = \sum_{i=1}^{t} \lambda_i E_i + N_i$.
(b) E_i are pairwise orthogonal idempotents and $\sum_{i=1}^{t} E_i = I$.
(c) N_i are nilpotent.
(d) $N_i = N_i E_i = E_i N_i$ and $N_i E_j = E_j N_i = 0$ if $i \neq j$.

Furthermore, show that there exist a diagonalizable linear map $D : V \to V$ and a nilpotent linear map $N : V \to V$ such that $T = D + N$ and $DN = ND$.

5.191. What are the possible Jordan normal forms of a linear map N satisfying $N^2 = 0$?

5.192. Let V be an n-dimensional vector space. Prove that there exists a linear map $N : V \to V$ such that $\operatorname{im} N = \ker N$ if and only if n is even.

5.193. Does there exist a linear map $T : \mathbb{C}^3 \to \mathbb{C}^3$ such that its centralizer $\{S \in \operatorname{End}_{\mathbb{C}}(\mathbb{C}^3) \mid ST = TS\}$ has dimension 6? What about 5?

5.194. Let $T, N : \mathbb{C}^n \to \mathbb{C}^n$ be linear maps such that $TN - NT = N$. Prove that N is nilpotent.

Chapter 6
Finite Groups

The success in classifying finite Abelian groups in the preceding chapter should not mislead us. The theory of finite groups is extremely complex and answers to general questions do not come easy. It would be far too much to expect that the structure of arbitrary finite groups can be completely determined. What we will show in this chapter is that every finite group contains certain subgroups that are relatively well understood. Thus, instead of tackling the whole group, which would be too ambitious, we will focus on its smaller pieces.

Besides giving some insight into the structure of finite groups, our goal is also to illustrate the applicability of the algebraic tools introduced in Part I.

6.1 Introductory Remarks

When considering the most basic group-theoretic concepts such as homomorphisms or quotient groups, there is no need to distinguish between finite and infinite groups. So far, therefore, only a small part of our study of groups has been devoted exclusively to finite groups. Still, it was an important part, so it seems appropriate to start this chapter by reviewing what we already know about them.

We begin with examples. First, there are finite cyclic groups. Any such group of order n is isomorphic to the additive group $\mathbb{Z}_n$ (Theorem 3.9). If finite cyclic groups are the simplest, the symmetric groups S_n are the most fundamental examples of finite groups. Cayley's Theorem tells us that every finite group can be viewed as a subgroup of an S_n (Corollary 3.104). This may be more interesting than useful, but still indicates the significance of symmetric groups. Their subgroups of particular importance are the alternating groups A_n. Further finite groups that were frequently mentioned in our discussions are the dihedral groups D_{2n} and the quaternion group Q. The general linear groups $\mathrm{GL}_n(F)$ and their various subgroups are also finite if F is a finite field (e.g., $F = \mathbb{Z}_p$).

The Fundamental Theorem of Finite Abelian Groups tells us that all finite Abelian groups are built out of finite cyclic groups. More precisely, written in the

© Springer Nature Switzerland AG 2019

M. Brešar, *Undergraduate Algebra*, Springer Undergraduate Mathematics Series, https://doi.org/10.1007/978-3-030-14053-3_6

language of multiplicative groups, this theorem states that every finite Abelian group G can be, essentially uniquely, decomposed into a direct product

$$G = \langle a_1 \rangle \times \cdots \times \langle a_t \rangle,$$

where $\langle a_i \rangle$ are cyclic subgroups of prime power order. We actually needed only a little group theory to prove this; the main computations were carried out in the ring of integers $\mathbb{Z}$. The proofs in the present chapter, however, will be genuinely group-theoretic.

In principle, we should now be interested only in non-Abelian finite groups. Nevertheless, Abelian groups, particularly cyclic groups, will often appear in our considerations. When dealing with a group element a, we usually also have to deal with the cyclic subgroup $\langle a \rangle$. A simple but extremely important fact is that the order of the element a is equal to $|\langle a \rangle|$, the order of the subgroup $\langle a \rangle$ (see Section 3.1). The next lemma, an extended version of Exercise 3.17, records some other basic facts concerning orders of elements.

Lemma 6.1. *Let a be an element of a group G.*

(a) *If a has order n, then, for every $m \in \mathbb{Z}$, $a^m = 1$ if and only if $n \mid m$.*
(b) *If $a \neq 1$ and $a^p = 1$ for some prime p, then a has order p.*
(c) *If a has order n and $N \triangleleft G$, then the order of aN divides n.*

Proof. (a) If $n \mid m$ then $m = qn$ for some $q \in \mathbb{Z}$, and hence

$$a^m = (a^n)^q = 1^q = 1.$$

Conversely, let $a^m = 1$. Using the division algorithm, we can write $m = qn + r$ where $q \in \mathbb{Z}$ and $0 \leq r < n$. Since n is the smallest natural number satisfying $a^n = 1$,

$$1 = a^m = (a^n)^q a^r = 1a^r = a^r$$

implies that $r = 0$. Thus, $n \mid m$.
   (b) Follows from (a).
   (c) Since $(aN)^n = a^n N = N$, this also follows from (a). $\square$

The last result we need to recall is Lagrange's Theorem (Theorem 4.9).

Lagrange's Theorem. *If H is a subgroup of a finite group G, then*

$$|G| = [G : H] \cdot |H|.$$

(Here, $[G : H]$ is the index of H in G, i.e., the number of distinct cosets of H in G.)

The proof is short and simple. The first step is to show that any two cosets aH and bH are either equal or disjoint (Lemma 4.7). We have derived this from Lemma 4.6, which describes when two cosets are equal, but it takes only a couple of lines to give a direct proof. The second step is to observe that $h \mapsto ah$ is a bijection from H to aH, which implies that every coset aH has the same cardinality as H. Since G is a disjoint union of all distinct cosets by the first step, the desired formula follows.

Lagrange's Theorem provides a foundation for the study of finite groups. We first state a trivial corollary, which, however, reveals the main point of the theorem.

Corollary 6.2. *The order of a subgroup of a finite group divides the order of the group.*

The rest of the section is devoted to elementary consequences of this corollary.

Corollary 6.3. *The order of an element of a finite group divides the order of the group.*

Proof. Since the order of a is equal to the order of the cyclic group $\langle a \rangle$, this follows from Corollary 6.2. $\square$

Corollary 6.4. *If G is a finite group, then $a^{|G|} = 1$ for all $a \in G$.*

Proof. This follows from Corollary 6.3 and the "if" part of Lemma 6.1 (a). $\square$

It is interesting to examine this corollary in the special case where $G = \mathbb{Z}_p^*$, the multiplicative group of all nonzero elements of the field $\mathbb{Z}_p$.

Corollary 6.5. (Fermat's Little Theorem) *If p is a prime and a is an integer, then*

$$a^p \equiv a \pmod{p}.$$

Proof. We will write elements of the field $\mathbb{Z}_p$ as cosets $x + p\mathbb{Z}$ where $x \in \mathbb{Z}$. Since the result of the corollary is trivial if $p \mid a$, we may assume that $p \nmid a$, meaning that $a + p\mathbb{Z}$ is a nonzero element of $\mathbb{Z}_p$. That is to say, $a + p\mathbb{Z}$ lies in the group $\mathbb{Z}_p^*$. Since $|\mathbb{Z}_p^*| = p - 1$, Corollary 6.4 implies that

$$(a + p\mathbb{Z})^{p-1} = 1 + p\mathbb{Z},$$

i.e.,

$$a^{p-1} + p\mathbb{Z} = 1 + p\mathbb{Z}.$$

Thus, $p \mid a^{p-1} - 1$, and hence also $p \mid a^p - a$. $\square$

In 1640, Pierre de Fermat mentioned this simple (but not obvious!) result in a letter to a friend. Compared with Fermat's Last Theorem (see Section 2.4), it is of course of very little depth. However, surprisingly, Fermat's Little Theorem has turned out to be useful in cryptography, the science studying techniques for secure communication (which is something that concerns our everyday lives—just think of the various applications of computers, phones, etc.). One of the most widely used methods of cryptography, called RSA, is based on the fact that while computers are capable of multiplying large numbers, factorizing them in a reasonable amount of time is beyond their capacity. The theoretical background of this method is exactly Fermat's Little Theorem. Let us picture a 17th century French lawyer, a mathematician by heart, perhaps sitting by candlelight while amusing himself through seemingly purposeless consideration of numbers. And then, in the 20th century, one of his discoveries is widely used on the Internet and other marvels of modern world. It seems unbelievable! In general, findings of pure mathematics sometimes turn out to be unexpectedly applicable to science and technology, usually decades or even centuries after their discovery.

Let us return to groups. Finding all groups of a given order n may be next to impossible. However, it is trivial if n is a prime.

Corollary 6.6. *Every group G of prime order is cyclic. Moreover, $G = \langle a \rangle$ holds for every $a \neq 1$ in G.*

Proof. Let $G = |p|$ where p is a prime. Take $a \neq 1$ in G. By Corollary 6.2, $|\langle a \rangle|$ divides p. Since $|\langle a \rangle| > 1$ it follows that $|\langle a \rangle| = p$ and hence $\langle a \rangle = G$. $\square$

The next corollary gives another description of groups of prime order.

Corollary 6.7. *A nontrivial group has no proper nontrivial subgroups if and only if it is a cyclic group of prime order.*

Proof. Let G be a nontrivial group having no proper nontrivial subgroups. Then $\langle a \rangle = G$ for every $a \in G \setminus \{1\}$. Therefore, G is a cyclic group. By Theorem 3.9, either $G \cong \mathbb{Z}$ or $G \cong \mathbb{Z}_n$ for some $n \in \mathbb{N}$. Now, the group $\mathbb{Z}$ contains proper nontrivial subgroups (e.g., $2\mathbb{Z}$), and if $d \mid n$ for some $1 < d < n$, then $d\mathbb{Z}_n$ is a proper nontrivial subgroup of $\mathbb{Z}_n$. Thus, the only possibility is that $G \cong \mathbb{Z}_p$ where p is a prime.

Conversely, let G be a cyclic group of prime order and let H be a nontrivial subgroup of G. By Corollary 6.6, for every $a \neq 1$ in H we have $H \supseteq \langle a \rangle = G$, and so $H = G$. $\square$

Recall that a group $G \neq \{1\}$ is said to be *simple* if $\{1\}$ and G are its only normal subgroups. If G is Abelian, then this is the same as saying that $\{1\}$ and G are its only subgroups. Corollary 6.7 therefore yields a characterization of simple Abelian groups.

Corollary 6.8. *An Abelian group is simple if and only if it is a cyclic group of prime order.*

A simple, memorable restatement of Corollary 6.6 is

$$|G| = p \implies G \cong \mathbb{Z}_p.$$

Thus, for every prime p, $\mathbb{Z}_p$ is the only group of order p (as usual, by "the only" we mean "the only up to isomorphism"). Consequently, the only groups of order at most 3 are the trivial group, $\mathbb{Z}_2$, and $\mathbb{Z}_3$. What about groups of order 4? We know that there are two non-isomorphic Abelian groups,

$$\mathbb{Z}_4 \quad \text{and} \quad \mathbb{Z}_2 \oplus \mathbb{Z}_2.$$

Let us show that these two groups are the only ones that exist. Take any group $G = \{1, a, b, c\}$ of order 4. If one of the elements a, b, c has order 4, then G is isomorphic to $\mathbb{Z}_4$. If none of them has order 4, then they all must have order 2 by Corollary 6.3. Thus,

$$a^2 = b^2 = c^2 = 1.$$

In view of the cancellation law, ab cannot be equal to any of a, b, or 1. Hence, $ab = c$. Since a, b, c appear symmetrically, we must have

$$ab = ba = c, \quad ac = ca = b, \quad bc = cb = a.$$

The reader should now be able to show that G is isomorphic to $\mathbb{Z}_2 \oplus \mathbb{Z}_2$ (as well as to D_4; compare Example 3.11). We remark that G is the smallest non-cyclic group. It is commonly called the **Klein four-group**.

It is slightly more difficult, but still elementary, to prove that

$$\mathbb{Z}_3 \oplus \mathbb{Z}_2 \quad \text{and} \quad S_3$$

are the only groups of order 6 (see Exercise 6.46), and that

$$\mathbb{Z}_2 \oplus \mathbb{Z}_2 \oplus \mathbb{Z}_2, \quad \mathbb{Z}_4 \oplus \mathbb{Z}_2, \quad \mathbb{Z}_8, \quad D_8, \quad \text{and} \quad Q$$

are the only groups of order 8. Since 5 and 7 are primes, the only groups of order at most 8 not yet mentioned are $\mathbb{Z}_5$ and $\mathbb{Z}_7$. Among the groups listed, exactly three are non-Abelian: S_3, D_8, and Q. The smallest one, S_3, has order 6.

There are other numbers n, either small enough or very special, for which it is possible to list all groups of order n. Some special cases will be encountered in the next chapters.

Exercises

6.9. Let G be a finite group and let H, K be subgroups of G such that $K \subseteq H$. Prove that $[G : K] = [G : H] \cdot [H : K]$.

6.10. Let K be a proper subgroup of a proper subgroup of a group G. What is a possible order of K if G has order 24?

6.11. Determine all finite groups G that contain a subgroup of order $|G| - 2$.

6.12. Suppose a group G contains elements $a \neq 1$ and $b \neq 1$ such that $a^{91} = 1$ and $b^{15} = 1$. Prove that $|G| \geq 21$.

6.13. Suppose $|G| = 55$. Prove:

(a) Every proper subgroup of G is cyclic.
(b) Either G is Abelian or has trivial center.

Hint. Exercise 4.61.

6.14. Explain why the number 55 in the preceding exercise can be replaced by any number which is a product of two primes. Can it be replaced by 8?

Comment. In Section 6.3, specifically in Examples 6.67 and 6.68, we will provide further information on groups whose order is a product of two primes.

6.15. If R is a ring of characteristic not 2, then the equation $x^2 = 1$ has at least two solutions in R, namely 1 and -1. Hence conclude that the group $\mathbb{Z}_n^*$ has even order for every $n \geq 3$.

Comment. This means that $\varphi(n)$ is always even for $n \geq 3$, where φ is Euler's phi function (see Exercise 2.35).

6.16. Let $n \geq 2$ and let a be an integer relatively prime to n. Prove that

$$a^{\varphi(n)} \equiv 1 \pmod{n}.$$

Comment. This generalization of Fermat's Little Theorem is called **Euler's Theorem**.

6.17. Use Fermat's Little Theorem to compute the remainder in the division of 4^{19} by 7 and by 17.

6.18. Use Euler's Theorem to compute the last digit of 23^8 and the last digit of 23^{23}.

6.19. Let p be a prime and let q be a prime divisor of $2^p - 1$. Prove that p is the order of the element $2 + q\mathbb{Z}$ in $\mathbb{Z}_q^*$, and hence conclude that $p \mid q - 1$ (and so, in particular, $q > p$).

Comment. This implies that there does not exist a largest prime, i.e., that there are infinitely many primes. Of course, this is not new to us (Theorem 2.8, Exercise 2.16). It is interesting, however, that we can deduce this fundamental mathematical statement from Lagrange's Theorem. Let us also mention that a number of the form $2^p - 1$ is called a **Mersenne number** (compare Exercise 2.15).

6.20. Let H and K be finite subgroups of a group G. Prove that each of the following two conditions implies that $H \cap K = \{1\}$:

(a) $|H|$ and $|K|$ are relatively prime numbers.
(b) $|H| = |K|$ is a prime and $H \neq K$.

6.21. Prove that if a group G has n subgroups of order p, then it has $n(p - 1)$ elements of order p.

6.22. Let H and K be finite subgroups of a group G. Prove that

$$|HK| = \frac{|H||K|}{|H \cap K|}.$$

Sketch of proof. Let m be the number of all distinct cosets hK where $h \in H$. First show that $|HK| = m|K|$. Then show that for any $h, h' \in H$, $hK = h'K$ if and only if $h(H \cap K) = h'(H \cap K)$. Hence deduce that $m = [H : H \cap K]$, and finally apply Lagrange's Theorem.

Comment. If one of H or K is a normal subgroup, this also follows from the Second Isomorphism Theorem.

6.23. The alternating group A_4 has order 12. Therefore, the order of a proper nontrivial subgroup of A_4 can be 2, 3, 4, or 6. Prove that subgroups of order 2, 3, and 4 exist, but a subgroup of order 6 does not.

Sketch of proof. Finding subgroups of order 2, 3, 4 is easy. Assume there is an $H \leq A_4$ such that $|H| = 6$. Then H is normal and $A_4/H \cong \mathbb{Z}_2$ (see Exercise 4.60), so $\sigma^2 \in H$ for all $\sigma \in A_4$. If $\pi \in A_4$ is a 3-cycle, we thus have $\pi = (\pi^2)^2 \in H$. Now, either apply the result of Exercise 2.86 or simply count the number of 3-cycles in A_4.

6.2 The Class Formula and Cauchy's Theorem

In this section, we will first derive a handy formula that relates the order of a finite group to the order of its center and indices of certain subgroups. As an application, we will prove Cauchy's Theorem, which states that a finite group G contains an element of order p for every prime p dividing $|G|$.

This section is intended to serve as a gentle transition to more advanced group theory. The results that will be obtained are just special cases of some of the results from the next section. The readers not intimidated by abstraction may therefore choose to skip this section, and look at some of its parts later if there is a need for clarification.

6.2.1 The Class Formula

Recall that two elements x and x' of a group G are said to be *conjugate* if there exists an $a \in G$ such that

$$x' = axa^{-1}.$$

We may consider *conjugacy* as a relation on the set G.

Lemma 6.24. *Conjugacy is an equivalence relation on the set of elements of any group.*

Proof. Reflexivity is obvious, since $x = 1x1^{-1}$. So is symmetry, since $x' = axa^{-1}$ yields $x = a^{-1}x'a$. As for transitivity, note that $x' = axa^{-1}$ and $x'' = bx'b^{-1}$ imply $x'' = (ba)x(ba)^{-1}$. $\qquad\square$

The conjugacy relation therefore partitions a group G into disjoint equivalence classes Cl_i, called **conjugacy classes**. Assuming that G finite, we thus have

$$|G| = \sum_i |Cl_i|. \qquad (6.1)$$

This is basically the formula we are looking for. We only have to express $|Cl_i|$ in a certain useful way.

For every $x \in G$, let $Cl(x)$ denote the conjugacy class containing x. It consists of all elements conjugate to x. Thus,

$$Cl(x) = \{axa^{-1} \mid a \in G\}.$$

Recall that the set

$$C_G(x) = \{g \in G \mid xg = gx\}$$

is called the **centralizer of** x. It is easy to see that $C_G(x)$ is a subgroup. Indeed, $C_G(x)$ is clearly closed under multiplication, and multiplying $xg = gx$ from the left and right by g^{-1} shows that $g \in C_G(x)$ implies $g^{-1} \in C_G(x)$.

Lemma 6.25. *If G is a group and $x \in G$, then $|Cl(x)| = [G : C_G(x)]$.*

Proof. For all $a, b \in G$, $aC_G(x) = bC_G(x)$ if and only if $a^{-1}b \in C_G(x)$ (Lemma 4.6), that is, if and only if $xa^{-1}b = a^{-1}bx$. Multiplying from the left by a and from the right by b^{-1} we see that this can be equivalently expressed as $axa^{-1} = bxb^{-1}$. Thus,

$$axa^{-1} = bxb^{-1} \iff aC_G(x) = bC_G(x).$$

This shows that

$$axa^{-1} \mapsto aC_G(x)$$

is a well-defined injective map from $Cl(x)$ to the set of all cosets of $C_G(x)$ in G. It is obviously also surjective, and so the cardinality of $Cl(x)$ is equal to the index of $C_G(x)$ in G.   □

We are now in a position to rewrite (6.1) in a suitable way. The equality in the next theorem is called the **class formula**.

Theorem 6.26. *If G is a finite group with center $Z(G)$, then there exist $x_1, \ldots, x_m \in G \setminus Z(G)$ such that*

$$|G| = |Z(G)| + \sum_{j=1}^{m} [G : C_G(x_j)].$$

Proof. If $x \notin Z(G)$, then $xa \neq ax$ for some $a \in G$, and so x and axa^{-1} are distinct elements of $Cl(x)$. Therefore, $|Cl(x)| > 1$. If, however, $x \in Z(G)$, then $Cl(x)$ is equal to $\{x\}$. Accordingly, $x \in Z(G)$ if and only if $|Cl(x)| = 1$. Hence, (6.1) can be written as

$$|G| = |Z(G)| + \sum_{j=1}^{m} |Cl_j|,$$

where Cl_j are conjugacy classes containing more than one element. For each j, choose $x_j \in Cl_j$. Then $x_j \notin Z(G)$ and $Cl_j = Cl(x_j)$. The desired formula therefore follows from Lemma 6.25.   □

The class formula is a very useful tool in studying finite non-Abelian groups. One important application will be given in the next subsection.

6.2.2 Cauchy's Theorem

Corollary 6.3 states that if r is the order of an element of a finite group G, then r divides the order of G. What about the converse: if r divides $|G|$, does G contain an element of order r? In general, the answer is negative. After all, if G is not cyclic, then it cannot contain an element of order $|G|$; or, as another example, the group $\mathbb{Z}_2 \oplus \mathbb{Z}_2 \oplus \mathbb{Z}_2$ does not contain elements of order 4. However, in the special case where r is a prime, the answer turns out to be positive. This is the content of the next theorem. It is named after *Augustin-Louis Cauchy* (1789–1857), one of the most prolific mathematicians of his time.

One more comment before stating the theorem. Our proof could be shortened by using the Fundamental Theorem of Finite Abelian Groups. However, using a profound theorem to establish something relatively simple may give the wrong picture. Besides, the argument that we will present nicely illustrates the usefulness of the concept of a quotient group.

Theorem 6.27. (Cauchy's Theorem) *Let G be a finite group. If a prime p divides $|G|$, then G contains an element of order p.*

Proof. We use induction on $n := |G|$. Since $p \mid n$, the first case to consider is when $n = p$. By Corollary 6.6, G is then a cyclic group and actually every element different from 1 has order p. We may thus assume that $n > p$ and that the theorem is true for all groups of order less than n. This, of course, includes all proper subgroups of G.

Assume first that G is non-Abelian. If $p \nmid |Z(G)|$, then we see from the class formula that $p \nmid [G : C_G(x_j)]$ for some $x_j \in G \setminus Z(G)$. Since

$$n = [G : C_G(x_j)] \cdot |C_G(x_j)|$$

by Lagrange's Theorem, it follows that $p \mid |C_G(x_j)|$. We have thus shown that p divides either $|Z(G)|$ or $|C_G(x_j)|$ with $x_j \notin Z(G)$. Since each of the two is a proper subgroup of G, the desired conclusion follows from the induction hypothesis.

It remains to consider the case where G is Abelian. Since $n > p$, n is not a prime. Therefore, by Corollary 6.7, G has a proper nontrivial subgroup N. As G is Abelian, N is a normal subgroup. Lagrange's Theorem applied to N can thus be written as

$$n = |G/N| \cdot |N|.$$

Hence, either $p \mid |G/N|$ or $p \mid |N|$. In the second case, we are done by the induction hypothesis. Since the order of G/N is less than n, in the first case we may use the induction hypothesis to conclude that there exists an $a \in G$ such that the coset aN has order p. According to Lemma 6.1 (c), a has order kp for some $k \in \mathbb{N}$. Hence, $a^k \neq 1$ and $(a^k)^p = 1$, so a^k has order p by Lemma 6.1 (b). $\qquad\square$

To give an example of applicability of Cauchy's Theorem, we first introduce the following important class of groups.

Definition 6.28. Let p be a prime. A finite group G is called a *p*-**group** if $|G| = p^m$ for some $m \geq 0$.

Corollary 6.29. *A finite group G is a p-group if and only if the order of every element in G is a power of p.*

Proof. Since the order of an element divides the order of the group (Corollary 6.3), the order of an element in a p-group is a power of p. Conversely, if the order of every element in G is a power of p, then, by Cauchy's Theorem, p is the only prime that divides $|G|$. Therefore, G is a p-group.   □

The condition that all elements in G have order a power of p still makes sense if G is an infinite group. We therefore use it to define the notion of a p-group for general, not necessarily finite groups. However, we will be interested only in finite p-groups. We have actually already encountered them in the treatment of Abelian groups, we just did not give them a name at that time. Note that the Fundamental Theorem of Finite Abelian Groups states that every finite Abelian group is a direct product of cyclic p-groups. In the next section, we will see that non-Abelian p-groups are also very important. The simplest examples are the quaternion group Q and the dihedral group D_8. Both have order 8, so they are 2-groups.

Exercises

6.30. It is easy to see that any two elements from the same conjugacy class have the same order. Do any two elements having the same order always lie in the same conjugacy class?

6.31. Show that the symmetric group S_3 has three conjugacy classes: the first one contains only 1, the second one all three transpositions, and the third one both 3-cycles.

6.32. Show that the quaternion group Q has five five conjugacy classes and describe them.

6.33. Show that the dihedral group D_8 has five conjugacy classes and describe them.

6.34. Prove that a nontrivial finite p-group has a nontrivial center.

6.35. Prove that a group G is Abelian if $|G| = p^2$ for some prime p.

6.36. The **exponent** of a group G is the smallest natural number m such that $x^m = 1$ for all $x \in G$; if no such m exists, the exponent of G is ∞. Prove that the exponent of a finite group G is the least common multiple of the orders of the elements in G, divides $|G|$, and has the same prime divisors as $|G|$.

6.37. What is the exponent of a finite cyclic group?

6.38. What is the exponent of S_3?

6.39. Give an example of a finite group whose exponent is not equal to its order.

6.40. For which finite p-groups is the exponent equal to the order?

6.41. For each $m \in \mathbb{N}$ and prime p, give an example of a p-group of order p^m and exponent p.

6.42. Give an example of an infinite group with finite exponent.

6.43. Give an example of a group with infinite exponent in which every element has finite order.

6.44. Let G be a finite group. Prove that the cardinality of the set $\{x \in G \mid x \neq x^{-1}\}$ is even. Hence derive Cauchy's Theorem for $p = 2$.

6.45. Suppose a group G has order mp, where p is a prime and $m < p$. Prove that G contains exactly one subgroup N of order p and that N is normal. As an illustration, examine the special case where $|G| = 6$ (so $G \cong \mathbb{Z}_3 \oplus \mathbb{Z}_2$ or $G \cong S_3$) and $p = 3$.

Hint. If you are unable to find a direct proof, employ the results of Exercises 6.20, 6.22, and 4.48.

6.46. Prove that a non-Abelian group of order 6 is isomorphic to S_3.

Sketch of proof. Cauchy's Theorem implies that G has an element a of order 3 and an element b of order 2. From Exercise 6.22 we deduce that $G = \langle a \rangle \langle b \rangle$. Show that the element bab can only be equal to a^2, and that this uniquely determines the multiplication in G.

6.47. Suppose a group G contains elements $a \neq 1$ and $b \neq 1$ such that $a^3 = b^2$ and $a^2 b = ba$. Prove that $|G| < 12$ implies $G \cong S_3$.

6.48. Prove that if a group G has order $2p$, where p is a prime different from 2, then $G \cong \mathbb{Z}_{2p}$ or $G \cong D_{2p}$.

Comment. This well-known result is proved in many textbooks. More interested readers looking for a good challenge should try to find a proof by themselves. Note that Exercise 6.46 is just a special case of this one (recall that $D_6 = S_3$).

6.49. Prove that a group G of order $4k + 2$, $k \in \mathbb{N}$, contains a normal subgroup of order $2k + 1$. In particular, G is not simple.

Hint. Let $\varphi : G \to S_{4k+2}$ be the embedding from the proof of Cayley's Theorem. Show that if $a \in G$ has order 2, then $\varphi(a)$ is an odd permutation.

6.3 Group Actions and the Sylow Theorems

In this section, we will generalize both the theory and techniques of the preceding section. First we will translate the notions and results around the class formula into the language of group actions, and then apply this to prove the Sylow theorems. These theorems, the first of which is a generalization of Cauchy's Theorem, open the doors to a finer analysis of finite groups.

6.3.1 Group Actions

Recall that a group G is said to **act on a set** X if there exists a map $(a, x) \mapsto a \cdot x$ from $G \times X$ to X, called a **group action** of G on X, that satisfies

$$(ab) \cdot x = a \cdot (b \cdot x) \quad \text{and} \quad 1 \cdot x = x$$

for all $a, b \in G$, $x \in X$. This definition was mentioned in Section 3.5 in connection with Cayley's Theorem, and we advise the reader to look at the brief discussion therein.

Let us look at some examples.

Example 6.50. Every group G acts on itself by **left multiplication**, i.e., $a \cdot x := ax$ for all $a, x \in G$, where ax is the usual product of a and x in G. This is the action on which the proof of Cayley's Theorem was based.

Example 6.51. Another standard action of a group G on itself is by **conjugation**: $a \cdot x := axa^{-1}$ for all $a, x \in G$.

Example 6.52. Let H be a subgroup of G. Then G acts on the set of all cosets of H in G by $a \cdot xH := axH$.

Example 6.53. Any group G acts on the set of all subgroups of G by $a \cdot K = aKa^{-1}$.

Example 6.54. The group S_n acts on the set $\{1, \ldots, n\}$ by $\sigma \cdot i := \sigma(i)$.

Example 6.55. From the discussion preceding Theorem 2.65 we see that S_n acts on the set of all polynomials in n variables over a ring R by permuting the variables:

$$(\sigma \cdot f)(X_1, \ldots, X_n) := f(X_{\sigma(1)}, \ldots, X_{\sigma(n)}).$$

Example 6.56. The general linear group $GL_n(\mathbb{R})$ acts on the set $\mathbb{R}^n$ in a natural way, i.e., by letting $A \cdot v$ be the usual product of the matrix A and the vector v (written as a column).

Example 6.57. The **trivial action** of G on X is given by $a \cdot x = x$ for all $a \in G, x \in X$. Here, G is an arbitrary group and X is an arbitrary set.

A variety of examples shows that the notion of a group action is quite general. We will not study it deeply. A few simple yet not obvious observations will be sufficient for our purposes. They concern the sets from the next definition.

Definition 6.58. Let a group G act on a set X. The **orbit** of an element $x \in X$ is the set

$$G \cdot x = \{a \cdot x \,|\, a \in G\} \subseteq X,$$

and the **stabilizer** of x is the set

$$G_x := \{g \in G \,|\, g \cdot x = x\} \subseteq G.$$

Example 6.59. Consider the action of G on itself by conjugation (Example 6.51). Then the orbit $G \cdot x$ is the conjugacy class $\mathrm{Cl}(x) = \{axa^{-1} \,|\, a \in G\}$, and the stabilizer G_x is the centralizer $C_G(x) = \{g \in G \,|\, xg = gx\}$.

We will show that the results concerning $\mathrm{Cl}(x)$ and $C_G(x)$ from the preceding section hold for general orbits and stabilizers. The proofs will be only notationally different. We start with two simple observations:

(a) The stabilizer G_x *is a subgroup of* G. Indeed, it is clearly closed under multiplication, and by multiplying $g \cdot x = x$ from the left by g^{-1} we see that $g \in G_x$ implies $g^{-1} \in G_x$.

(b) The rule

$$x \sim y \iff y = a \cdot x \ \text{ for some } a \in G$$

defines an *equivalence relation* on X, and the equivalence class containing x is the orbit $G \cdot x$. The proof is straightforward and is left to the reader.

We continue with a generalization of Lemma 6.25.

Lemma 6.60. *Let a group G act on a set X. Then $|G \cdot x| = [G : G_x]$ for every $x \in X$.*

Proof. Note that for all $a, b \in G$,

$$a \cdot x = b \cdot x \iff b^{-1}a \cdot x = x \iff b^{-1}a \in G_x \iff aG_x = bG_x.$$

Therefore, $a \cdot x \mapsto aG_x$ is a well-defined injective map from $G \cdot x$ to the set of all cosets of G_x in G. Since it is obviously also surjective, the lemma follows. □

In view of Lagrange's Theorem, the result of the lemma can also be written as $|G| = |G \cdot x| \cdot |G_x|$. In this form, the lemma is known as the **Orbit-Stabilizer Theorem**.

A generalization of the class formula is now at hand.

Theorem 6.61. *Let a group G act on a finite set X. Denote by Z the set of all elements $x \in X$ satisfying $a \cdot x = x$ for all $a \in G$. Then there exist $x_1, \ldots, x_m \in X \setminus Z$ such that*

$$|X| = |Z| + \sum_{j=1}^{m} [G : G_{x_j}].$$

Furthermore, if G is a finite p-group, then $|X| \equiv |Z| \pmod{p}$.

Proof. Note that $x \in Z$ if and only if the orbit $G \cdot x$ contains exactly one element. Since, in light of (b), X is a disjoint union of orbits $G \cdot x$, it follows that there exist $x_1, \ldots, x_m \in X \setminus Z$ such that

$$|X| = |Z| + \sum_{j=1}^{m} |G \cdot x_j|.$$

The desired formula thus follows from Lemma 6.60.

Suppose G is a finite p-group. Since $x_j \notin Z$, $[G : G_{x_j}] = |G \cdot x_j| > 1$. Lagrange's Theorem implies that $[G : G_{x_j}]$ is a power of p. Hence, p divides $|X| - |Z|$. $\qquad\square$

6.3.2 The Sylow Theorems

Let p be a prime. A group G contains an element of order p if and only if it contains a subgroup of order p. Indeed, if $a \in G$ has order p, then the cyclic subgroup $\langle a \rangle$ has order p. Conversely, a subgroup of order p is cyclic (Corollary 6.6) and thus contains an element of order p. Cauchy's Theorem can therefore be equivalently stated as follows: if a prime p divides the order $|G|$ of the group G, then G has a subgroup of order p. Now, one may ask whether this can be generalized to non-prime numbers. That is, if $r \in \mathbb{N}$ divides $|G|$, does G contain a subgroup of order r? By Lagrange's Theorem, we know that this question makes no sense without assuming that $r \mid |G|$. What we are asking is whether the converse to Lagrange's Theorem holds. Unfortunately, the answer is negative in general. For example, the alternating group A_4 has order 12, but does not contain a subgroup order 6 (see Exercise 6.23). We will show, however, that the answer is positive not only when r is a prime, but, more generally, when r is a power of a prime. Moreover, we will provide a deeper insight into subgroups of prime power order.

To avoid any possible confusion, let us mention that $p^{\ell} \mid |G|$, where p is a prime and $\ell \geq 2$, does not necessarily imply that G contains an element of order p^{ℓ}. We will only show that G has a subgroup H of order p^{ℓ}, but there is no reason why H should be cyclic (see the comments before Cauchy's Theorem).

We need some definitions.

Definition 6.62. Let H be a subgroup of a group G. The set

$$N_G(H) := \{a \in G \mid aHa^{-1} = H\}$$

is called the **normalizer** of H.

It is clear that $N_G(H) = G$ if and only if H is a normal subgroup of G, and an easy exercise to show that

$$N_G(H) \leq G \quad \text{and} \quad H \triangleleft N_G(H).$$

(One way to prove that $N_G(H)$ is a subgroup is to observe that $N_G(H)$ is the stabilizer G_H with respect to the action from Example 6.53.)

Definition 6.63. Let p be a prime. A subgroup H of a finite group G is called a *p*-**subgroup** of G if $|H| = p^\ell$ for some $\ell \geq 0$ (i.e., if H is a p-group). If $|H| = p^k$ and $p^{k+1} \nmid |G|$, then we call H a **Sylow *p*-subgroup** of G.

Suppose $|G| = p^k t$ where p does not divide t. The p-subgroups of G may only be of order p^ℓ with $\ell \leq k$ by Lagrange's Theorem. The Sylow p-subgroups are those of maximal order p^k. The existence of Sylow p-subgroups and other p-subgroups of orders p^ℓ with $\ell \leq k$ is the first statement of what is commonly called the Sylow theorems. They were discovered by the Norwegian mathematician *Ludwig Sylow* (1832–1918) in 1872.

Theorem 6.64. (Sylow theorems) *Let G be a finite group and let p be a prime that divides $|G|$.*

(a) *If p^ℓ divides $|G|$, then G has at least one p-subgroup of order p^ℓ.*
(b) *Every p-subgroup of G is contained in some Sylow p-subgroup of G.*
(c) *Any two Sylow p-subgroups are conjugate.*
(d) *The number of Sylow p-subgroups of G divides $|G|$.*
(e) *The number of Sylow p-subgroups of G is of the form $mp + 1$ for some $m \geq 0$.*

Proof. (a) The proof is similar to that of Cauchy's Theorem. We proceed by induction on $n := |G|$. The base case is trivial, so we assume that the statement is true for all groups of order less than n.

We first consider the case where $p \nmid |Z(G)|$. From the class formula we see that then $p \nmid [G : C_G(x_j)]$ for some $x_j \in G \setminus Z(G)$. Since p^ℓ divides

$$n = |G| = [G : C_G(x_j)] \cdot |C_G(x_j)|,$$

it follows that $p^\ell \mid |C_G(x_j)|$. However, $C_G(x_j)$ is a proper subgroup of G, and so we are done by the induction hypothesis.

Assume now that $p \mid |Z(G)|$. Cauchy's Theorem (for Abelian groups) implies that $Z(G)$ contains an element c of order p. Since c lies in the center, $\langle c \rangle$ is a normal subgroup of G. Lagrange's Theorem implies that $p^{\ell-1}$ divides the order of the quotient group $G/\langle c \rangle$. Therefore, $G/\langle c \rangle$ has a subgroup of order $p^{\ell-1}$ by the induction hypothesis. According to Theorem 4.38, this subgroup can be written as $H/\langle c \rangle$ for some $H \leq G$. Hence,

$$|H| = |H/\langle c \rangle| \cdot |\langle c \rangle| = p^{\ell-1} \cdot p = p^\ell.$$

(b)–(c) By (a), G has at least one Sylow p-subgroup, which we denote by S. Take an arbitrary p-subgroup H of G. Define an action of the group H on the set X of all cosets xS of S in G by

$$a \cdot xS := axS$$

(as in Example 6.52). Let Z denote the set of all xS in X such that $a \cdot xS = xS$ for every $a \in H$. Since H is a p-group, Theorem 6.61 tells us that $|X| \equiv |Z| \pmod{p}$. Now, $|X| = \frac{|G|}{|S|}$ by Lagrange's Theorem and hence, as S is a Sylow p-subgroup, p does not divide $|X|$. Consequently, p does not divide $|Z|$. In particular, $Z \neq \emptyset$. Take $x \in G$ such that $xS \in Z$. For every $a \in H$ we then have $axS = a \cdot xS = xS$, and therefore $x^{-1}ax \in S$. That is to say, $x^{-1}Hx \subseteq S$, or equivalently,

$$H \subseteq xSx^{-1}. \tag{6.2}$$

Since $|xSx^{-1}| = |S|$, xSx^{-1} is also a Sylow p-subgroup. Thus, (6.2) proves (b). So far, H was an arbitrary p-subgroup. Assuming that it is a Sylow p-subgroup, we have $|H| = |xSx^{-1}|$ and so (6.2) yields $H = xSx^{-1}$, proving (c).

(d) Pick a Sylow p-subgroup S. By (c),

$$Y := \{xSx^{-1} \mid x \in G\}$$

is the set of all Sylow p-subgroups of G. The rule

$$a \cdot T := aTa^{-1} \tag{6.3}$$

defines an action of G on Y (as in Example 6.53). The orbit of S is Y and the stabilizer of S is $N_G(S)$, the normalizer of S. Lemma 6.60 therefore shows that

$$|Y| = [G : N_G(S)] = \frac{|G|}{|N_G(S)|}. \tag{6.4}$$

This proves that $|Y|$, the number of Sylow p-subgroups, divides $|G|$.

(e) Now consider the action of S on Y by conjugation. By this we mean that we take the same rule (6.3) as before, but a can be taken only from S. Let W be the set of all elements T in Y such that $a \cdot T = T$ for all $a \in S$. Note that this condition can be read as

$$S \subseteq N_G(T).$$

Obviously, S lies in W. Take any T in W. Both S and T are Sylow p-subgroups of G and hence also of its subgroup $N_G(T)$ (indeed, $p^{k+1} \nmid |G|$ implies $p^{k+1} \nmid |N_G(T)|$ by Lagrange's Theorem). By (c), S and T are conjugate in $N_G(T)$. However, by the very definition, $aTa^{-1} = T$ for every $a \in N_G(T)$, so this actually means that T and S are equal. We have thus proved that $W = \{S\}$. Since S is a p-group and $|W| = 1$, Theorem 6.61 shows that $|Y| \equiv 1 \pmod{p}$. That is to say, $|Y| = mp + 1$ for some $m \geq 0$. $\qquad\square$

Remark 6.65. A consequence of (c) is that a Sylow p-subgroup of G is a normal subgroup if and only if it is the only Sylow p-subgroup of G.

The proofs of statements (a)–(e) were not easy, but they all followed the same pattern. The main principle was always to apply the elementary observations from the preceding subsection to some concrete group actions.

If the Fundamental Theorem of Finite Abelian Groups completely determines the structure of finite Abelian groups of a given order n, the Sylow theorems tell us quite a bit about the structure of general groups of order n. We will try to illustrate this through a few examples. For convenience of exposition, we set:

$$n_p := \text{the number of Sylow } p\text{-subgroups of } G.$$

Example 6.66. Let $|G| = 12$. Since $12 = 2^2 \cdot 3$, the Sylow 2-subgroups have order 4 and the Sylow 3-subgroups have order 3. By (d), n_2 divides 12 and so $n_2 \in \{1, 2, 3, 4, 6, 12\}$. Since it is of the form $2m + 1$ by (e), the only possibilities are $n_2 = 1$ and $n_2 = 3$. Both can occur. For example, the alternating group A_4 has only one Sylow 2-subgroup and the dihedral group D_{12} has three. Checking this is an elementary exercise. Similarly, we see that $n_3 = 1$ or $n_3 = 4$. Again, both can occur: A_4 has four Sylow 3-subgroups and D_{12} has only one.

Example 6.67. Suppose $|G| = pq$ where p and q are primes and $p < q$. Since n_q divides pq and is of the form $mq + 1$, the only possibility is that $n_q = 1$. According to Remark 6.65, this means that G has a normal Sylow q-subgroup. In particular, G has a proper nontrivial normal subgroup and therefore is not a simple group.

This was just one simple example of applying the Sylow theorems to show that a group of a certain order cannot be simple. With some effort one can for instance prove that none of the non-Abelian groups of order less than 60 is simple. The group A_5 has order 60 and, as we will prove in Section 6.5, is simple.

Example 6.68. As in the preceding example, let $|G| = pq$ where $p < q$ are primes, and assume additionally that p does not divide $q - 1$. We claim that in this case G is a cyclic group (and hence isomorphic to $\mathbb{Z}_{pq}$). We already know that G has a normal Sylow q-subgroup N. From $n_p \mid pq$, $n_p = mp + 1$, and $p \nmid q - 1$, it follows that n_p is also equal to 1. Thus, G also has a normal Sylow p-subgroup M. Lagrange's Theorem shows that $M \cap N = \{1\}$. We may now use Theorem 4.30 (d) to conclude that the elements of M commute with the elements of N. Having prime orders, M and N are cyclic groups. Thus, there are $a, b \in G$ such that $M = \langle a \rangle$, $N = \langle b \rangle$, and $ab = ba$. From $a^p = 1$ and $b^q = 1$ we get $(ab)^{pq} = 1$, which readily implies that ab has order pq. Hence, $G = \langle ab \rangle$.

Example 6.69. This last example is similar to, but slightly more involved than the preceding one. Suppose $|G| = p^2q$ where $p < q$ are primes and p does not divide $q - 1$. Let us show that G is necessarily Abelian (and hence, as we know, isomorphic to either $\mathbb{Z}_{p^2} \oplus \mathbb{Z}_q$ or $\mathbb{Z}_p \oplus \mathbb{Z}_p \oplus \mathbb{Z}_q$). We leave it to the reader to show that $n_p = n_q = 1$. Thus, G has a normal Sylow p-subgroup M (of order p^2) and a normal Sylow q-subgroup N (of order q). By Theorem 4.30 (c), MN is a subgroup of G. A standard application of Lagrange's Theorem shows that $MN = G$ and $M \cap N = \{1\}$. Thus, G is the internal direct product of M and N. In particular, the elements of M commute with the elements of N. Moreover, both M and N are Abelian groups; namely, N is even cyclic since its order is a prime, while M is Abelian since its order is a square of a prime (see Exercise 6.35). But then G itself is Abelian.

Exercises

6.70. Let a group G act on a set X. Prove that for any $x, y \in X$, $x \sim y$ implies that G_x and G_y are conjugate subgroups.

6.71. Let G be the group of real numbers under addition, and let X be the set of all linear functions from $\mathbb{R}$ to $\mathbb{R}$. Verify that the following rules define group actions of G on X, and describe the corresponding orbits and stabilizers:

(a) For $t \in G$ and $f \in X$, $t \cdot f$ is the function defined by $(t \cdot f)(x) = f(x) + t$.
(b) For $t \in G$ and $f \in X$, $t \cdot f$ is the function defined by $(t \cdot f)(x) = f(x + t)$.

6.72. Let $1 \leq k \leq n$. Describe the orbit and stabilizer of the monomial $X_1 \cdots X_k$ with respect to the action from Example 6.55.

6.73. An action of a group G on a nonempty set X is said to be **transitive** if $G \cdot x = X$ for every $x \in X$ (i.e., there is only one orbit). Show that the actions from Examples 6.50, 6.52, 6.54, and the proof of (d) are transitive. What about the action from Example 6.56?

6.74. Suppose a finite group G acts transitively on a set X. Prove that X is a finite set and that $|X|$ divides $|G|$.

6.75. Let $N \triangleleft G$. Prove that if G acts transitively on a set X, then $|N \cdot x| = |N \cdot y|$ for all $x, y \in X$.

6.76. Let $H \leq G$. The set $C_G(H) := \{c \in G \mid ch = hc \text{ for all } h \in H\}$ is called the **centralizer of** H. Prove that $C_G(H)$ is a normal subgroup of the normalizer $N_G(H)$ and that the quotient group $N_G(H)/C_G(H)$ is isomorphic to a subgroup of $\mathrm{Aut}(H)$, the group of all automorphisms of H. (This is known as the **N/C Theorem**.)
Hint. Example 4.121.

6.77. Let G be the multiplicative group of all nonzero quaternions (i.e., $G = \mathbb{H}^*$), and let H be its subgroup $\{1, -1, i, -i\}$. Determine the normalizer $N_G(H)$ and the centralizer $C_G(H)$, and show that

$$N_G(H)/C_G(H) \cong \mathbb{Z}_2 \cong \mathrm{Aut}(H).$$

6.78. Let H be a p-subgroup of a finite group G. Prove that

$$[N_G(H) : H] \equiv [G : H] \pmod{p}.$$

Hint. Consider the action of H on the set of all cosets of H in G defined in the usual way (as in Example 6.52).

6.79. Using the following steps, give an alternative proof of Cauchy's Theorem.

(a) Let p be a prime dividing the order of a group G. Set
$$X := \{(a_1, \ldots, a_p) \mid a_i \in G \text{ and } a_1 \cdots a_p = 1\}.$$

Show that the rule
$$k \cdot (a_1, \ldots, a_p) := (a_{k+1}, \ldots, a_p, a_1, \ldots, a_k)$$

defines an action of the group $(\mathbb{Z}_p, +)$ on X.

(b) Define Z as in Theorem 6.61. Show that $|X| = |G|^{p-1}$ (hint: $a_p = (a_1 \cdots a_{p-1})^{-1}$) and hence conclude that $p \mid |Z|$.

(c) Show that $(a_1, \ldots, a_p)$ lies in Z if and only if $a_1 = \cdots = a_p$. Since $(1, \ldots, 1) \in Z$, and hence $Z \neq \emptyset$, conclude from $p \mid |Z|$ that there exists an $a \neq 1$ in G such that $(a, \ldots, a) \in Z$. Note that a has order p.

6.80. Let a finite group G act on a finite set X. For each $g \in G$, let fix(g) denote the set of all elements in X that are fixed by g, i.e., fix$(g) = \{x \in X \mid g \cdot x = x\}$. Note that $\sum_{g \in G} |\text{fix}(g)|$ is equal to the cardinality of the set $\{(g, x) \in G \times X \mid g \cdot x = x\}$, which in turn is equal to $\sum_{x \in X} |G_x|$. Using this along with the Orbit-Stabilizer Theorem, derive that the number of orbits is equal to $\frac{1}{|G|} \sum_{g \in G} |\text{fix}(g)|$.

Comment. This result, sometimes called **Burnside's Lemma**, has many applications in combinatorics.

6.81. Let $n, N \in \mathbb{N}$, $n < N$. Consider the following question: in how many ways we can write N as a sum of n natural numbers? Here, it should be understood that the order of the summands is irrelevant, so we are not interested in the cardinality of the set
$$X := \{(k_1, \ldots, k_n) \in \mathbb{N}^n \mid k_1 + \cdots + k_n = N\},$$

but in the number of orbits with respect to the natural action of the group S_n on X. Use the preceding exercise to answer the given question for, say, $n = 3$ and $N = 9$, and then verify your result by simply counting the number of all possibilities. (To gain a better feeling for the strength of this approach, you may also consider some larger numbers like $n = 4$ and $N = 100$ where direct methods do not seem to work.)

6.82. Show that in the case of Abelian groups, the Sylow theorems follow from the Fundamental Theorem of Finite Abelian Groups.

6.83. Prove that a normal p-subgroup of a finite group G is contained in every Sylow p-subgroup of G.

6.84. Suppose $|G| = p^k t$ where p is a prime and $p \nmid t$. Prove that $n_p \mid t$.

Comment. This simple observation slightly simplifies calculations, so it is advisable to use it in what follows.

6.85. Prove that a group of order 88 cannot be simple.

6.86. Prove that a group of order 441 cannot be simple.

6.87. From the definition of the dihedral group D_8 in Section 2.6, it can be extracted that D_8 can be identified with the following Sylow 2-subgroup of S_4:

$$\{1, (1\,2\,3\,4), (1\,4\,3\,2), (1\,2)(3\,4), (1\,3)(2\,4), (1\,4)(2\,3), (1\,3), (2\,4)\}.$$

Show that this subgroup is not normal. How many Sylow 2-subgroups does S_4 then have?

6.88. Note that every 3-cycle generates a Sylow 3-subgroup of S_4. Hence deduce, either directly or with the help of the Sylow theorems, that S_4 has four Sylow 3-subgroups. Together with the preceding exercise, this shows that S_4 has no normal Sylow p-subgroups. Does this mean that S_4 is simple?

6.89. Use the Sylow theorems to explain why the quaternion group Q cannot be embedded into the symmetric group S_5.

6.90. In Example 6.66, we saw that $|G| = 12$ implies that $n_2 = 1$ or $n_2 = 3$ and $n_3 = 1$ or $n_3 = 4$. Prove that at least one of n_2 and n_3 is equal to 1 (i.e., at least one of the Sylow p-subgroups of G is normal).

Hint. Exercise 6.21.

6.91. Give examples showing that the conclusions in Examples 6.68 and 6.69 do not hold without assuming that $p \nmid q - 1$.

6.92. Determine all $n \le 20$ with the property that all groups of order n are isomorphic.

6.93. Suppose $|G| = p^2 q^2$ where p and q are distinct primes such that $p \nmid q^2 - 1$ and $q \nmid p^2 - 1$. Prove that G is Abelian.

6.94. Prove that a group of order 3993 contains an element of order 33.

Hint. Formula (6.4), Example 6.68.

6.95. Let G be a p-group. Then G itself is a Sylow p-subgroup and therefore the Sylow theorems do not tell us much. In fact, the only nontrivial statement is (a). Generalize it by showing that G has at least one *normal* subgroup of order p^ℓ.

6.96. Prove that a group of order pqr, where p, q, r are not necessarily distinct primes, cannot be simple.

6.4 Solvable Groups

Even after being exposed to the Sylow theorems, we are still far from a complete understanding of the structure of finite non-Abelian groups. On the other hand, the structure of finite Abelian groups is simple and clear. In this section, we will consider groups that are close to Abelian groups—maybe not very close, but close enough for some purposes. Let us look at two simple examples before giving the definition.

Example 6.97. Let $G = S_3$. Although G is non-Abelian, it has a normal subgroup $N = A_3$ with the property that both N and G/N are Abelian. More precisely, N has order 3 and is hence isomorphic to $\mathbb{Z}_3$, and G/N has order 2 and is hence isomorphic to $\mathbb{Z}_2$.

Understanding the structure of a normal subgroup N and the quotient group G/N does not exactly imply understanding the structure of G. But it can be very helpful. It is often the case that information on N and G/N can be transferred to information on G. The proof of the Abelian case of Cauchy's Theorem is quite illustrative in this regard.

Example 6.98. It can be checked that the only nontrivial Abelian normal subgroup of $G = S_4$ is
$$N_1 = \{1, (1\,2)(3\,4), (1\,3)(2\,4), (1\,4)(2\,3)\}.$$
However, the group G/N_1 is not Abelian (for example, the cosets $(1\,2)N_1$ and $(1\,3)N_1$ do not commute) and is therefore, as a non-Abelian group of order 6, isomorphic to S_3. Our point here is that G has no normal subgroup N such that both N and G/N are Abelian. Yet the situation is not so bad, just slightly more involved than in the previous example. Denoting the alternating subgroup A_4 by N_2, we have a chain of normal subgroups
$$\{1\} = N_0 \subseteq N_1 \subseteq N_2 \subseteq N_3 = G$$
such that N_{i+1}/N_i is an Abelian group for each i. More precisely, $N_1/N_0 \cong N_1 \cong \mathbb{Z}_2 \oplus \mathbb{Z}_2$, and since $|N_2/N_1| = 3$ and $|N_3/N_2| = 2$ by Lagrange's Theorem, $N_2/N_1 \cong \mathbb{Z}_3$ and $N_3/N_2 \cong \mathbb{Z}_2$.

We have provided enough motivation for the following definition.

Definition 6.99. A group G is said to be **solvable** if there exist normal subgroups $N_0, N_1, \ldots, N_m$ of G such that
$$\{1\} = N_0 \subseteq N_1 \subseteq \cdots \subseteq N_m = G$$
and N_{i+1}/N_i is an Abelian group for $i = 0, 1, \ldots, m - 1$.

The $m = 1$ case shows that Abelian groups are solvable. We can thus focus on non-Abelian groups. Incidentally, the definition makes sense regardless of whether G is finite or not, but of course we are more interested in finite groups. The above two examples show that S_3 and S_4 are solvable. It turns out that actually "most" finite groups are solvable. **Burnside's Theorem** from 1904 states that every group of order $p^k q^\ell$, where p and q are primes, is solvable. This is a deep theorem with a long proof. Even more fascinating is the **Feit–Thompson Theorem** from 1963—its proof takes over 250 pages—stating that every group of odd order is solvable. So, how to find a non-solvable group? At least theoretically, this should be easy. If G has no normal subgroups different from $\{1\}$ and G, then the condition from the definition is fulfilled only when G is Abelian. That is to say, a non-Abelian simple group is not solvable. However, what are concrete examples of such groups? This question will be discussed in the next section. It turns out that the simplest examples are the alternating groups A_n where $n \geq 5$. As we shall see, from this it can be deduced that the symmetric groups S_n are not solvable for $n \geq 5$.

Solvable groups play an important role in group theory. Moreover, in Section 7.9 we will use them as a tool for solving a classical problem on polynomial equations. We thus have good reasons for devoting a part of this chapter to solvable groups. However, studying them in detail would be beyond the scope of this book. In the rest of the section, we will derive only some of their most basic properties.

Let G be a group and $x, y \in G$. Recall that the element

$$[x, y] = xyx^{-1}y^{-1}$$

is called the *commutator* of x and y. Obviously, x and y commute if and only if $[x, y] = 1$. The subgroup of G generated by all commutators in G is called the **commutator subgroup** of G. We will denote it by G'. Since the inverse of a commutator is again a commutator, in fact

$$[x, y]^{-1} = [y, x],$$

G' consists of all products of commutators in G.

Lemma 6.100. *If $N \triangleleft G$, then $N' \triangleleft G$. In particular, $G' \triangleleft G$.*

Proof. For each $a \in G$, let φ_a denote the corresponding inner automorphism of G, i.e., $\varphi_a(x) = axa^{-1}$ for every $x \in G$. The condition that a subgroup N is normal can be expressed as $\varphi_a(N) \subseteq N$ for every $a \in G$.

Note that φ_a, just as any homomorphism, satisfies

$$\varphi_a\big([x_1, y_1] \cdots [x_n, y_n]\big) = [\varphi_a(x_1), \varphi_a(y_1)] \cdots [\varphi_a(x_n), \varphi_a(y_n)].$$

This shows that $\varphi_a(N) \subseteq N$ implies $\varphi_a(N') \subseteq N'$. That is to say, N' is normal if N is normal. □

Lemma 6.101. *Let $N \triangleleft G$. Then G/N is Abelian if and only if $G' \subseteq N$. In particular, G/G' is Abelian.*

Proof. Let $x, y \in G$. The cosets xN and yN commute if and only if $xyN = yxN$, or, equivalently, if and only if $[x^{-1}, y^{-1}] \in N$. Therefore, G/N is Abelian if and only if N contains all commutators, i.e., if and only if $G' \subseteq N$. □

We define the **derived subgroups** of G by $G^{(0)} := G$ and inductively

$$G^{(i+1)} := (G^{(i)})', \quad i \geq 0.$$

Thus, $G^{(1)} = G', G^{(2)} = (G')'$, etc.

Theorem 6.102. *A group G is solvable if and only if $G^{(m)} = \{1\}$ for some $m \in \mathbb{N}$.*

Proof. By Lemma 6.100, $G^{(i)} \triangleleft G$, and by Lemma 6.101, $G^{(i)}/G^{(i+1)}$ is an Abelian group. Therefore, if $G^{(m)} = \{1\}$, then the chain

$$\{1\} = G^{(m)} \subseteq G^{(m-1)} \subseteq \cdots \subseteq G^{(1)} \subseteq G^{(0)} = G$$

shows that G is solvable.

Conversely, suppose G is solvable. Let $N_0, N_1, \ldots, N_m$ be as in Definition 6.99. Since N_{i+1}/N_i is Abelian, Lemma 6.101 implies that $N'_{i+1} \subseteq N_i$. Thus,

$$G^{(1)} \subseteq N_{m-1},$$

hence

$$G^{(2)} = (G^{(1)})' \subseteq N'_{m-1} \subseteq N_{m-2},$$

and continuing in this way, we obtain $G^{(j)} \subseteq N_{m-j}$ for $0 \leq j \leq m$. In particular, $G^{(m)} \subseteq N_0 = \{1\}$. $\qquad\square$

Remark 6.103. In many books, the definition of a solvable group requires only $N_i \triangleleft N_{i+1}$ rather than $N_i \triangleleft G$. Note that this change does not affect the proof of Theorem 6.102. Therefore, this definition is equivalent to ours.

Corollary 6.104. *A subgroup of a solvable group is solvable.*

Proof. It is clear that $H \leq G$ implies $H^{(1)} \leq G^{(1)}$, and, in general, $H^{(i)} \leq G^{(i)}$ for every i. Hence, $G^{(m)} = \{1\}$ implies $H^{(m)} = \{1\}$. $\qquad\square$

Corollary 6.105. *Let $N \triangleleft G$. Then G is solvable if and only if both N and G/N are solvable.*

Proof. Observe that $(G/N)^{(1)} = G^{(1)}N/N$ (see Remark 4.39), and inductively, $(G/N)^{(i)} = G^{(i)}N/N$. Hence, $G^{(m)} = \{1\}$ implies $(G/N)^{(m)} = \{1\}$. Thus, G/N is solvable if G is solvable. By Corollary 6.104, the same is true for N.

Now assume that N and G/N are solvable. In particular, $(G/N)^{(m)} = \{1\}$ for some m. Let $\pi : G \rightarrow G/N$ be the canonical epimorphism. Note that $\pi(G^{(i)}) = \pi(G)^{(i)}$ for every i, and so

$$\pi(G)^{(m)} = (G/N)^{(m)} = \{1\}$$

implies

$$G^{(m)} \subseteq \ker \pi = N.$$

Corollary 6.104 tells us $G^{(m)}$ is solvable. Hence,

$$G^{(m+m')} = (G^{(m)})^{(m')} = \{1\}$$

for some m'. $\qquad\square$

Exercises

6.106. Let G and N be as in Example 6.97. Show that the group $G_1 = \mathbb{Z}_6$ has a subgroup N_1 such that $N_1 \cong N$ and $G_1/N_1 \cong G/N$. However, $G \not\cong G_1$.

Comment. The point we wish to make here is that the groups N and G/N unfortunately do not contain all information about the group G. The approach suggested in the comment following Example 6.97 thus has limitations. On the other hand, Corollary 6.105 speaks in its favor.

6.107. Determine S_3'.

6.108. Determine Q'.

6.109. Determine D_8'.

6.110. Determine A_4'.

6.111. Recall that a subgroup K of G is *characteristic* if $\varphi(K) \subseteq K$ for every automorphism φ of G. Prove that the derived subgroups $G^{(i)}$ are characteristic.

6.112. Let $N \triangleleft G$. Prove that $[G : N] < 6$ implies $G' \subseteq N$. Does this also hold for subgroups that are not normal?

6.113. Let G be any group. Define $G^1 := G'$, and, inductively, G^{i+1} as the subgroup of G generated by all commutators $[g, x]$ where $g \in G^i$ and $x \in G$. We say that G is **nilpotent** if $G^m = \{1\}$ for some $m \in \mathbb{N}$.

(a) Show that $G^{(i)} \leq G^i$ and hence conclude that every nilpotent group is solvable.
(b) Show that every Abelian group is nilpotent.
(c) Show that a nontrivial nilpotent group has a nontrivial center.
(d) Give an example of a nilpotent non-Abelian group.
(e) Give an example of a solvable group that is not nilpotent.

6.114. Prove that a finite p-group is solvable.

Comment. Using Corollary 6.105 along with the fact that a nontrivial finite p-group has a nontrivial center (Exercise 6.34), this should be easy to prove. The more interested reader may try to prove the more general statement that a finite p-group is nilpotent.

6.115. Prove that D_{2n} is a solvable group.

6.116. Prove that a group of order pq, where p and q are primes, is solvable.

6.5 Simple Groups

We first make some general comments about the notion of simplicity. Besides simple groups, we have also introduced simple rings and simple modules. We may think of these simple objects as building blocks for constructing more complex objects. For example, by taking direct products (or sums) of at least two simple objects we obtain a new object that is no longer simple. In the case of vector spaces, this straightforward construction already yields general objects. Indeed, regarding vector spaces as modules over fields, the simple modules are the 1-dimensional vector spaces (see Exercise 5.127), and every nontrivial vector space can be written as their direct sum, with summands being the 1-dimensional subspaces generated by vectors from a basis. The structure of groups, rings, and modules over general rings is, of course, much more complicated, and using direct products (or sums) we can obtain only some of the objects. For example, even a cyclic group of order p^2, where p is a prime, cannot be written as a direct product of simple groups.

There is, however, another, more sophisticated way to associate certain simple groups to a general finite group $G \neq \{1\}$. We start by choosing a proper normal subgroup G_1 of G of maximal order (if G is simple, then $G_1 = \{1\}$). Using the Correspondence Theorem it is easy to see that G/G_1 is a simple group (this is Exercise 4.59). In the next step, we choose a proper normal subgroup G_2 of G_1 of maximal order, after that a proper normal subgroup G_3 of G_2 of maximal order, and so on. Since G is finite, this process must terminate with $G_m = \{1\}$. Thus, there exist subgroups $G_0, G_1, \ldots, G_m$ of G such that

$$\{1\} = G_m \subseteq G_{m-1} \subseteq \cdots \subseteq G_1 \subseteq G_0 = G,$$

$G_{i+1} \lhd G_i$ and G_i/G_{i+1} is a simple group for $i = 0, 1, \ldots, m - 1$. Such a sequence of subgroups is called a **composition series** for G. Usually G has more than just one composition series. However, each of them gives rise to the same simple groups G_i/G_{i+1}. More precisely, the following holds: if

$$\{1\} = H_k \subseteq H_{k-1} \subseteq \cdots \subseteq H_0 = G$$

is another composition series for G, then $k = m$ and there exists a permutation σ of $\{1, \ldots, m\}$ such that
$$H_i/H_{i+1} \cong G_{\sigma(i)}/G_{\sigma(i)+1}.$$

This result is called the **Jordan–Hölder Theorem**. The proof is not very difficult, but we shall not give it here. The purpose of this introduction is only to provide additional motivation for considering simple groups. Let us conclude with a rough summary: to every nontrivial finite group G we can associate a unique collection of simple groups, and although G is not completely determined by them (see Exercise 6.106), they can be very helpful in understanding the structure of G.

On a more down-to-earth level, let us provide some examples of simple groups. Corollary 6.8 states that among Abelian groups, the simple ones are precisely the cyclic groups of prime order. Establishing the simplicity of a non-Abelian group is never entirely easy. Among our standard examples of non-Abelian groups, the only simple ones turn out to be the alternating groups A_n where $n \geq 5$ (Example 6.98 shows that A_4 is not simple, A_3 is simple but Abelian, and A_1 and A_2 are trivial groups). We will prove only that A_5 is simple. The general case is not that much more difficult, but requires more computation.

Theorem 6.117. *The group A_5 is simple.*

Proof. Our first step is to show that A_5 (and in fact every A_n with $n \geq 3$) is generated by the 3-cycles (which of course are even permutations). The reader was asked to prove this in Exercise 2.86, but we give the proof here for completeness. Since every element in an alternating group is a product of an even number of transpositions, it is enough to show that the product $\tau_1 \tau_2$ of two distinct transpositions τ_1 and τ_2 lies in the subgroup generated by the 3-cycles. If τ_1 and τ_2 are not disjoint, then $\tau_1 \tau_2 = (i_1 \, i_2)(i_1 \, i_3) = (i_1 \, i_3 \, i_2)$ is a 3-cycle. If they are disjoint, then

$$\tau_1 \tau_2 = (i_1 \, i_2)(i_3 \, i_4) = (i_1 \, i_2)(i_2 \, i_3)(i_2 \, i_3)(i_3 \, i_4) = (i_2 \, i_3 \, i_1)(i_3 \, i_4 \, i_2),$$

which completes the proof.

In the second step, we will prove that any two 3-cycles $(i_1 \, i_2 \, i_3)$ and $(j_1 \, j_2 \, j_3)$ are conjugate in A_5. All we have to show is that there exists a $\sigma \in A_n$ satisfying $\sigma(i_1) = j_1$, $\sigma(i_2) = j_2$, and $\sigma(i_3) = j_3$, since then

$$(j_1 \, j_2 \, j_3) = \sigma(i_1 \, i_2 \, i_3)\sigma^{-1}$$

follows. There certainly exists a permutation $\sigma \in S_n$ with the desired properties. If σ happens to be an odd permutation, we simply replace it by $(j_4 \, j_5)\sigma$ which then lies in A_5 and has the same properties (here, of course, j_4 and j_5 are distinct and do not belong to $\{j_1, j_2, j_3\}$).

We are ready for the final step of the proof. Take a normal subgroup $N \neq \{1\}$ of A_5. It suffices to prove that N contains a 3-cycle. Indeed, by the second step N then contains all the 3-cycles, and is hence equal to A_5 by the first step. Pick $\sigma \neq 1$ in N and write it as a product of disjoint cycles. Since σ is an even permutation, there are only three possibilities:

$$\sigma = (i_1 \, i_2 \, i_3), \quad \sigma = (i_1 \, i_2 \, i_3 \, i_4 \, i_5), \quad \text{or} \quad \sigma = (i_1 \, i_2)(i_3 \, i_4)$$

(where $i_u \neq i_v$ if $u \neq v$). We want to prove that N has a 3-cycle, so we may assume that σ is not of the first type. Suppose $\sigma = (i_1 \, i_2 \, i_3 \, i_4 \, i_5)$. Denote the 3-cycle $(i_1 \, i_2 \, i_3)$ by π. Since N is normal, it contains the element

$$\left(\pi \sigma \pi^{-1}\right)\sigma^{-1} = (i_1 \, i_2 \, i_3)(i_1 \, i_2 \, i_3 \, i_4 \, i_5)(i_3 \, i_2 \, i_1)(i_5 \, i_4 \, i_3 \, i_2 \, i_1) = (i_1 \, i_2 \, i_4),$$

so we are done in this case. Finally, suppose $\sigma = (i_1 \, i_2)(i_3 \, i_4)$. Let i_5 be different from i_1, i_2, i_3, i_4 and let us write ρ for $(i_1 \, i_2)(i_3 \, i_5)$. Then

$$\left(\rho \sigma \rho^{-1}\right)\sigma^{-1} = (i_1 \, i_2)(i_3 \, i_5)(i_1 \, i_2)(i_3 \, i_4)(i_1 \, i_2)(i_3 \, i_5)(i_1 \, i_2)(i_3 \, i_4) = (i_3 \, i_5 \, i_4),$$

so N contains a 3-cycle in this case too. □

Corollary 6.118. *For $n \geq 5$, the symmetric group S_n is not solvable.*

Proof. Since A_5 is simple and non-Abelian, it is not solvable. According to Corollary 6.104, any group containing A_5 as a subgroup is not solvable either. Now, since $n \geq 5$, we may view S_5, and hence also A_5, as a subgroup of S_n. □

What about groups of matrices over finite fields? It turns out, for example, that the general linear group $GL_3(\mathbb{Z}_2)$ is simple and has order 168. Moreover, it is the second smallest non-Abelian simple group, right after A_5 whose order is 60. In general, however, the group $GL_n(F)$ is rarely simple since $SL_n(F)$, the special linear group, is a normal subgroup (if $F = \mathbb{Z}_2$ then of course $SL_n(F) = GL_n(F)$). The group $SL_n(F)$ is also only exceptionally simple since its center is usually nontrivial (see Exercise 6.129). However, the **projective special linear group**, defined as the quotient group of $SL_n(F)$ by its center,

$$PSL_n(F) := SL_n(F)/Z(SL_n(F)),$$

is simple for any $n \geq 2$ and any field F, except when $n = 2$ and F is either $\mathbb{Z}_2$ or $\mathbb{Z}_3$. We mention as a curiosity that the aforementioned group $GL_3(\mathbb{Z}_2)$ is isomorphic to $PSL_2(\mathbb{Z}_7)$.

By relatively elementary means we can thus find three infinite families of finite simple groups: the cyclic groups $\mathbb{Z}_p$ where p is a prime, the alternating groups A_n where $n \geq 5$, and the projective special linear groups $PSL_n(F)$ where $n \geq 2$ and, except for two special cases, F can be any finite field (we can, for example, take $\mathbb{Z}_p$ for F, but there are more possibilities as we will see in Section 7.6). It turns out that, besides these three, there are 15 more infinite families of finite simple groups. Additionally, there are 26 special finite simple groups not belonging to these families, called **sporadic groups**. The largest of these, known as the **Monster group**, has order approximately 8×10^{53}. However, what is truly amazing is that these 18 infinite families together with 26 sporadic groups form a *complete list* of finite simple groups! This **classification of finite simple groups** is a landmark result, one of the triumphs of modern mathematics. The proof is spread out over thousands of pages in hundreds of papers, written by many different authors. One can thus say that simple groups are anything but simple.

Exercises

6.119. Find a composition series for S_3.

6.120. Find a composition series for S_4.

6.121. Find a composition series for S_5.

6.122. Find two composition series for $\mathbb{Z}_6$.

6.123. Find all composition series for Q.

6.124. The dihedral group D_8 has seven different composition series. Find all of them.

6.125. Prove that an infinite Abelian group does not have a composition series.

6.126. Recall that S_n' denotes the commutator subgroup of S_n.

(a) Prove, either directly or by using Lemma 6.101, that $S_n' \subseteq A_n$.
(b) Prove that every 3-cycle is a commutator.
(c) Deduce from (a) and (b) that $S_n' = A_n$.

6.127. Let $n \geq 5$. Using the fact that A_n is simple, prove that the only normal subgroup of S_n are $\{1\}$, A_n, and S_n.

Comment. This is not true if $n = 4$, see Example 6.98.

6.128. Let $G \leq S_n$. Prove that if G is simple and $|G| > 2$, then $G \leq A_n$.

6.129. Prove that $Z(\mathrm{SL}_n(F))$ consists of scalar matrices λI where $\lambda \in F$ and $\lambda^n = 1$.

6.130. For $i \neq j$, define the **elementary matrices** $A_{ij}(\lambda) = I + \lambda E_{ij}$ where $\lambda \in F$ and E_{ij} is a matrix unit (see Exercise 4.88). Note that $A_{ij}(\lambda) \in \mathrm{SL}_n(F)$.

(a) Prove that the matrices $A_{ij}(\lambda)$ generate $\mathrm{SL}_n(F)$.
(b) Prove that $A_{ij}(\lambda)$ are commutators in $\mathrm{SL}_n(F)$ except when $n = 2$ and $|F| \leq 3$.
(c) Prove that $\mathrm{SL}_n(F)' = \mathrm{SL}_n(F)$ except when $n = 2$ and $|F| \leq 3$.
(d) Prove that $\mathrm{GL}_n(F)' = \mathrm{SL}_n(F)$ except when $n = 2$ and $|F| \leq 3$.

Comment. It is easy to see that $|F| \leq 3$ if and only if $F \cong \mathbb{Z}_2$ or $F \cong \mathbb{Z}_3$.

Chapter 7
Field Extensions

This last chapter is the closest to the origin of algebra. Field extensions are intimately connected with polynomial equations, the subject of investigation of classical algebra. In the first section, we present a historical overview which should help the reader to better understand the content of this chapter, and, in fact, of the whole book. The theory developed in subsequent sections is mathematically rich with several impressive results, culminating in Galois theory, which, along with its applications, often strikes mathematical souls with its harmony and perfection.

7.1 A Peek into the History

When we think of mathematics, we first think of numbers. By numbers, we usually mean elements of one of the numbers sets

$$\mathbb{N}, \mathbb{Z}, \mathbb{Q}, \mathbb{R}, \mathbb{C}.$$

We first get acquainted with the *natural numbers*, which occur in our daily lives. The *integers* are nearly as "natural" as the natural numbers; after all, we can own as well as owe. Nevertheless, the negative integers and the number zero were introduced relatively late in the development of mathematics, especially when compared to positive *rational numbers*, which were known to the Babylonians as far back as four millennia ago. Of course, we need to represent parts of a whole, so we need the rational numbers. Why should we need any other numbers? This question may not have a single answer. Irrational numbers are not too useful when dealing with practical problems. Their decimal notation consists of an infinite, non-repeating sequence of digits, so we are forced to use their rational approximations. However, the rational numbers are easily seen to be insufficient from the theoretical viewpoint. For example, we know from the Pythagorean theorem that the length of the diagonal of a unit square is $\sqrt{2}$. However, this number is not rational. That is to say, the square

© Springer Nature Switzerland AG 2019
M. Brešar, *Undergraduate Algebra*, Springer Undergraduate Mathematics Series,
https://doi.org/10.1007/978-3-030-14053-3_7

of a rational number cannot equal 2. This has been known since ancient times. Let us recall the standard proof.

Proof of irrationality of $\sqrt{2}$. Suppose there exists a rational number $q = \frac{m}{n}$ such that $q^2 = 2$. We may assume that $\frac{m}{n}$ is simplified to lowest terms. In particular, m and n are not both even. Writing $q^2 = 2$ as $m^2 = 2n^2$, we see that m^2 is even. Hence m is even too, and, consequently, n is odd. Writing $m = 2k$ with $k \in \mathbb{Z}$ it follows from $m^2 = 2n^2$ that $2k^2 = n^2$. But this is a contradiction, since the square of an odd number cannot be even.

Thus, the length of the diagonal of a unit square cannot be written as a fraction. It turns out that the same holds for π, i.e., the ratio of a circle's circumference to its diameter. The proof is not too difficult, but requires some knowledge of calculus. There are, of course, further examples. One thus has to realize that in order to describe some fundamental geometric observations in precise terms, we need numbers that are not rational. Eventually this led to the introduction of *real numbers*. At first, mathematicians dealt with them without worrying about mathematical rigor. The question of the precise definition of real numbers became an issue in the latter half of the 19th century. The axiomatic definition that is commonly used today was formulated at the beginning of the 20th century. Should things change, our view of real numbers may not remain the same in the future. Anyhow, the current approach enables the development of beautiful mathematical theories with a wide spectrum of applications.

Why should we also extend the real numbers? It is a little bit more difficult to give a short, convincing answer to this question. We usually start from the fact that the equation $x^2 = -1$ has no solution in real numbers. The *complex numbers* are defined in such a way that this equation is solvable. We first introduce a "fictitious" number i, called the imaginary unit, that satisfies $i^2 = -1$. After that, the definition of general complex numbers along with their addition and multiplication is almost self-explanatory. As we know, the so-defined numbers form a field. However, what are they good for? Even though they had appeared in different forms since the 16th century or even earlier, mathematicians were skeptical towards them for a long time. This is understandable. The study of abstract algebraic structures, such as rings and fields, is a feature of modern mathematics. The fact that a certain set is a field may be interesting to us since this aspect is emphasized in our education, but we cannot expect mathematicians to have thought in similar terms centuries ago. Complex numbers were eventually accepted since they have proved to be extremely useful. It often turns out that a problem that concerns real numbers can be successfully solved by using complex numbers as the main tool. For example, look at Theorem 5.6 which describes irreducible polynomials over $\mathbb{R}$: the statement itself does not involve complex numbers, but the proof depends on them. More precisely, it is based on the *Fundamental Theorem of Algebra*, which states that for any complex numbers $a_0, a_1, \ldots, a_n$, the polynomial equation

$$a_n x^n + a_{n-1} x^{n-1} + \cdots + a_1 x + a_0 = 0$$

has at least one solution in $\mathbb{C}$, as long as $a_n \neq 0$ and $n \geq 1$. This theorem is often attributed to one of the greatest mathematicians of all time, *Carl Friedrich Gauss*. However, by today's standards, Gauss' proof from 1799 was not entirely rigorous. The first complete proof was given some years later by *Jean-Robert Argand*, who was a bookkeeper in Paris and only an amateur mathematician. Many different proofs are known today, but, paradoxically, they all use some non-algebraic tools such as continuity. One must look at the name "Fundamental Theorem of Algebra" from the historical perspective outlined in the next paragraphs. In fact, algebra is now such a broad subject that calling any of its theorems "fundamental" sounds like an overstatement. At any rate, the Fundamental Theorem of Algebra is one of the most famous theorems of mathematics and is behind many applications of complex numbers. And there are indeed many applications, both within mathematics and outside of mathematics. Even though the complex numbers arose from a seemingly controversial idea to extract square roots of negative real numbers, nowadays there is absolutely no doubt about their meaningfulness. What would modern mathematics be without complex numbers?

The introductions of rational, real, and complex numbers share a common feature. They were all motivated by the inability to solve polynomial equations, or, in other words, by the nonexistence of roots of polynomials in existing numbers sets. Linear polynomials $nX - m$ with integer coefficients do not necessarily have roots in $\mathbb{Z}$, but they do have them in $\mathbb{Q}$. Some polynomials, such as $X^2 - 2$, do not have roots in $\mathbb{Q}$, which led to the introduction of real numbers. Similarly, some polynomials, notably $X^2 + 1$, do not have roots in $\mathbb{R}$, which led to the introduction of complex numbers. Polynomial equations have thus played one of the key roles in the history of mathematics. Classical algebra was essentially the study of polynomial equations. Only in the 19th century, with the introduction of the first abstract algebraic structures, did its transition to modern (or abstract) algebra occur.

Let us take a short stroll through the history of solving polynomial equations.

The way to solve *linear and quadratic equations* was known in several ancient civilizations. The Babylonians were able to solve quadratic equations in a manner that does not essentially differ from ours. This does not mean that they understood the concept of an equation in the same way as we do now. In today's language, we would say that they were considering only "word problems" involving positive rational numbers. The history of mathematical symbolism and terminology is another interesting subject, but we will not discuss it here. Let us just say that it took a long time before mathematicians started to write the general quadratic equation and its solution in the form

$$ax^2 + bx + c = 0$$

and

$$x = \frac{-b \pm \sqrt{b^2 - 4ac}}{2a}. \tag{7.1}$$

Cubic equations were also known in ancient times. However, it was not until the Renaissance when their solutions were found (roughly three millennia after solving quadratic equations!). The 16th-century Italian mathematicians discovered that the equation

$$x^3 = px + q$$

has the solution

$$x = \sqrt[3]{\frac{q}{2} + \sqrt{\left(\frac{q}{2}\right)^2 - \left(\frac{p}{3}\right)^3}} + \sqrt[3]{\frac{q}{2} - \sqrt{\left(\frac{q}{2}\right)^2 - \left(\frac{p}{3}\right)^3}}. \qquad (7.2)$$

Moreover, the general *cubic equation*, as well as the general equation of the fourth degree (*quartic equation*), can be reduced to this special cubic equation. In this way, one can derive explicit formulas for solutions of cubic and quartic equations, which, however, are very long and complicated. This is documented in the 1545 book *The Great Art* by *Gerolamo Cardano*. Today, (7.2) is known as **Cardano's formula** (the fact is, however, that in his book Cardano presented the discoveries of his contemporaries *Scipione del Ferro*, *Niccolò F. Tartaglia*, and *Ludovico Ferrari*). It was not yet known in Cardano's time that one can use this formula for complex coefficients p and q to get complex solutions x. What is perhaps even more interesting, and Cardano was also not aware of, is that even if both p and q as well as all solutions x are real numbers, the complex numbers may still appear in this formula. Indeed, even if $\left(\frac{q}{2}\right)^2 - \left(\frac{p}{3}\right)^3$ is negative, the imaginary part of the solution x may still be 0 (see Exercise 7.5). This is a classical example of the usefulness of complex numbers.

The next challenge was the equation of the fifth degree (*quintic equation*). According to the Fundamental Theorem of Algebra, such an equation certainly has solutions, but how to find them? Solutions of polynomial equations of lower degrees can be expressed by the coefficients of a polynomial using the operations of addition, subtraction, multiplication, and division, together with the extraction of nth roots. Moreover, we have explicit formulas (like (7.1) and (7.2)) for them. One would expect that there should be a similar general formula for solutions of quintic equations, but probably an extremely complicated one. All the attempts to discover it were unsuccessful, until it finally became clear that such a formula cannot exist! This fascinating result was published in 1799 by the Italian mathematician *Paolo Ruffini*. His proof, however, was incomplete. Somewhat later, in 1824, a correct proof was given by *Niels Henrik Abel*. We have already mentioned Abel, along with *Évariste Galois*, in Section 1.3. It was Galois who made the next major breakthrough. He developed a theory which, in modern terminology, describes a profound connection between field extensions and finite groups. For him, a group was what we would call a subgroup of S_n. By Cayley's Theorem, this definition is essentially equivalent to our definition of a finite group. *Galois theory* provides a deep insight into the discovery by Ruffini and Abel. It shows not only that there cannot exist a general formula for roots of *arbitrary* polynomials of degree 5 (and more), but provides a group-theoretic criterion for deciding whether or not the roots of a *specific* polynomial can be expressed by combining its coefficients using only addition, subtraction, multiplication, division, and taking nth roots.

Some consider Galois' work to be the birth of abstract algebra. Even now, nearly 200 years later, it is regarded as one of the greatest and most beautiful achievements of mathematics.

Exercises

7.1. Let n be an integer that is not a perfect square. Prove that $\sqrt{n} \notin \mathbb{Q}$.

7.2. Prove that $\sqrt{2} + \sqrt{3} \notin \mathbb{Q}$.

7.3. Prove that $\log_2 3 \notin \mathbb{Q}$.

7.4. Prove that for any $s \in \mathbb{N}$,

$$\frac{1}{s+1} + \frac{1}{(s+1)(s+2)} + \frac{1}{(s+1)(s+2)(s+3)} + \cdots < 1.$$

Hence, derive that $e = \sum_{n=0}^{\infty} \frac{1}{n!}$ cannot be written as $\frac{r}{s}$ with $r, s \in \mathbb{N}$, i.e., $e \notin \mathbb{Q}$.

Comment. The two most famous mathematical constants, π and e, are thus irrational numbers. So are many other constants. Proving the irrationality of a number, however, can often be difficult. For example, it is still not known whether or not $\pi + e$ and πe are irrational (we do know, however, that at least one of them is irrational—see Exercise 7.103 below).

7.5. Use Cardano's formula to find an integer solution of the equation $x^3 = 15x + 4$.

Hint. $(2+i)^3 = 2 + 11i$ and $(2-i)^3 = 2 - 11i$.

Comment. This equation, as well as the indicated method for solving it, are from the 1572 book *Algebra* by the Italian mathematician *Rafael Bombelli*. He is often considered the inventor of complex numbers. The further development was slow, however. It was not until the 19th century that complex numbers were generally accepted.

7.2 Algebraic and Transcendental Elements

Thus far we have not considered fields systematically. Still, we have encountered them throughout the book, so it seems appropriate to begin with a brief summary of what we already know about them.

First, we list some examples of fields. We start with our "old friends" $\mathbb{Q}$, $\mathbb{R}$, and $\mathbb{C}$. There are actually many fields lying between $\mathbb{Q}$ and $\mathbb{C}$, not only $\mathbb{R}$. One simple example is the field $\mathbb{Q}(i) = \{p + qi \mid p, q \in \mathbb{Q}\}$ which has already been mentioned a couple of times. A similar example is $\mathbb{Q}(\sqrt{2}) = \{p + q\sqrt{2} \mid p, q \in \mathbb{Q}\}$. There are, on the other hand, fields whose elements are not numbers. For example, $\mathbb{Z}_p$ is a (finite) field for every prime number p. In Section 3.6, we introduced the field of fractions of an integral domain. An interesting special case is the field of rational functions $F(X)$. Here, F can be an arbitrary field.

We say that a field K is an *extension (or extension field)* of a field F if F is a subfield of K. For example, $\mathbb{R}$ is an extension of $\mathbb{Q}$, $\mathbb{C}$ is an extension of $\mathbb{R}$ (as well as of $\mathbb{Q}$), and $F(X)$ is an extension of F. When considering extensions of a fixed field F, we sometimes call F the **base field**. Especially in the exercises, the role of the base field will often be played by $\mathbb{Q}$.

The *characteristic* of a field F is either 0 (meaning that $n \cdot 1 \neq 0$ for every $n \in \mathbb{N}$) or a prime p (meaning that p is the smallest natural number satisfying $p \cdot 1 = 0$). For example, $\mathbb{Q}$ has characteristic 0 and $\mathbb{Z}_p$ has characteristic p. Obviously, every extension of F has the same characteristic as F. According to Theorem 3.137, any field of characteristic 0 can be viewed as an extension of $\mathbb{Q}$, and any field of characteristic p can be viewed as an extension of $\mathbb{Z}_p$.

The study of field extensions is closely tied to the study of the polynomial ring $F[X]$. We will therefore also make use of some results from Chapter 5. They will be recalled when needed. It goes without saying that we will also need elementary facts concerning polynomials from Section 2.4.

After these preliminary remarks, we turn to the topic of this section, which is algebraic and transcendental elements. They will accompany us throughout the chapter. Let us first give some motivation for the definition.

We divide the real numbers into rational and irrational numbers. As mentioned in the preceding section, two of the basic examples of irrational numbers are $\sqrt{2}$ and π. However, π is, in some sense, further from the rational numbers than $\sqrt{2}$. Indeed, it can be proved (but it is not easy!) that for any $q_0, q_1 \ldots, q_n \in \mathbb{Q}$ that are not all 0,

$$q_0 + q_1\pi + \cdots + q_{n-1}\pi^{n-1} + q_n\pi^n \neq 0.$$

In other words, π is not a root of a nonzero polynomial in $\mathbb{Q}[X]$. Being an irrational number, $\sqrt{2}$ is not a root of a linear polynomial in $\mathbb{Q}[X]$, but of course it is a root of $X^2 - 2 \in \mathbb{Q}[X]$. Numbers like $\sqrt{2}$ are called algebraic, and numbers like π are called transcendental. We state these definitions in a more general context.

Definition 7.6. Let K be an extension of a field F. We say that $a \in K$ is **algebraic over** F if there exists a nonzero polynomial $f(X) \in F[X]$ such that $f(a) = 0$. If a is not algebraic over F, then a is said to be **transcendental over** F.

An algebraic element is a root of many polynomials, but one of them is of special interest.

Definition 7.7. Let $a \in K$ be algebraic over F. The monic polynomial of lowest degree having a as a root is called the **minimal polynomial** of a over F. We say that a is **algebraic of degree** n (over F) if its minimal polynomial has degree n.

The *existence* of the minimal polynomial is clear. We just take any nonzero polynomial $f(X)$ of lowest degree such that $f(a) = 0$, and if λ is its leading coefficient, then $p(X) := \lambda^{-1}f(X)$ has the desired properties. The *uniqueness* of the minimal polynomial is also immediate. Indeed, if $p(X)$ and $p_1(X)$ were two different minimal polynomials of a, then a would be a root of $p(X) - p_1(X)$. But this is a contradiction since this polynomial has lower degree than $p(X)$ and $p_1(X)$.

The next theorem gives a more complete picture of the notion of a minimal polynomial. Incidentally, the minimal polynomial of a linear map on a finite-dimensional vector space is defined in the same way (see Exercise 5.182), but does not always satisfy condition (ii).

Theorem 7.8. *Let $a \in K$ be algebraic over F and let $p(X) \in F[X]$ be a monic polynomial such that $p(a) = 0$. The following conditions are equivalent:*

(i) *$p(X)$ is the minimal polynomial of a.*

(ii) *$p(X)$ is irreducible over F.*

(iii) *$p(X)$ divides every polynomial $f(X) \in F[X]$ such that $f(a) = 0$.*

Proof. (i) $\Longrightarrow$ (ii). If the minimal polynomial $p(X)$ can be written as the product $g(X)h(X)$, then $g(a)h(a) = p(a) = 0$ and hence $g(a) = 0$ or $h(a) = 0$. Suppose the former holds. The minimality of $p(X)$ implies that $\deg(g(X)) \geq \deg(p(X))$. However, since $g(X)$ divides $p(X)$, the converse inequality also holds. Thus, $g(X)$ and $p(X)$ are of the same degree, which proves that $p(X)$ is irreducible.

(ii) $\Longrightarrow$ (iii). It is easy to see that

$$\mathscr{I} := \{f(X) \in F[X] \mid f(a) = 0\}$$

is an ideal of $F[X]$. Since $F[X]$ is a principal ideal domain (Theorem 5.71), there exists a $p_1(X) \in F[X]$ such that $\mathscr{I} = (p_1(X))$. In particular, $p(X) = q(X)p_1(X)$ for some $q(X) \in F[X]$. As $p(X)$ is irreducible and $p_1(X)$ is obviously non-constant, $q(X)$ has to be a constant polynomial. Therefore,

$$(p(X)) = (p_1(X)) = \mathscr{I},$$

from which (iii) follows.

(iii) $\Longrightarrow$ (i). This is immediate from the obvious fact that $p(X) \mid f(X)$ implies $f(X) = 0$ or $\deg(p(X)) \leq \deg(f(X))$. $\qquad\square$

The notions "irreducible polynomial" and "minimal polynomial" are thus closely connected. In order to prove that a certain polynomial is the minimal polynomial of a, it is often easier to show that it is irreducible than checking that a is not a root of polynomials of lower degree.

Let us proceed to some examples.

Example 7.9. The elements of F are algebraic of degree 1 over F. Indeed, every $a \in F$ is a root of $X - a \in F[X]$. Conversely, if a is algebraic of degree 1 over F, then a is a root of a polynomial of the form $a_1 X + a_0$ where $a_1 \neq 0$ and a_0 belong to F, and hence $a = -a_1^{-1} a_0 \in F$.

Example 7.10. Every $z \in \mathbb{C}$ is algebraic over $\mathbb{R}$, either of degree 2 (if $z \notin \mathbb{R}$) or 1 (if $z \in \mathbb{R}$). Indeed, z is a root of the polynomial

$$(X - z)(X - \bar{z}) = X^2 - (z + \bar{z})X + z\bar{z}.$$

Since $z + \bar{z}$ and $z\bar{z}$ are real numbers, this polynomial lies in $\mathbb{R}[X]$.

Example 7.11. Consider the field of rational functions $F(X)$ as an extension of the field F. Note that $X \in F(X)$ is transcendental over F. (This is obvious, it only requires understanding the definitions.)

The classical case where $F = \mathbb{Q}$ and $K = \mathbb{C}$ is particularly interesting. In this context, we call algebraic elements **algebraic numbers** and transcendental elements **transcendental numbers**. Thus, a complex number a is an algebraic number if there exists a nonzero polynomial $f(X) \in \mathbb{Q}[X]$ such that $f(a) = 0$. By clearing denominators if necessary, we may assume that $f(X)$ has integer coefficients.

Example 7.12. The imaginary unit i is a root of $X^2 + 1 \in \mathbb{Q}[X]$ and is therefore an algebraic number of degree 2.

Example 7.13. For any prime p, $\sqrt[n]{p}$ is an algebraic number of degree n. Indeed, the polynomial $X^n - p$ is irreducible over $\mathbb{Q}$ by the Eisenstein Criterion (see Example 5.10), and is hence the minimal polynomial of $\sqrt[n]{p}$ by Theorem 7.8.

Example 7.14. We already mentioned that π is transcendental, and it turns out that so is e. However, proving this is difficult and beyond the scope of this book. Generally speaking, transcendence is not a simple notion. How do we know that transcendental numbers exist at all? The reader having some knowledge of cardinal numbers should be able to prove that the set of algebraic numbers is countable (hint: the set of all polynomials with rational coefficients is countable, and each of them has only finitely many roots). This not only implies that transcendental numbers exist, but that "most" complex numbers are transcendental. In precise terms, the cardinality of the set of transcendental numbers is the same as that of the set of complex numbers. In spite of this, it is never easy to prove that some concrete number is transcendental. For many numbers it is already difficult to prove that they are irrational, which only means that they are not algebraic of degree 1. Still, there are numbers for which the proof of their transcendence requires only elementary tools. The most well-known example is **Liouville's constant** $\sum_{n=1}^{\infty} 10^{-n!}$. Finding a proof may be a challenge for the more ambitious reader.

So far, we have mentioned only some of the most obvious examples of algebraic numbers. Providing further examples is not particularly difficult. For example, take $a := \sqrt{2} + \sqrt{3}$. By squaring, we obtain $a^2 - 5 = 2\sqrt{6}$. Squaring again, it follows that a is a root of

$$p(X) := X^4 - 10X^2 + 1.$$

Therefore, a is algebraic (and in Example 7.126 below we will see that $p(X)$ is its minimal polynomial). What about, e.g., $\sqrt{2} + \sqrt[3]{3}$? It is easy to see that this number is also algebraic, and we leave it to the reader to find an appropriate polynomial. In the next section, we will prove a much more general fact: the sum, difference, product, and quotient of algebraic elements are again algebraic. In other words, the algebraic elements form a field. We will not show this by finding an appropriate polynomial for every element, but by another, more subtle approach.

Exercises

7.15. Let K be an extension of F, and let $a \in K$. Suppose there exists a non-constant polynomial $f(X) \in F[X]$ such that $f(a)$ is algebraic over F. Prove that then a itself is algebraic over F.

7.16. Let $a \in K \setminus \{0\}$ be algebraic of degree n over F. Prove that there exists a polynomial $g(X) \in F[X]$ of degree $n - 1$ such that $a^{-1} = g(a)$.

7.17. Let $a \in K$ be algebraic of degree n over F. Prove that for any $\alpha, \beta \in F$, $\beta \neq 0$, $\alpha + \beta a$ is also algebraic of degree n.

7.18. Explain why an element of some extension of $\mathbb{C}$ (resp. $\mathbb{R}$) cannot be algebraic of degree 2 (resp. 3) or more over $\mathbb{C}$ (resp. $\mathbb{R}$).

7.19. Let z be an algebraic number. Prove that $\bar{z}$ is also algebraic and has the same minimal polynomial as z.

7.20. Prove that $\sqrt{2 + \sqrt{3 + \sqrt{5}}}$ is an algebraic number.

7.21. Prove that $\sqrt{2} + i\sqrt{3}$ is an algebraic number.

7.22. Prove that $\sqrt{10} + \sqrt[4]{6}$ is an algebraic number.

7.23. Let p be a prime. Explain why the cyclotomic polynomial $\Phi_p(X)$ (see Example 5.11) is the minimal polynomial of the algebraic number $e^{\frac{2\pi i}{p}}$.

7.24. What is the minimal polynomial of the algebraic number $e^{\frac{2\pi i}{6}}$?

7.25. What is the minimal polynomial of the algebraic number $e^{\frac{2\pi i}{8}}$?

7.3 Finite Extensions

Every extension field K of a field F can be viewed as a *vector space* over F. Perhaps this sounds paradoxical—how can elements of a field be vectors? But actually this is a very natural viewpoint, and crucial for our approach to field extensions. Let us explain. Being a field, K is an additive group, and for any $\lambda \in F$ and $x \in K$, their product λx is an element of K. We now simply interpret the operation $(\lambda, x) \mapsto \lambda x$ as scalar multiplication. The vector space axioms are obviously fulfilled: $\lambda(x + y) = \lambda x + \lambda y$ and $(\lambda + \mu)x = \lambda x + \mu x$ follow from the distributive law in K, $\lambda(\mu x) = (\lambda\mu)x$ is a consequence of the associative law in K, and $1x = x$ holds since F and K share the same unity. Incidentally, K is even an algebra over F, but this will not be particularly helpful to us. What is important is that we can talk about the dimension of K over F.

Definition 7.26. Let K be an extension of a field F. We say that K is a **finite extension** of F if K is a finite-dimensional vector space over F. The dimension of K over F is called the **degree of the extension** K and is denoted by $[K : F]$.

In linear algebra, we would write $\dim_F K$ rather than $[K : F]$. It is often the case, usually due to historical reasons, that different notations are used in different areas.

Example 7.27. An obvious example of a basis of the vector space $\mathbb{C}$ over the field $\mathbb{R}$ is $\{1, i\}$. Hence, $\mathbb{C}$ is a finite extension of $\mathbb{R}$ and $[\mathbb{C} : \mathbb{R}] = 2$.

Example 7.28. The reader with a basic knowledge of cardinal numbers can easily verify that every finite extension of $\mathbb{Q}$ is a countable set. Therefore, $\mathbb{R}$ is not a finite extension of $\mathbb{Q}$. In Example 7.42, we will prove this (and more) by using the theory that we are about to develop.

More examples will be given toward the end of the section. We continue with a theorem which, in particular, states that a finite extension of a finite extension is itself finite.

Theorem 7.29. *Let L be a finite extension of F, and let K be a finite extension of L. Then K is a finite extension of F, and*

$$[K : F] = [K : L] \cdot [L : F]. \tag{7.3}$$

Proof. Let $\{a_1, \ldots, a_m\}$ be a basis of L over F, and $\{b_1, \ldots, b_n\}$ a basis of K over L. Thus, $m = [L : F]$ and $n = [K : L]$. Set

$$B := \{a_i b_j \mid j = 1, \ldots, n, i = 1, \ldots, m\}.$$

Observe that $a_i b_j = a_{i'} b_{j'}$ implies $i = i'$ and $j = j'$. Hence, $|B| = nm$. The theorem will be proved by showing that B is a basis of K over F.

Take $x \in K$. Then there exist elements $\ell_j \in L$ such that $x = \sum_{j=1}^{n} \ell_j b_j$. Writing each ℓ_j as $\sum_{i=1}^{m} \lambda_{ij} a_i$ with $\lambda_{ij} \in F$, we thus obtain

$$x = \sum_{j=1}^{n} \left(\sum_{i=1}^{m} \lambda_{ij} a_i \right) b_j = \sum_{j=1}^{n} \sum_{i=1}^{m} \lambda_{ij} a_i b_j.$$

This proves that B spans K over F.

To show that B is a linearly independent set, suppose that

$$\sum_{j=1}^{n} \sum_{i=1}^{m} \lambda_{ij} a_i b_j = 0$$

for some $\lambda_{ij} \in F$. Writing this as

$$\sum_{j=1}^{n} \left(\sum_{i=1}^{m} \lambda_{ij} a_i \right) b_j = 0$$

and taking into account that $b_1, \ldots, b_n$ are linearly independent over L, it follows that

$$\sum_{i=1}^{m} \lambda_{ij} a_i = 0$$

for each j. Since $a_1, \ldots, a_m$ are linearly independent over F, each λ_{ij} has to be 0. $\quad\square$

This theorem is of fundamental importance. Its role in field theory may be compared with that of Lagrange's Theorem in group theory. The following simple corollary is particularly useful.

Corollary 7.30. *Let K be a finite extension of F. If L is subfield of K that contains F, then $[L : F]$ divides $[K : F]$.*

Proof. Since K is a finite extension of F, it is also a finite extension of L. Indeed, if $\{c_1, \ldots, c_r\}$ spans the vector space K over the field F, then it also spans the vector space K over the field L. Of course, L is also a finite extension of F (a subspace of a finite-dimensional vector space is finite-dimensional). Hence, (7.3) holds. $\quad\square$

With the next definition we return to the theme of the preceding section. The difference, however, is that now we will not consider particular elements of a field extension, but all elements simultaneously. Such an approach is characteristic for algebra—often the most efficient way to obtain information on a single element is to study an algebraic structure to which this element belongs.

Definition 7.31. An extension K of a field F is said to be an **algebraic extension** of F if every element of K is algebraic over F. An extension that is not algebraic is called a **transcendental extension**.

The condition that $a \in K$ is algebraic over F, i.e., that

$$\lambda_0 + \lambda_1 a + \cdots + \lambda_n a^n = 0$$

for some $n \in \mathbb{N}$ and some $\lambda_i \in F$ that are not all 0, can be equivalently described as the linear dependence of the elements $1, a, \ldots, a^n \in K$ over F. This interpretation can be quite useful. We have already used it indirectly in the proof of Lemma 5.173, and now we will use it again.

Lemma 7.32. *Every finite extension is algebraic.*

Proof. Let $[K : F] = n$. For every $a \in K$, the elements $1, a, \ldots, a^n$ are then linearly dependent over F, since their number exceeds the dimension of the vector space K. This means that a is algebraic over F. $\quad\square$

Remark 7.33. The proof actually shows that $[K : F] = n$ implies that every element of K is algebraic of degree at most n over F. Since $[\mathbb{C} : \mathbb{R}] = 2$, this gives another proof of the fact that every $z \in \mathbb{C} \setminus \mathbb{R}$ is algebraic of degree 2 over $\mathbb{R}$ (compare Example 7.10).

In Example 7.42, we will show that the converse of Lemma 7.32 does not hold, i.e., not all algebraic extensions are finite.

We have to introduce some notation. Let us point out in advance the similarity with the notation used to denote the polynomial ring $F[X]$ (the square brackets) and the field of rational functions $F(X)$ (the round brackets).

Let K be an extension of F, and let $a_1, \ldots, a_n$ be elements in K. By

$$F[a_1, \ldots, a_n]$$

we denote the *subring* of K generated by F and $a_1, \ldots, a_n$, and by

$$F(a_1, \ldots, a_n)$$

we denote the *subfield* of K generated by F and $a_1, \ldots, a_n$. We call it the **field obtained by adjoining** $a_1, \ldots, a_n$ to F. We claim that for any $1 \le k \le n - 1$,

$$F(a_1, \ldots, a_n) = \big(F(a_1, \ldots, a_k)\big)(a_{k+1}, \ldots, a_n). \tag{7.4}$$

Indeed, $F(a_1, \ldots, a_n)$ contains the field $F(a_1, \ldots, a_k)$ and the elements $a_{k+1}, \ldots, a_n$, and so it also contains the field $\big(F(a_1, \ldots, a_k)\big)(a_{k+1}, \ldots, a_n)$; this field, on the other hand, contains F and all a_i, and so it also contains $F(a_1, \ldots, a_n)$. In what follows, (7.4) will be frequently (and often tacitly) used, so the reader should keep it in mind.

We say that K is a **simple extension** of F if there exists an $a \in K$ such that $K = F(a)$. Every such element a is called a **primitive element** of the extension K. Either by referring to Section 1.7 or by a direct verification, we see that $F(a)$ consists of elements of the form xy^{-1} with $x, y \in F[a]$, $y \ne 0$, and $F[a]$ consists of elements of the form

$$\lambda_0 + \lambda_1 a + \cdots + \lambda_r a^r$$

with $\lambda_i \in F$. That is to say,

$$F[a] = \{f(a) \mid f(X) \in F[X]\} \tag{7.5}$$

and

$$F(a) = \{f(a)g(a)^{-1} \mid f(X), g(X) \in F[X], g(a) \ne 0\}. \tag{7.6}$$

The description of $F[a_1, \ldots, a_n]$ and $F(a_1, \ldots, a_n)$ is similar, but involves polynomials in several variables. So, $F[a_1, \ldots, a_n]$ consists of elements of the form $f(a_1, \ldots, a_n)$ where $f(X_1, \ldots, X_n) \in F[X_1, \ldots, X_n]$.

The next theorem shows that the description (7.6) of $F(a)$ becomes much simpler if a is algebraic. Let us first give an example.

Example 7.34. The imaginary unit i is an algebraic number of degree 2. Since $i^n \in \{1, -1, i, -i\}$ for all $n \in \mathbb{N}$, for every polynomial $f(X) \in \mathbb{Q}[X]$ there exist $\lambda_0, \lambda_1 \in \mathbb{Q}$ such that $f(i) = \lambda_0 + \lambda_1 i$. We may therefore restrict to linear polynomials $f(X)$ in (7.5). Since the ring $\mathbb{Q}[i] = \{\lambda_0 + \lambda_1 i \mid \lambda_i \in \mathbb{Q}\}$ is actually a field (as noticed already in Example 1.149), we have $\mathbb{Q}[i] = \mathbb{Q}(i)$ and $[\mathbb{Q}(i) : \mathbb{Q}] = 2$. Replacing the role of

$\mathbb{Q}$ by $\mathbb{R}$, we see that, similarly, $\mathbb{R}[i] = \mathbb{R}(i) = \mathbb{C}$. Thus, $\mathbb{C}$ is a simple extension of $\mathbb{R}$, and i is an example of a primitive element.

Theorem 7.35. *Let K be an extension of a field F. If $a \in K$ is algebraic of degree n over F, then*

$$F(a) = F[a] = \{\lambda_0 + \lambda_1 a + \cdots + \lambda_{n-1} a^{n-1} \mid \lambda_i \in F\} \tag{7.7}$$

is a finite extension of F and $[F(a) : F] = n$.

Proof. Every element of $F[a]$ can be written as $f(a)$ for some polynomial $f(X) \in F[X]$. Suppose $f(a) \neq 0$. Let $p(X)$ be the minimal polynomial of a. Since $p(X)$ is irreducible (Theorem 7.8) and does not divide $f(X)$ (as $f(a) \neq 0$), $p(X)$ and $f(X)$ are relatively prime polynomials. Therefore, by Theorem 5.74, there exist $h(X), k(X) \in F[X]$ such that

$$p(X)h(X) + f(X)k(X) = 1.$$

Since $p(a) = 0$, this yields $f(a)k(a) = 1$. Thus, $f(a)^{-1} = k(a) \in F[a]$. This proves that $F[a]$ is a field, and so $F[a] = F(a)$.

Our next goal is to show that $f(a)$ can be written as the evaluation of some polynomial of degree less than n at a. This is easy: if $q(X), r(X) \in F[X]$ are polynomials satisfying

$$f(X) = q(X)p(X) + r(X)$$

with $r(X) = 0$ or $\deg(r(X)) < n$, then $f(a) = r(a)$. With this, (7.7) is proved.

From (7.7) we see that the set $\{1, a, \ldots, a^{n-1}\}$ spans the vector space $F(a)$ over F. Since the degree of the minimal polynomial of a is equal to n, this set is linearly independent. Hence, it is a basis of $F(a)$, and so $[F(a) : F] = n$. $\qquad\square$

Example 7.36. If p is a prime and n is any natural number, $\sqrt[n]{p}$ is an algebraic number of degree n (see Example 7.13). Therefore,

$$[\mathbb{Q}(\sqrt[n]{p}) : \mathbb{Q}] = n,$$

and $\{1, \sqrt[n]{p}, \sqrt[n]{p^2}, \ldots, \sqrt[n]{p^{n-1}}\}$ is a basis of the vector space $\mathbb{Q}(\sqrt[n]{p})$ over $\mathbb{Q}$. Thus,

$$\mathbb{Q}(\sqrt{2}) = \{\lambda_0 + \lambda_1 \sqrt{2} \mid \lambda_i \in \mathbb{Q}\},$$
$$\mathbb{Q}(\sqrt[3]{2}) = \{\lambda_0 + \lambda_1 \sqrt[3]{2} + \lambda_2 \sqrt[3]{4} \mid \lambda_i \in \mathbb{Q}\}, \text{ etc.}$$

Remark 7.37. For every element $a \in K$, algebraic or transcendental, the map

$$\varepsilon : F[X] \to F[a], \quad \varepsilon(f(X)) = f(a),$$

is a ring epimorphism, called the *evaluation homomorphism at a* (see Example 3.75).

(a) Let a be algebraic over F. Then $\ker \varepsilon = (p(X))$ where $p(X)$ is the minimal polynomial of a (Theorem 7.8). Thus, by the Isomorphism Theorem,

$$F[a] \cong F[X]/(p(X)).$$

Applying Corollary 5.73, we see that this implies that $F[a]$ is a field, thus giving an alternative proof of the equality $F[a] = F(a)$.

(b) Now let a be transcendental over F. Then

$$F[a] \cong F[X] \quad \text{and} \quad F(a) \cong F(X).$$

Indeed, the epimorphism ε is injective and hence an isomorphism, and

$$f(X)g(X)^{-1} \mapsto f(a)g(a)^{-1}$$

defines its extension to an isomorphism from $F(X)$ to $F(a)$. Note that $F[a]$ is not a field and so $F[a] \subsetneq F(a)$.

Remark 7.38. One of the important consequences of Theorem 7.35 is that for every algebraic element a over F,

$$[F(a) : F] = \text{degree of the minimal polynomial of } a.$$

If L is a field lying between F and K, then a is also algebraic over L, and the degree of the minimal polynomial over L is less than or equal to the degree of the minimal polynomial over F. We can write this as $[L(a) : L] \le [F(a) : F]$.

The next theorem considers a more general situation where we adjoin any finite number of algebraic elements to F.

Theorem 7.39. *Let K be an extension of a field F. If $a_1, \ldots, a_n \in K$ are algebraic over F, then $F(a_1, \ldots, a_n) = F[a_1, \ldots, a_n]$ is a finite extension of F.*

Proof. The $n = 1$ case follows from Theorem 7.35. We may therefore assume that the theorem holds for fewer than n elements. In particular, $L := F(a_1, \ldots, a_{n-1})$ is a finite extension of F and is equal to $F[a_1, \ldots, a_{n-1}]$. Since a_n is algebraic over F, it is also algebraic over L. Therefore, $L(a_n)$—which is, by (7.4), equal to $F(a_1, \ldots, a_n)$—is a finite extension of L by Theorem 7.35, and hence also of F by Theorem 7.29. Moreover, Theorem 7.35 tells us that the elements of $L(a_n)$ can be written as sums of elements of the form ℓa_n^k with $\ell \in L$ and $k \ge 0$. Since $L = F[a_1, \ldots, a_{n-1}]$, we can write each ℓ as

$$f(a_1, \ldots, a_{n-1}), \quad \text{where } f(X_1, \ldots, X_{n-1}) \in F[X_1, \ldots, X_{n-1}].$$

Accordingly, $L(a_n)$ consists of precisely the elements of the form

$$g(a_1, \ldots, a_{n-1}, a_n), \quad \text{where } g(X_1, \ldots, X_{n-1}, X_n) \in F[X_1, \ldots, X_{n-1}, X_n].$$

This means that $L(a_n) = F[a_1, \ldots, a_{n-1}, a_n]$, which is the desired conclusion. $\square$

Example 7.40. Consider the field $\mathbb{Q}(\sqrt{2}, \sqrt{3})$. By Theorem 7.39, its elements can be written as $f(\sqrt{2}, \sqrt{3})$, where $f(X_1, X_2) \in \mathbb{Q}[X_1, X_2]$. Hence,

$$\mathbb{Q}(\sqrt{2}, \sqrt{3}) = \{\lambda_0 + \lambda_1\sqrt{2} + \lambda_2\sqrt{3} + \lambda_3\sqrt{6} \,|\, \lambda_i \in \mathbb{Q}\}.$$

We now show that

$$[\mathbb{Q}(\sqrt{2}, \sqrt{3}) : \mathbb{Q}] = 4. \tag{7.8}$$

Every element of $\mathbb{Q}(\sqrt{2}, \sqrt{3})$ is a linear combination of the four elements $1, \sqrt{2}, \sqrt{3}$, and $\sqrt{6}$, so $[\mathbb{Q}(\sqrt{2}, \sqrt{3}) : \mathbb{Q}] \le 4$. Since $\mathbb{Q}(\sqrt{2}) \subseteq \mathbb{Q}(\sqrt{2}, \sqrt{3})$ and $[\mathbb{Q}(\sqrt{2}) : \mathbb{Q}] = 2$, it follows from Corollary 7.30 that $[\mathbb{Q}(\sqrt{2}, \sqrt{3}) : \mathbb{Q}]$ can be either 2 or 4. If it was 2, then $1, \sqrt{2}$, and $\sqrt{3}$ would be linearly dependent, implying that $\sqrt{3} = \lambda + \mu\sqrt{2}$ for some $\lambda, \mu \in \mathbb{Q}$. Squaring both sides and using $\sqrt{2} \notin \mathbb{Q}$ and $\sqrt{3} \notin \mathbb{Q}$ we easily reach a contradiction. Therefore, (7.8) must hold.

We close this section by proving the result announced at the end of the preceding section.

Corollary 7.41. *Let K be an extension of a field F. The set L of all elements in K that are algebraic over F is a subfield of K.*

Proof. Take $a, b \in L$. By Theorem 7.39, $F(a, b)$ is a finite extension of F. Lemma 7.32 therefore tells us that $F(a, b) \subseteq L$. In particular, L contains the elements $a - b$ and ab, as well as a^{-1} provided that $a \ne 0$. This means that L is a subfield. $\square$

Example 7.42. The set of all algebraic numbers is thus a field. It is an example of an algebraic extension of $\mathbb{Q}$ which is not finite. In fact, any extension K of $\mathbb{Q}$ that contains $\sqrt[n]{2}$ for every $n \in \mathbb{N}$ is not finite. Indeed, if it was finite, then, by Corollary 7.30 and Example 7.36, $n = [\mathbb{Q}(\sqrt[n]{2}) : \mathbb{Q}]$ would divide $[K : \mathbb{Q}]$ for every n, which is of course impossible.

Exercises

7.43. Prove that $\mathbb{R}(a) = \mathbb{C}$ for every $a \in \mathbb{C} \setminus \mathbb{R}$.

7.44. Suppose $[K : F] = 12$. Explain why an element of K cannot be algebraic of degree 8 over F. Can it be algebraic of degree 4?

7.45. Determine all $k \in \mathbb{N}$ such that $\mathbb{Q}(\sqrt[12]{2})$ contains an element that is algebraic of degree k over $\mathbb{Q}$. For each k find an example of such an element.

7.46. Let n be odd and let $a \in \mathbb{Q}(\sqrt[n]{2})$. Prove that $a^2 \in \mathbb{Q}$ implies $a \in \mathbb{Q}$.

Hint. Do not rush into calculations but rather use the theory. The same advice applies to many exercises in this section.

7.47. Suppose $[K : F] = p$ where p is a prime. Prove that every element in $K \setminus F$ is algebraic of degree p over F.

7.48. Determine $[\mathbb{Q}(3 - \sqrt{7}) : \mathbb{Q}]$ and $[\mathbb{Q}(3 - 5\sqrt[3]{7} + 4\sqrt[3]{49}) : \mathbb{Q}]$.

7.49. Determine $[\mathbb{Q}(\sqrt[3]{3+\sqrt{3}}) : \mathbb{Q}]$.

Hint. Find an irreducible polynomial having $\sqrt[3]{3+\sqrt{3}}$ as a root.

7.50. Let a and b be algebraic elements over F. Suppose the degrees $[F(a) : F]$ and $[F(b) : F]$ are relatively prime. Prove that $[F(a, b) : F] = [F(a) : F] \cdot [F(b) : F]$.

7.51. Determine $[\mathbb{Q}(\sqrt{3}, \sqrt[3]{5}) : \mathbb{Q}]$ and $[\mathbb{Q}(\sqrt[3]{3}, \sqrt[3]{5}) : \mathbb{Q}]$.

7.52. Prove that $F(a) = F(a^2)$ if $[F(a) : F]$ is odd.

7.53. Prove that $F(a^k, a^\ell) = F(a^d)$ where $d = \gcd(k, \ell)$.

7.54. With the help of the preceding exercise, determine $[\mathbb{Q}(\sqrt{2}, \sqrt[3]{2}) : \mathbb{Q}]$ and $[\mathbb{Q}(\sqrt[4]{2}, \sqrt[6]{2}) : \mathbb{Q}]$.
Hint. $(\sqrt[6]{2})^2 = \sqrt[3]{2}$ and $(\sqrt[6]{2})^3 = \sqrt{2}$.

7.55. Determine $[\mathbb{Q}(\sqrt{2} + \sqrt[3]{2}) : \mathbb{Q}]$.

7.56. Determine $[\mathbb{Q}(\sqrt{2} + \sqrt[4]{2}) : \mathbb{Q}]$.

7.57. Determine $[\mathbb{Q}(\sqrt[6]{2}) : \mathbb{Q}(\sqrt{2})]$.

7.58. Determine $[\mathbb{Q}(\sqrt[3]{3}, i) : \mathbb{Q}(\sqrt{3} + i)]$.

7.59. Let a be a transcendental number and let $n \in \mathbb{N}$. Determine $[\mathbb{Q}(a) : \mathbb{Q}(a^n)]$.

7.60. Prove that $\mathbb{Q}(\sqrt{2}) \not\cong \mathbb{Q}(\sqrt{3})$.

7.61. Let $a_1, \ldots, a_n$ be algebraic over F. Prove that

$$[F(a_1, \ldots, a_n) : F] \le [F(a_1) : F] \cdots [F(a_n) : F].$$

7.62. Let $a_1, \ldots, a_n$ be algebraic of degree 2 over F. Prove that there is a $k \le n$ such that $[F(a_1, \ldots, a_n) : F] = 2^k$.

7.63. Let $p_1, \ldots, p_n$ be distinct primes. Prove that $[\mathbb{Q}(\sqrt{p_1}, \ldots, \sqrt{p_n}) : \mathbb{Q}] = 2^n$.

7.4 Straightedge and Compass Constructions

The mathematics of ancient Greece was amazingly original and sophisticated. It is not only that its discoveries are astonishing. Its major contribution was the introduction of mathematical proof, and, generally speaking, the transformation of mathematics into the abstract science we know today.

The algebraic way of thinking, however, was rather unfamiliar to the Greek mathematicians. From today's perspective it looks curious that they expressed even some of the simplest algebraic statements in geometric language. Indeed, the Greeks were primarily geometers. They are particularly famous for ingenious constructions by straightedge and compass. Yet they also encountered geometric problems that they were unable to solve. Especially renowned are the following three.

Fig. 7.1 Using a straightedge

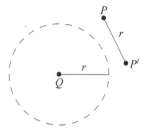

Fig. 7.2 Using a compass

1. **Doubling the cube:** Using only a straightedge and compass, construct a cube having double the volume of a given cube.
2. **Trisecting an angle:** Using only a straightedge and compass, construct an angle equal to one third of a given angle.
3. **Squaring the circle:** Using only a straightedge and compass, construct a square having the same area as a given circle.

It took over two millennia before the answers were finally revealed. Each of the three problems actually has the same answer: such a construction cannot exist! For the first two problems, doubling the cube and trisecting an angle, this was proved in 1837 by *Pierre Wantzel*, and for the squaring the circle problem in 1882 by *Ferdinand von Lindemann* ("squaring the circle" is now a metaphor for trying to do the impossible). The main idea for solving these problems is translating them from geometric to algebraic language. As we will see, this enables one to effectively apply the field theory from the previous two sections.

Let us first make precise what we mean by using a straightedge and compass. Our straightedge is not marked with numbers. We can only use it to draw the line through given points P and Q in the plane (Figure 7.1). Given points P, P', and Q, we can use a compass to draw the circle with center Q and radius equal to the length of the line segment with endpoints P and P' (Figure 7.2). So, these are our "rules of the game". At the beginning, we are given at least two points. The points of intersections of lines and circles, obtained from the initial points by using a straightedge and compass, are our new points which can be used for further constructions. The usual goal is to arrive at a certain point in a finite number of steps. Let us look at what the initial points are and what the goal is in the above problems.

1. We are given a cube. To ease the discussion, we assume that the length of its edge is 1. Its volume is then also 1. We are asked to construct a cube of volume

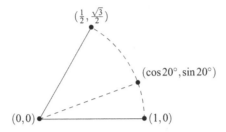

Fig. 7.3 Trisection of 60°

2, i.e., a cube whose edge has length $\sqrt[3]{2}$. Our problem is thus equivalent to the following one: given a line segment of length 1, construct a line segment of length $\sqrt[3]{2}$. For convenience, we consider it in the usual Cartesian plane $\mathbb{R}^2$. Then it can be stated as follows: given the points

$$P_1 = (0, 0), \quad P_2 = (1, 0),$$

construct the point

$$Z = (\sqrt[3]{2}, 0).$$

2. It is actually possible to trisect some angles. For example, it is quite easy to construct a 30° angle trisecting a 90° angle. Therefore, it matters which angle we consider. Choosing 60°, our problem reads as follows: given the points

$$P_1 = (0, 0), \quad P_2 = (1, 0), \quad P_3 = (\cos 60°, \sin 60°) = \left(\frac{1}{2}, \frac{\sqrt{3}}{2}\right),$$

construct the point

$$Z = (\cos 20°, \sin 20°)$$

(see Figure 7.3).

3. The area of a circle of radius 1 is π. A square with the same area has side of length $\sqrt{\pi}$. The problem of squaring the circle can thus be stated as follows: given the points

$$P_1 = (0, 0), \quad P_2 = (1, 0),$$

construct the point

$$Z = (\sqrt{\pi}, 0).$$

We now consider the general situation. Let $\mathscr{P} = \{P_1, P_2, \dots\}$ be a given set of points in the plane with at least two elements. We are interested in properties of points that can be constructed by straightedge and compass from $\mathscr{P}$. Every such point is obtained in a finite number of steps. In the first step, we obtain a point, let us call it A, in one of the three ways:

(ll) A is the point of intersection of two (distinct) lines obtained from $\mathscr{P}$;

(lc) A is a point of intersection of a line and a circle obtained from $\mathscr{P}$;

(cc) A is a point of intersection of two (distinct) circles obtained from $\mathscr{P}$.

In the second step, we obtain a point, let us call it B, constructed in the same way, only the role of the set $\mathscr{P}$ is now taken by the set $\mathscr{P} \cup \{A\}$. In the third step, we take the set $\mathscr{P} \cup \{A, B\}$, etc. We will call the points obtained in this way, i.e., the points A, B, etc., the **points constructed from** $\mathscr{P}$.

How to connect these geometric objects with field theory? The essence of our approach is that, rather than considering particular points P_i of $\mathscr{P}$, we will study a subfield F of $\mathbb{R}$ such that $\mathscr{P} \subseteq F \times F$, i.e., a subfield that contains all components of the points of $\mathscr{P}$. Such a field F is not unique. Although this will not be required in the next theorem, we suggest the reader to think of F as being the smallest possible subfield. If $\mathscr{P}$ consists of the points $P_i = (\lambda_i, \mu_i)$, then this is the subfield of $\mathbb{R}$ generated by all λ_i in μ_i.

Example 7.64. In the problems of doubling the cube and squaring the circle, $\mathscr{P} = \{(0, 0), (1, 0)\}$ and hence $F = \mathbb{Q}$. This is because $\mathbb{Q}$ is the smallest subfield (the prime subfield) of $\mathbb{R}$. In the problem of trisecting $60°$, $F = \mathbb{Q}(\sqrt{3})$.

The next theorem provides the key to solving all three problems.

Theorem 7.65. *Let $\mathscr{P}$ be a set of points in the plane and let F be a subfield of $\mathbb{R}$ such that $\mathscr{P} \subseteq F \times F$. If a point $Z = (a, b)$ is constructed from $\mathscr{P}$, then a and b are algebraic over F of degree a power of 2.*

Proof. As above, let A be the point constructed from $\mathscr{P}$ in the first step. We will show that there exists a field L such that

$$A \in L \times L \quad \text{and} \quad [L : F] \in \{1, 2\}. \tag{7.9}$$

From this the theorem follows easily. All we have to do is to use this result repeatedly, for appropriate sets of points and appropriate fields, as follows. If B is as above, then we take the set $\mathscr{P} \cup \{A\}$ (instead of $\mathscr{P}$) and the field L (instead of F). Hence, there exists a field M such that $B \in M \times M$ and $[M : L] \in \{1, 2\}$. Theorem 7.29 shows that

$$[M : F] = [M : L] \cdot [L : F] \in \{1, 2, 4\}.$$

By repeating this argument, we see that for any point $Z = (a, b)$ constructed from $\mathscr{P}$ there exists a field K such that $Z \in K \times K$ and $[K : F] = 2^r$ for some $r \geq 0$. Now it is easy to complete the proof. Being elements of a finite extension of F, a and b are algebraic over F by Lemma 7.32. Since $F(a)$ is a subfield of K, $[F(a) : F]$ divides $[K : F] = 2^r$ by Corollary 7.30. But then $[F(a) : F]$ is also a power of 2. Of course, the same holds for $[F(b) : F]$. According to Theorem 7.35, this means that a and b are algebraic over F of degree a power of 2.

Thus, from now on we focus on the point A and finding a field L satisfying (7.9).

We will consider separately the three cases **(ll)**, **(lc)**, and **(cc)**. But first we state two general remarks. The first one is that any line joining two points from $\mathscr{P}$ has equation

$$y = \alpha x + \beta \quad \text{or} \quad x = \gamma, \quad \text{where } \alpha, \beta, \gamma \in F. \tag{7.10}$$

Indeed, the equation of the line through the points (a_1, b_1) and (a_2, b_2) of $\mathscr{P}$ is

$$y - b_1 = \frac{b_2 - b_1}{a_2 - a_1}(x - a_1),$$

provided that $a_1 \neq a_2$ (and $x = a_1$ otherwise). Since a_i, b_i belong to F, so do $\alpha := \frac{b_2 - b_1}{a_2 - a_1}$ and $\beta := b_1 - \alpha a_1$ (and $\gamma := a_1$). The second remark is that a circle with center $Q \in \mathscr{P}$ and radius equal to the length of the line segment with endpoints $P \in \mathscr{P}$ and $P' \in \mathscr{P}$ has equation

$$x^2 + y^2 = \delta x + \varepsilon y + \zeta, \quad \text{where } \delta, \varepsilon, \zeta \in F. \tag{7.11}$$

This follows from the standard equation of a circle $(x - c_1)^2 + (y - c_2)^2 = r^2$.

We now consider the separate cases.

(ll) Let A be the point of intersection of two (distinct) lines obtained from $\mathscr{P}$. Suppose that the equation of the first line is given by (7.10), and the equation of the second line by

$$y = \alpha' x + \beta' \quad \text{or} \quad x = \gamma', \quad \text{where } \alpha', \beta', \gamma' \in F.$$

The components of the point A are thus the solutions of the system of two linear equations over F, and therefore they both lie in F. Hence, we can simply take F for L (and so $[L : F] = 1$).

(lc) Let now A be a point of intersection of the line with equation (7.10) and the circle with equation (7.11). Writing A as (x, y), x and y thus satisfy (7.10) and (7.11). We claim that $L := F(x, y)$ satisfies $[L : F] \in \{1, 2\}$. Assume first that $x \notin F$. Then $y = \alpha x + \beta$ for some $\alpha, \beta \in F$. Using this in (7.11), we see that x is a root of a quadratic polynomial with coefficients in F. Since $x \notin F$, this shows that x is algebraic of degree 2 over F, and so $[F(x) : F] = 2$ by Theorem 7.35. As $y = \alpha x + \beta$, $F(x)$ is actually equal to L, so this proves our claim. Assume now that $x \in F$. Then $L = F(y)$ and (7.11) shows that y is a root of a quadratic polynomial with coefficients in F. Hence, our claim holds in this case too.

(cc) Finally, let A be a point of intersection of the circle with equation (7.11) and the circle with equation

$$x^2 + y^2 = \delta' x + \varepsilon' y + \zeta', \quad \text{where } \delta', \varepsilon', \zeta' \in F.$$

Of course, we are assuming that the circles are distinct and have a point of intersection, which implies that $\delta \neq \delta'$ or $\varepsilon \neq \varepsilon'$. By subtracting one equation from the other we obtain

$$(\delta - \delta')x + (\varepsilon - \varepsilon')y + (\zeta - \zeta') = 0.$$

Since $\delta - \delta', \varepsilon - \varepsilon'$, and $\zeta - \zeta'$ belong to F and at least one of the first two is nonzero, we can express this equation in the same way as the equation (7.10). We have thereby reduced the case **(cc)** to the case **(lc)**. $\qquad\square$

The ancient geometric problems can now be solved.

Corollary 7.66. *Using only a straightedge and compass, it is impossible to construct a cube having double the volume of a given cube.*

Proof. As explained above, the problem of doubling the cube can be phrased as follows: is it possible to construct the point $Z = (\sqrt[3]{2}, 0)$ from the set $\mathscr{P} = \{(0, 0), (1, 0)\}$? If the answer was positive, then, by Theorem 7.65, $\sqrt[3]{2}$ would be algebraic of degree a power of 2 over $\mathbb{Q}$. However, it is algebraic of degree 3. Indeed, either directly or by applying the Eisenstein Criterion (see Example 7.13) we see that the polynomial $X^3 - 2$ is irreducible over $\mathbb{Q}$, and is therefore (by Theorem 7.8) the minimal polynomial of $\sqrt[3]{2}$. □

It was not that difficult. Why did we have to wait for the solution for over two millennia? This question, of course, does not have a clear answer. We can say, however, that the solution required great leaps of thought. The idea of proving the impossibility of a geometric construction by introducing and studying various abstract algebraic concepts is far from obvious.

The problem of trisecting an angle will be solved in a similar manner, but will take up a bit more space.

Corollary 7.67. *Using only a straightedge and compass, it is impossible to trisect the angle 60°.*

Proof. We have to prove that the point $Z = (\cos 20°, \sin 20°)$ cannot be constructed from the set $\mathscr{P} = \left\{ (0, 0), (1, 0), \left(\frac{1}{2}, \frac{\sqrt{3}}{2} \right) \right\}$. In light of Theorem 7.65, it suffices to show that $u := \cos 20°$ is not algebraic of degree a power of 2 over $\mathbb{Q}(\sqrt{3})$. We will prove that it is actually algebraic of degree 3.

Calculating the real part of

$$\cos 3\varphi + i \sin 3\varphi = (\cos \varphi + i \sin \varphi)^3$$

we arrive at the formula

$$\cos 3\varphi = 4 \cos^3 \varphi - 3 \cos \varphi.$$

Taking 20° for φ we obtain $\frac{1}{2} = 4u^3 - 3u$. That is, u is a root of the polynomial

$$p(X) := 8X^3 - 6X - 1 \in \mathbb{Q}[X] \subseteq \mathbb{Q}(\sqrt{3})[X].$$

Our goal now is to show that $p(X)$ is irreducible over $\mathbb{Q}(\sqrt{3})$. Suppose this were not true. Then $p(X)$ would have a root $a \in \mathbb{Q}(\sqrt{3})$. The number $b := 2a$, which also lies in $\mathbb{Q}(\sqrt{3})$, would then satisfy $b^3 - 3b - 1 = 0$. Writing $b = q + r\sqrt{3}$ with $q, r \in \mathbb{Q}$, and using the fact that 1 and $\sqrt{3}$ are linearly independent over $\mathbb{Q}$, it follows that

$$q^3 + 9qr^2 - 3q - 1 = 0 \quad \text{and} \quad r(q^2 + r^2 - 1) = 0.$$

The second equality implies that $r = 0$ or $q^2 + r^2 - 1 = 0$, and hence, by the first equality, either

$$q^3 - 3q - 1 = 0 \quad \text{or} \quad q'^3 - 3q' + 1 = 0,$$

where $q' = 2q$. However, each of these two possibilities leads to a contradiction. Indeed, write $q = \frac{m}{n}$ in lowest terms. The first possibility implies that $m^3 - 3mn^2 - n^3 = 0$. In particular, $m | n^3$ and $n | m^3$. Since m and n are relatively prime, it follows that they are equal to 1 or -1, which is clearly impossible. Similarly, we see that no rational number q' satisfies $q'^3 - 3q' + 1 = 0$. $\square$

We have to add a comment to this proof. By elementary geometric considerations it can be shown that the point $\left(\frac{1}{2}, \frac{\sqrt{3}}{2}\right)$ can be constructed from the points $(0, 0)$ and $(1, 0)$. Had we proved this, we could exclude this point from $\mathscr{P}$, and hence deal with $\mathbb{Q}$ rather than with $\mathbb{Q}(\sqrt{3})$. The computational part of the proof of Corollary 7.67 would then be slightly shorter. We have chosen our approach to emphasize more clearly that the problem can be solved by exclusively algebraic means.

Here we should also mention the notion of a **constructible point**. This is any point that can be constructed from the points $(0, 0)$ and $(1, 0)$. It is easy to see that the point (k, ℓ) is constructible if and only if $(k, 0)$ and $(0, \ell)$ are constructible. We say that k is a **constructible number** if $(k, 0)$ is a constructible point (or, equivalently, k is one of the components of a constructible point). It turns out that the set K of all constructible numbers is a field with the property that $\sqrt{k} \in K$ for every positive $k \in K$ (the point $\left(\frac{1}{2}, \frac{\sqrt{3}}{2}\right)$ is thus constructible). Proving this is a nice exercise in elementary geometry, which the interested reader may wish to tackle.

Finally, we consider the problem of squaring the circle.

Corollary 7.68. *Using only a straightedge and compass, it is impossible to construct a square having the same area as a given circle.*

Proof. We now consider the case where $Z = (\sqrt{\pi}, 0)$ and $\mathscr{P} = \{(0, 0), (1, 0)\}$. In view of Theorem 7.65, it is enough to prove that $\sqrt{\pi}$ is not an algebraic number. If it were, then its square, π, would also be algebraic (Corollary 7.41). However, this is not true. $\square$

To be honest, we should not call this a "proof" because we have used the fact that π is a transcendental number without proving it. What we really did is reduce the problem of squaring the circle to the problem of showing that π is a transcendental number. To solve the latter, one has to use tools that are outside the scope of this book.

Exercises

7.69. Trisect the angle $90°$ by straightedge and compass.

7.70. Determine all $n \in \mathbb{N}$ such that $\sqrt[n]{2}$ is a constructible number.

7.5 Splitting Fields

After an excursion into geometry, we return to developing the theory of field extensions. From now on, our main topic will be roots of polynomials. One might say that this topic is not new since algebraic elements are nothing but roots of polynomials. However, we will approach it from a different perspective. Rather than treating a fixed algebraic element, we will study the existence and properties of roots of a fixed polynomial.

7.5.1 The Multiplicity of a Root of a Polynomial

A basic result concerning roots of polynomials is Corollary 5.2, which states that $a \in F$ is a root of $f(X) \in F[X]$ if and only if $X - a$ divides $f(X)$. In this chapter, we are interested in roots that may lie in some extension K of F, not necessarily in F itself. That is why we will now state Corollary 5.2 in a slightly different, but equivalent way.

Lemma 7.71. *Let K be an extension of a field F. An element $a \in K$ is a root of the polynomial $f(X) \in F[X]$ if and only if there exists a polynomial $g(X) \in K[X]$ such that $f(X) = (X - a)g(X)$.*

Proof. Since K is an extension of F, $F[X]$ is a subring of the ring $K[X]$. We may therefore consider $f(X)$ as an element of $K[X]$ and apply Corollary 5.2 to this situation. □

In other words, $a \in K$ is a root of $f(X) \in F[X]$ if and only if $X - a$ divides $f(X)$ in $K[X]$. It is possible that some higher power of $X - a$ divides $f(X)$ too, i.e., a may also be a root of the polynomial $g(X)$.

Definition 7.72. If $a \in K$ is a root of $f(X) \in F[X]$ and there exists an $h(X) \in K[X]$ such that

$$f(X) = (X - a)^k h(X) \quad \text{and} \quad h(a) \neq 0, \tag{7.12}$$

then we say that a is a **root of multiplicity** k of $f(X)$. A root of multiplicity 1 is called a **simple root**.

Since (7.12) implies that $\deg(f(X)) = k + \deg(h(X))$, the multiplicity of a root cannot exceed the degree of the polynomial. Actually, more can be said.

Theorem 7.73. *A polynomial $f(X) \in F[X]$ of degree n has at most n roots, counting multiplicity, in any extension K of F.*

Proof. We claim that $f(X)$ can be written as

$$f(X) = (X - a_1)^{k_1} \cdots (X - a_r)^{k_r} f_0(X) \tag{7.13}$$

for some distinct $a_1, \ldots, a_r \in K$, $k_1, \ldots, k_r \in \mathbb{N}$, and a polynomial $f_0(X) \in K[X]$ having no root in K. This is trivially true for $n = 0$, so we may assume that $n > 0$ and that this claim holds for all polynomials of degree less than n (and with coefficients in any subfield of K). We may also assume that $f(X)$ has at least one root a_1 (otherwise we take $f_0(X) = f(X)$). If the multiplicity of a_1 is k_1, then there exists an $h(X) \in K[X]$ such that $f(X) = (X - a_1)^{k_1} h(X)$ and $h(a_1) \neq 0$. Since $\deg(h(X)) = n - k_1 < n$, the claim holds for $h(X)$. Hence it holds for $f(X)$ too.

It is obvious from (7.13) that each a_i is a root of $f(X)$. Moreover, $f(X)$ has no other root in K. Indeed, $f(b) = 0$ with $b \in K$ can be written as

$$(b - a_1)^{k_1} \cdots (b - a_r)^{k_r} f_0(b) = 0,$$

and since $f_0(b) \neq 0$, b must be equal to a_i for some i. The multiplicity of a_i is k_i. Therefore, $f(X)$ has $k_1 + \cdots + k_r$ roots in K, counting multiplicity. Since (7.13) implies that

$$n = k_1 + \cdots + k_r + \deg(f_0(X)),$$

the theorem follows. $\square$

Example 7.74. Consider the polynomial

$$f(X) = X^6 - 3X^4 + 4 \in \mathbb{Q}[X].$$

Note that we can write it as

$$f(X) = (X^2 - 2)^2(X^2 + 1).$$

How to express $f(X)$ in the form (7.13)? The answer depends on the choice of the field K. Let us look at three possibilities.

(a) Taking $\mathbb{Q}$ for K, $f(X)$ has no root in K, and so $f(X) = f_0(X)$.
(b) Taking $\mathbb{R}$ for K, we have $a_1 = \sqrt{2}$, $a_2 = -\sqrt{2}$, $k_1 = k_2 = 2$, and $f_0(X) = X^2 + 1$. Thus, $f(X)$ has 4 roots in K, counting multiplicity.
(c) Taking $\mathbb{C}$ for K, we have $a_1 = \sqrt{2}$, $a_2 = -\sqrt{2}$, $a_3 = i$, $a_4 = -i$, $k_1 = k_2 = 2$, $k_3 = k_4 = 1$, and $f_0(X) = 1$. In this case, the number of roots, counted with multiplicity, is equal to $\deg(f(X)) = 6$.

We are particularly interested in extensions in which we can factor the given polynomial into linear factors (like in (c)). This is the topic of the next subsection.

7.5.2 Splitting Fields: Existence

As we know, not every polynomial with coefficients in a field F has a root in F. However, sometimes we can find a root in some extension of F. *Does such an extension always exist?* In other words, does every polynomial equation have a solution in some larger field? The meaning and importance of this question is evident from the outline of the history of classical algebra from the beginning of the chapter. Our goal is to prove that it has an affirmative answer. If F is a subfield of $\mathbb{C}$, then this of course follows from the Fundamental Theorem of Algebra. But how to construct an appropriate extension of an abstract field? The next example, in which we will "rediscover" the field of complex numbers, indicates a possible approach.

Example 7.75. Let $\mathscr{I}$ be the principal ideal of $\mathbb{R}[X]$ generated by the polynomial $X^2 + 1$ (i.e., $\mathscr{I} = (X^2 + 1)$). As a quadratic polynomial without real roots, this polynomial is irreducible over $\mathbb{R}$. Corollary 5.73 therefore tells us that $\mathbb{R}[X]/\mathscr{I}$ is a field. We claim that it is isomorphic to the field of complex numbers. The usual way of proving this is indicated in Exercise 7.95. However, let us instead verify directly that $\varphi : \mathbb{C} \to \mathbb{R}[X]/\mathscr{I}$ defined by

$$\varphi(\lambda + \mu i) = \lambda + \mu X + \mathscr{I}, \quad \text{where } \lambda, \mu \in \mathbb{R},$$

is an isomorphism. It is clear that φ preserves addition. To show that φ also preserves multiplication, consider

$$\varphi\big((\lambda + \mu i)(\lambda' + \mu' i)\big) = (\lambda\lambda' - \mu\mu') + (\lambda\mu' + \mu\lambda')X + \mathscr{I}$$

and

$$\varphi(\lambda + \mu i)\varphi(\lambda' + \mu' i) = \big(\lambda + \mu X + \mathscr{I}\big)\big(\lambda' + \mu' X + \mathscr{I}\big)$$
$$= \lambda\lambda' + (\lambda\mu' + \mu\lambda')X + \mu\mu' X^2 + \mathscr{I}.$$

Since

$$\mu\mu' X^2 + \mathscr{I} = -\mu\mu' + \mu\mu'(X^2 + 1) + \mathscr{I} = -\mu\mu' + \mathscr{I},$$

the two expressions coincide and so φ is a homomorphism. The ideal $\mathscr{I}$ does not contain polynomials of degree 0 or 1, implying that φ has trivial kernel. Take an arbitrary $f(X) \in \mathbb{R}[X]$. By the division algorithm for polynomials,

$$f(X) = q(X)(X^2 + 1) + \lambda + \mu X$$

for some $q(X) \in \mathbb{R}[X]$ and $\lambda, \mu \in \mathbb{R}$. Therefore,

$$f(X) + \mathscr{I} = (\lambda + \mu X) + \mathscr{I},$$

which proves that φ is surjective and hence an isomorphism. Thus, we may indeed identify $\mathbb{R}[X]/\mathscr{I}$ with the field of complex numbers. The role of real numbers in $\mathbb{R}[X]/\mathscr{I}$ is played by the cosets $\varphi(\lambda) = \lambda + \mathscr{I}$ where $\lambda \in \mathbb{R}$, and the role of the imaginary unit i by the coset $\varphi(i) = X + \mathscr{I}$. Since φ is an isomorphism and i is a root of the polynomial $X^2 + 1$, so is $X + \mathscr{I}$. Let us also check this by a direct calculation. We have

$$(X + \mathscr{I})^2 = X^2 + \mathscr{I} = -1 + \mathscr{I}. \tag{7.14}$$

Here we used that $X^2 + 1 \in \mathscr{I}$. Since $-1 + \mathscr{I}$ is the additive inverse of the unity of $\mathbb{R}[X]/\mathscr{I}$, (7.14) is exactly what we wanted to show.

So how did we arrive at a field containing a root of the given polynomial in this example? We took the principal ideal generated by this polynomial and introduced the desired field as the quotient ring of the polynomial ring by this ideal. In the next proof, we will see that this construction works for any irreducible polynomial over any field.

Theorem 7.76. *Let F be a field and let $f(X) \in F[X]$ be a non-constant polynomial. Then there exists an extension K of F in which $f(X)$ has a root.*

Proof. Let $p(X) \in F[X]$ be an irreducible polynomial that divides $f(X)$ (this exists by Theorem 5.80). Then $f(X)$ belongs to the principal ideal $\mathscr{I} := (p(X))$. Let $K := F[X]/\mathscr{I}$. By Corollary 5.73, K is a field.

Define
$$\varphi : F \to K, \quad \varphi(\lambda) = \lambda + \mathscr{I},$$

where λ on the right-hand side is considered to be a constant polynomial. It is obvious that φ is a ring homomorphism. Since $p(X)$ is a non-constant polynomial, so are the nonzero polynomials in $\mathscr{I}$. Hence φ is an embedding. We may therefore consider K as an extension of F (identifying $\lambda \in F$ with $\lambda + \mathscr{I} \in K$).

Take any $h(X) \in F[X]$ and let us show that its evaluation at $X + \mathscr{I} \in K$ equals $h(X) + \mathscr{I}$, i.e.,
$$h(X + \mathscr{I}) = h(X) + \mathscr{I}. \tag{7.15}$$

Let $h(X) = \sum_i \lambda_i X^i$. In view of our goal it is more convenient to write λ_i, which is an element of F, as $\lambda_i + \mathscr{I} \in K$. Accordingly,
$$h(X + \mathscr{I}) = \sum_i (\lambda_i + \mathscr{I})(X + \mathscr{I})^i = \sum_i (\lambda_i + \mathscr{I})(X^i + \mathscr{I}) = \sum_i \lambda_i X^i + \mathscr{I},$$

proving (7.15). Taking $f(X)$ for $h(X)$, we obtain
$$f(X + \mathscr{I}) = f(X) + \mathscr{I} = 0,$$

since $f(X) \in \mathscr{I}$. So, $X + \mathscr{I}$ is a root of $f(X)$. $\qquad\square$

Every polynomial thus has a root—if not "at home," in the base field, then in one of its extensions. This is one of the fundamental theorems of field theory. Its proof is a nice illustration of the power of the abstract approach. Although the theorem speaks about only one polynomial, its proof depends on the consideration of the ring of all polynomials. Besides, it involves abstract notions such as irreducibility, quotients rings, and (indirectly) maximal ideals, which seemingly have no connection with the classical mathematical problem of finding a root of a polynomial.

We will now derive two sharpened versions of Theorem 7.76, which apparently tell us much more. However, from their proofs it will be evident that they are rather simple consequences of the basic theorem.

Theorem 7.77. *Let F be a field and let $f(X) \in F[X]$ be a non-constant polynomial with leading coefficient c. Then there exists an extension K of F containing (not necessarily distinct) elements $a_1, \ldots, a_n$ such that $f(X) = c(X - a_1) \cdots (X - a_n)$.*

Proof. We proceed by induction on $n := \deg(f(X))$. The case where $n = 1$ is obvious, so let $n > 1$. We may assume that the desired conclusion holds for all polynomials of degree less than n that have coefficients in any field (not necessarily in F). Theorem 7.76 tells us that there exists an extension K_1 of F containing a root a_1 of $f(X)$. By Lemma 7.71, there exists a polynomial $g(X) \in K_1[X]$ such that $f(X) = (X - a_1)g(X)$. Since $\deg(g(X)) = n - 1$, by assumption there exists an extension K of K_1 (and thereby also an extension of F) such that $g(X)$ is equal to a product of linear polynomials with coefficients in K. Hence, the same holds for $f(X) = (X - a_1)g(X)$. □

Theorem 7.73 states that a polynomial in $F[X]$ of degree n has at most n roots, counting multiplicity, in any extension of F. Now we know that there exists an extension K of F such that the number of roots in K, counted with multiplicity, is equal to n.

Another way of stating Theorem 7.77 is as follows: every non-constant polynomial with coefficients in F can be written as a product of linear polynomials with coefficients in some extension of F. There are several such extensions. We are particularly interested in the smallest one.

Definition 7.78. Let K be an extension of a field F. We say that a polynomial $f(X) \in F[X]$ **splits over** K if $f(X)$ is equal to a product of linear polynomials in $K[X]$. If $f(X)$ splits over K but does not split over any proper subfield of K containing F, then we say that K is a **splitting field for** $f(X)$ **over** F.

Theorem 7.77 shows that every non-constant polynomial $f(X) \in F[X]$ splits over some extension of F. The existence of a splitting field now easily follows.

Theorem 7.79. *Let F be a field and let $f(X) \in F[X]$ be a non-constant polynomial. Then there exists a splitting field for $f(X)$ over F.*

Proof. Let K and $a_1, \ldots, a_n \in K$ be as in Theorem 7.77. We claim that $F(a_1, \ldots, a_n)$ is a splitting field for $f(X)$ over F. Obviously, $f(X)$ splits over this field. If K_0 is a subfield of K containing F such that $f(X)$ splits over K_0, then K_0 must contain all roots of $f(X)$ in K. That is, $a_1, \ldots, a_n \in K_0$ and hence $K_0 \supseteq F(a_1, \ldots, a_n)$. In particular, K_0 is not a proper subfield of $F(a_1, \ldots, a_n)$, and so $F(a_1, \ldots, a_n)$ is indeed a splitting field for $f(X)$ over F. □

Remark 7.80. As we see from the proof, a splitting field for $f(X) \in F[X]$ is simply $F(a_1, \ldots, a_n)$ where $a_1, \ldots, a_n$ are the roots of $f(X)$. Thus, it is the smallest field containing F and all the roots of the polynomial in question. Note that, in view of Theorem 7.39, a splitting field is a *finite* extension of F.

Example 7.81. What is a splitting field for the polynomial $X^2 + 1$? This question has no answer without specifying the field F.

(a) The polynomial $X^2 + 1 \in \mathbb{Q}[X]$ of course splits over $\mathbb{C}$, as well as over any
subfield of $\mathbb{C}$ containing all rational numbers along with i and $-i$. Therefore, a
splitting field for $X^2 + 1$ over $\mathbb{Q}$ is $\mathbb{Q}(i) = \{\lambda_0 + \lambda_1 i \mid \lambda_i \in \mathbb{Q}\}$.
(b) A splitting field for $X^2 + 1$ over $\mathbb{R}$ is $\mathbb{R}(i) = \mathbb{C}$.
(c) A splitting field for $X^2 + 1$ over $\mathbb{C}$ is $\mathbb{C}$ itself.
(d) Similarly, a splitting field for $X^2 + 1$ over $\mathbb{Z}_2$ is $\mathbb{Z}_2$ itself. This is because $X^2 + 1 = (X + 1)^2$ in $\mathbb{Z}_2[X]$.

Remark 7.82. If K is a splitting field for $f(X)$ over F, then K is also a splitting
field for $f(X)$ over any field L such that $F \subseteq L \subseteq K$. This is immediate from the
definition of a splitting field.

7.5.3 Splitting Fields: Uniqueness

We have proved that a splitting field for every non-constant polynomial $f(X)$ *exists*.
What about its *uniqueness*? We cannot expect that there is literally only one splitting
field. Our definitions concern arbitrary extensions of the given field which may be
different as sets. The best we can hope for is that any two splitting fields for $f(X)$
are isomorphic—and this is indeed true, as we will now show. We will actually prove
slightly more than that, but not because we truly need this generalization. It is just
that sometimes a method of proof yields a more general result than one wishes to
prove.

Before starting, we warn the reader that the proofs in this subsection may seem
intimidating at first because of the large amount of notation. Once we extract their
main idea, however, they seem natural to us.

First we record two general remarks on isomorphisms. The first one is a sharpened
version of Remark 7.37 (a).

Remark 7.83. Let K be an extension of F, let $a \in K$ be a root of an irreducible
polynomial $p(X) \in F[X]$, and let $\varepsilon : F[X] \to F(a)$ be the *evaluation homomor-
phism at* a, i.e., $\varepsilon(f(X)) = f(a)$. Note that ε is surjective and $\ker \varepsilon = (p(X))$. The
Isomorphism Theorem therefore tells us that there is an isomorphism

$$\overline{\varepsilon} : F[X]/(p(X)) \to F(a).$$

Recall that $\overline{\varepsilon}$ is defined by $\overline{\varepsilon} \circ \pi = \varepsilon$, where π is the canonical epimorphism from
$F[X]$ to $F[X]/(p(X))$ (Figure 7.4). In particular,

$$\overline{\varepsilon}\big(X + (p(X))\big) = a \quad \text{and} \quad \overline{\varepsilon}\big(\lambda + (p(X))\big) = \lambda \quad \text{for all } \lambda \in F. \tag{7.16}$$

Remark 7.84. Let F and F' be fields. An isomorphism $\varphi : F \to F'$ can be extended
to an isomorphism between the rings $F[X]$ and $F'[X]$. Indeed, for every polynomial

$$f(X) = \lambda_0 + \lambda_1 X + \cdots + \lambda_n X^n \in F[X]$$

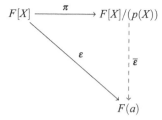

Fig. 7.4 Commutative diagram corresponding to Remark 7.83

define the polynomial

$$f_\varphi(X) = \varphi(\lambda_0) + \varphi(\lambda_1)X + \cdots + \varphi(\lambda_n)X^n \in F'[X].$$

One immediately verifies that

$$f(X) \mapsto f_\varphi(X)$$

is an isomorphism from $F[X]$ to $F'[X]$ (we encountered a similar homomorphism in the proof of Gauss' Lemma). Note that $p(X)$ is irreducible in $F[X]$ if and only if $p_\varphi(X)$ is irreducible in $F'[X]$.

Lemma 7.85. *Let F be a field, let $p(X) \in F[X]$ be an irreducible polynomial, and let a be a root of $p(X)$ in some extension K of F. If $\varphi : F \to F'$ is an isomorphism of fields and a' is a root of $p_\varphi(X)$ in some extension K' of F', then there exists a unique isomorphism $\Phi : F(a) \to F(a')$ extending φ and satisfying $\Phi(a) = a'$.*

Proof. Let

$$\overline{\varepsilon} : F[X]/(p(X)) \to F(a)$$

be an isomorphism satisfying (7.16). Since $p_\varphi(X)$ is also irreducible, there is an analogous isomorphism

$$\overline{\varepsilon}' : F'[X]/(p_\varphi(X)) \to F'(a')$$

satisfying

$$\overline{\varepsilon}'\big(X + (p_\varphi(X))\big) = a' \quad \text{and} \quad \overline{\varepsilon}'\big(\lambda' + (p_\varphi(X))\big) = \lambda' \quad \text{for all } \lambda' \in F'. \quad (7.17)$$

Next, we define

$$\widetilde{\varphi} : F[X]/(p(X)) \to F'[X]/(p_\varphi(X))$$

by

$$\widetilde{\varphi}\big(f(X) + (p(X))\big) = f_\varphi(X) + (p_\varphi(X)).$$

It is easy to check that $\widetilde{\varphi}$ is a well-defined isomorphism. Now we can define

$$\Phi := \overline{\varepsilon}' \circ \widetilde{\varphi} \circ \overline{\varepsilon}^{-1} : F(a) \to F'(a')$$

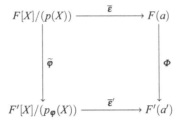

Fig. 7.5 Commutative diagram corresponding to Lemma 7.85

(see Figure 7.5). Since $\bar{\varepsilon}'$, $\widetilde{\varphi}$, and $\bar{\varepsilon}^{-1}$ are isomorphisms, so is Φ. From (7.16) and (7.17), we infer that $\Phi(a) = a'$ and $\Phi(\lambda) = \varphi(\lambda)$ for all $\lambda \in F$, as desired. Moreover, as every element of $F(a)$ can be written as $\sum_i \lambda_i a^i$ with $\lambda_i \in F$, Φ is a unique isomorphism with these properties.   □

We will use this lemma twice, first in the proof of the next theorem and later in the proof of Theorem 7.128.

Theorem 7.86. *Let F be a field, let $f(X) \in F[X]$ be a non-constant polynomial, and let K be a splitting field for $f(X)$ over F. If $\varphi : F \to F'$ is an isomorphism of fields and K' is a splitting field for $f_\varphi(X)$ over F', then φ can be extended to an isomorphism from K to K'.*

Proof. The proof is by induction on $n := [K : F]$. If $n = 1$, then $f(X)$ splits over F, and hence $f_\varphi(X)$ splits over F'. Thus, $K = F$ and $K' = F'$, so there is nothing to prove. Let $n > 1$. Then $f(X)$ does not split over F. Therefore, there exists an irreducible polynomial $p(X) \in F[X]$ that divides $f(X)$ and has degree greater than 1. Being a splitting field for $f(X)$, K contains a root, let us call it a, of $p(X)$. Similarly, $p_\varphi(X)$ has a root $a' \in K'$. Let $\Phi : F(a) \to F(a')$ be the isomorphism from Lemma 7.85. Since K is a splitting field for $f(X)$ over F, it is also a splitting field for $f(X)$ over the larger field $F(a)$ (see Remark 7.82). Similarly, K' is a splitting field for $f_\varphi(X)$ over $F'(a')$. Note that

$$[K : F] = [K : F(a)][F(a) : F]$$

and

$$[F(a) : F] = \deg(p(X)) > 1.$$

Therefore, $[K : F(a)] < n$ and we may apply the induction hypothesis to $\Phi : F(a) \to F(a')$. Therefore, there is an isomorphism from K to K' that agrees with Φ on $F(a)$, and hence also with φ on F.   □

In the simplest case where $F = F'$ and $\varphi = \mathrm{id}_F$, Theorem 7.86 reduces to the following corollary. The reader may wish to examine why the proof of the theorem does not work if we consider only this case.

Corollary 7.87. *Let F be a field and let $f(X) \in F[X]$ be a non-constant polynomial. Then any two splitting fields for $f(X)$ over F are isomorphic.*

To summarize, a splitting field exists and is essentially unique. It is therefore common to speak of "the" splitting field.

7.5.4 Algebraic Closures

Recall that a field A is said to be *algebraically closed* if every non-constant polynomial $f(X) \in A[X]$ has a root in A. By a standard argument based on Lemma 7.71, one shows that this is equivalent to the condition that every non-constant polynomial $f(X) \in A[X]$ splits over A, i.e., it can be written as

$$f(X) = c(X - a_1) \cdots (X - a_n)$$

for some not necessarily distinct $a_i \in A$ (here, c is its leading coefficient).

The Fundamental Theorem of Algebra states that the field of complex numbers $\mathbb{C}$ is algebraically closed. The next lemma will lead us to further examples.

Lemma 7.88. *Let L be an algebraic extension of a field F, and let K be an extension of L. If $x \in K$ is algebraic over L, then x is also algebraic over F.*

Proof. By assumption, there are $a_i \in L$, not all 0, such that

$$a_0 + a_1 x + \cdots + a_n x^n = 0.$$

Therefore, x is also algebraic over the subfield $F(a_0, \ldots, a_n)$ of L. Theorem 7.35 shows that $F(a_0, \ldots, a_n)(x)$ is a finite extension of $F(a_0, \ldots, a_n)$. The latter field, on the other hand, is a finite extension of F. Indeed, this follows from Theorem 7.39 since every a_i lies in L and is therefore algebraic over F. By Theorem 7.29, $F(a_0, \ldots, a_n)(x)$ is a finite extension of F. Now, Lemma 7.32 implies that every element of $F(a_0, \ldots, a_n)(x)$, and thus also x, is algebraic over F. $\qquad \square$

We remark that this lemma shows that an algebraic extension of an algebraic extension is itself algebraic. This is analogous to the frequently used fact that a finite extension of a finite extension is itself finite.

Being algebraically closed is a rather exceptional property of a field. If a given field F does not have it, one of its extensions A might. For example, if $F = \mathbb{Q}$, then one can take $A = \mathbb{C}$. However, the complex numbers are a "huge" extension of the rational numbers. Is there any algebraically closed extension that is more closely tied to $\mathbb{Q}$? Let us make precise what we mean by this.

Definition 7.89. Let F be a field. A field $\overline{F}$ is called an **algebraic closure** of F if $\overline{F}$ is algebraically closed and is an algebraic extension of F.

Example 7.90. The field $\mathbb{C}$ is an algebraic extension of $\mathbb{R}$ and is therefore an algebraic closure of $\mathbb{R}$. On the other hand, since there exist transcendental numbers, $\mathbb{C}$ is not an algebraic closure of $\mathbb{Q}$.

It is easy to describe algebraic closures of fields that are contained in some algebraically closed field. This, of course, includes all fields lying between $\mathbb{Q}$ and $\mathbb{C}$.

Theorem 7.91. *Let F be a subfield of an algebraically closed field A. Then the set $\overline{F}$ of all elements in A that are algebraic over F is an algebraic closure of F.*

Proof. As we know, $\overline{F}$ is a field (see Corollary 7.41), and, by the very definition, it is an algebraic extension of F. We must prove that $\overline{F}$ is algebraically closed. Take a non-constant polynomial $f(X) \in \overline{F}[X]$. Since A is algebraically closed and $\overline{F} \subseteq A$, there exists an $x \in A$ such that $f(x) = 0$. Hence, x is algebraic over $\overline{F}$. By Lemma 7.88, x is also algebraic over F, meaning that $x \in \overline{F}$. Thus, $f(X)$ has a root in $\overline{F}$. □

Corollary 7.92. *The field of all algebraic numbers is an algebraic closure of $\mathbb{Q}$.*

It turns out that every field is a subfield of an algebraically closed field. In light of Theorem 7.91, this means that *every field F has an algebraic closure*. Moreover, an *algebraic closure of F is unique up to isomorphism*. The proofs of these statements are somewhat sophisticated applications of Zorn's lemma and will be omitted. Note that the existence of an algebraic closure of a field is a much deeper result than the existence of a splitting field for a polynomial. If $\overline{F}$ is an algebraic closure of F, then *all* polynomials with coefficients in F split over $\overline{F}$. A splitting field, on the other hand, concerns only one polynomial.

Exercises

7.93. Determine the multiplicity of the root a of the polynomial $f(X) \in F[X]$ in the case where:

(a) $a = 1$, $f(X) = X^5 - 5X^3 + 5X^2 - 1$, and $F = \mathbb{Q}$.
(b) $a = 1$, $f(X) = X^4 + X^3 + X + 1$, and $F = \mathbb{Z}_2$.
(c) $a = -1$, $f(X) = X^{10} - 1$, and $F = \mathbb{Z}_5$.
(d) $a = 1$, $f(X) = X^{2^n} + 1$, and $F = \mathbb{Z}_2$.

7.94. Let p be a prime. Apply Fermat's Little Theorem to prove that

$$X^{p-1} - 1 = (X - 1)(X - 2) \cdots (X - (p - 1))$$

holds in $\mathbb{Z}_p[X]$. Use this to derive another proof of *Wilson's Theorem* (see Exercise 3.27).

7.95. In Example 7.75, we saw that $\mathbb{R}[X]/(X^2 + 1) \cong \mathbb{C}$. Give a different (and shorter) proof by making use of the evaluation homomorphism $f(X) \mapsto f(i)$ from $\mathbb{R}[X]$ to $\mathbb{C}$.

7.96. Prove that $\mathbb{Q}(\sqrt{2})$ is the splitting field over $\mathbb{Q}$ for each of the following polynomials:

(a) $X^2 - 2$.
(b) $3X^3 - 6X$
(c) $X^4 - 3X^2 + 2$.

7.97. Prove that $\mathbb{Q}(\sqrt{2}, \sqrt{3})$ is the splitting field for $X^4 - 5X^2 + 6$ over $\mathbb{Q}$.

7.98. Prove that $\mathbb{Q}(\sqrt[3]{2}, -\frac{1}{2} + i\frac{\sqrt{3}}{2})$ is the splitting field for $X^3 - 2$ over $\mathbb{Q}$.

7.99. Prove that $\mathbb{Q}(i)$ is the splitting field for $X^4 + 4$ over $\mathbb{Q}$.

7.100. Prove that $\mathbb{Q}(\sqrt[4]{2}, i)$ is the splitting field for $X^4 + 2$ over $\mathbb{Q}$.

7.101. Prove that $\mathbb{Q}(\sqrt{2}, i)$ is the splitting field for $X^4 + 1$ over $\mathbb{Q}$.

7.102. Let $f(X) \in F[X]$ be of degree n and let K be the splitting field for $f(X)$ over F. Prove that $[K : F] \leq n!$ and that $n \mid [K : F]$ if $f(X)$ is irreducible. Determine $[K : F]$ in the preceding exercises.

7.103. Let a and b be transcendental numbers. Prove that at least one of $a + b$ and ab is transcendental too.

7.104. Let $a_1, \ldots, a_n$ be elements of a field F. Show that there exists a polynomial $f(X) \in F[X]$ such that $f(a_1) = \cdots = f(a_n) = 1$. Hence deduce that a finite field cannot be algebraically closed.

7.105. Let A be a field. Prove that the following conditions are equivalent:

(i) A is algebraically closed.
(ii) Every finite-dimensional algebra over A of dimension at least 2 has zero-divisors.
(iii) There does not exist a finite extension $K \supsetneq A$.
(iv) Every irreducible polynomial over A is linear.

Hint. If $p(X) \in A[X]$ is irreducible, then we can regard $A[X]/(p(X))$ as an extension of A (see the proof of Theorem 7.76).

7.6 Finite Fields

This section is devoted to the classification of finite fields, i.e., fields having a finite number of elements. Although the result that we will establish is quite deep, the proof will not be too long. However, this is mainly because we will employ the theory established earlier in this chapter. It is especially interesting that theoretical results will help us greatly when searching for concrete examples. In this way, we will be rewarded for our efforts in the previous sections.

What do we already know about finite fields? We are familiar with fields of prime cardinality $\mathbb{Z}_p$ (Theorem 2.24), and we know that we can embed $\mathbb{Z}_p$ into any field of characteristic p (Theorem 3.137 (b)). Every finite field K has a prime characteristic p, since otherwise it would contain a field isomorphic to $\mathbb{Q}$ (Theorem 3.137 (a)). Thus, we may regard K as a finite extension of $\mathbb{Z}_p$. This enables us to make the first, easy but important step towards the description of finite fields.

Lemma 7.106. *If K is a finite field of characteristic p, then $|K| = p^n$ for some $n \in \mathbb{N}$.*

Proof. Let $\{b_1, \ldots, b_n\}$ be a basis of the vector space K over $\mathbb{Z}_p$. Every element in K can be *uniquely* written as

$$\lambda_1 b_1 + \cdots + \lambda_n b_n, \quad \text{where } \lambda_i \in \mathbb{Z}_p.$$

The number of elements in K is therefore equal to the number of all distinct n-tuples $(\lambda_1, \ldots, \lambda_n)$, which in turn is equal to $|\mathbb{Z}_p|^n = p^n$. $\square$

The cardinality of a finite field is thus a power of its characteristic, which is necessarily a prime number p. We know that there are fields with p elements, but do there also exist fields of cardinality p^2, p^3, etc.? Incidentally, we can forget about the rings $\mathbb{Z}_{p^2}$, $\mathbb{Z}_{p^3}$, etc., which have zero-divisors and are not fields. A more subtle approach is needed. From the proof of Lemma 7.106 it is evident that

$$|K| = p^n \iff [K : \mathbb{Z}_p] = n.$$

We can therefore reword our question and ask whether there exist finite extensions of $\mathbb{Z}_p$ of degree 2, 3, etc. Recall that splitting fields are finite extensions of the base field (see Remark 7.80). Thus, in principle, we can take any non-constant polynomial in $\mathbb{Z}_p[X]$, and its splitting field, which we know to exist by Theorem 7.79, will be an example of a field of cardinality p^n for some $n \in \mathbb{N}$. However, the next lemma shows that it is enough to consider only specific polynomials.

Lemma 7.107. *If K is a field with p^n elements, then K is the splitting field for the polynomial $f(X) = X^{p^n} - X$ over $\mathbb{Z}_p$.*

Proof. The set K^* of all nonzero, and hence invertible, elements of K is a multiplicative group of order $p^n - 1$. Corollary 6.4 therefore tells us that $x^{p^n-1} = 1$ for all $x \in K^*$. Multiplying by x we obtain $x^{p^n} = x$. Since this equality also holds for $x = 0$, every element of K is a root of $f(X)$. Thus, the number of distinct roots of $f(X)$ in K is equal to its degree. Therefore, $f(X)$ splits over K. Since K happens to be the set of all its roots, $f(X)$ cannot split over a proper subfield of K. $\square$

The lemma states that *if* there exists a field of p^n elements, then it must be the splitting field for the polynomial $f(X) = X^{p^n} - X$ over $\mathbb{Z}_p$. Except for $n = 1$, we do not yet know whether such a field exists. The lemma only reveals the only possible candidate for such a field. On the other hand, since any two splitting fields for $f(X)$ are isomorphic (Corollary 7.87), the lemma also implies that any two finite fields of the same cardinality are isomorphic.

Our goal now is to prove that the splitting field for $f(X)$ over $\mathbb{Z}_p$ indeed has p^n elements. But first we have to get familiar with a special endomorphism peculiar to prime characteristic.

Lemma 7.108. *If the characteristic of a commutative ring R is a prime p, then the map $\varphi : R \to R$, $\varphi(x) = x^p$, is an endomorphism of R.*

Proof. It is clear that

$$\varphi(xy) = (xy)^p = x^p y^p = \varphi(x)\varphi(y).$$

We have to show that φ also preserves addition. Note first that the standard induction proof shows that the usual binomial formula,

$$(x + y)^p = x^p + \binom{p}{1}x^{p-1}y + \binom{p}{2}x^{p-2}y^2 + \cdots + \binom{p}{p-1}xy^{p-1} + y^p,$$

holds in any commutative ring. However, since $p \mid \binom{p}{k}$, $1 \le k \le p - 1$, (see Example 5.11) and R has characteristic p, all terms in the above summation, except for the first and last, vanish. Accordingly, the following friendly formula (sometimes called the *freshman's dream*) holds in R:

$$(x + y)^p = x^p + y^p.$$

This means that $\varphi(x + y) = \varphi(x) + \varphi(y)$. $\qquad\qquad\qquad\qquad\qquad\qquad$ □

We call φ the **Frobenius endomorphism**. Write

$$\varphi^n := \underbrace{\varphi \circ \cdots \circ \varphi}_{n \text{ times}}.$$

For example, $\varphi^2(x) = (x^p)^p = x^{p^2}$. In general,

$$\varphi^n(x) = x^{p^n}. \qquad\qquad\qquad (7.18)$$

Since φ is an endomorphism, so is φ^n.

The proof of Lemma 7.107 shows that every element of the field K is a root of $f(X)$. This gives us an idea of how to approach the proof of our final lemma.

Lemma 7.109. *The splitting field for the polynomial $f(X) = X^{p^n} - X$ over $\mathbb{Z}_p$ has p^n elements.*

Proof. Denote this field by $\mathbb{F}_{p^n}$. Thus, $\mathbb{F}_{p^n}$ is the smallest field containing all roots of $f(X)$. Let K be the set of its roots, i.e.,

$$K = \{x \in \mathbb{F}_{p^n} \mid x^{p^n} = x\}.$$

By (7.18), we can write K as

$$K = \{x \in \mathbb{F}_{p^n} \mid \varphi^n(x) = x\}.$$

Hence we see that K itself is a field. Indeed, $1 \in K$ and since φ^n is an endomorphism, $x, y \in K$ implies $x - y \in K$, $xy \in K$, and $x^{-1} \in K$ provided that $x \ne 0$. Therefore,

$$K = \mathbb{F}_{p^n}.$$

As $f(X)$ has degree p^n, it has at most p^n distinct roots. Thus, $|\mathbb{F}_{p^n}| = |K| \le p^n$. On the other hand, if we count the roots of $f(X)$ with their multiplicity, then their number is exactly p^n. In order to prove that $|\mathbb{F}_{p^n}| = p^n$, it is therefore enough to show that all roots of $f(X)$ are simple. From

$$f(X) = (X^{p^n-1} - 1)X$$

we see that 0 is a simple root. Take a root $a \ne 0$. Then $a^{p^n-1} = 1$ and hence

$$f(X) = (X^{p^n-1} - a^{p^n-1})X = (X - a)g(X),$$

where

$$g(X) = X^{p^n-1} + aX^{p^n-2} + a^2X^{p^n-3} + \cdots + a^{p^n-3}X^2 + a^{p^n-2}X.$$

We have

$$g(a) = (p^n - 1)a^{p^n-1} = (p^n - 1) \cdot 1.$$

Of course, $\mathbb{F}_{p^n}$ is an extension of $\mathbb{Z}_p$ and thus has characteristic p. Therefore,

$$g(a) = -1 \ne 0,$$

and so a is a simple root of $f(X)$.   □

The field $\mathbb{F}_{p^n}$ is sometimes called the **Galois field of order** p^n, and is alternatively denoted by $\mathrm{GF}(p^n)$. A field with p elements can thus be denoted by $\mathbb{Z}_p$, $\mathbb{F}_p$, or $\mathrm{GF}(p)$.

Lemmas 7.106, 7.107, and 7.109, along with Theorem 7.79 (the existence of a splitting field) and Corollary 7.87 (the uniqueness of a splitting field) yield the following theorem.

Theorem 7.110. *For each prime p and each natural number n, there exists a field, denoted $\mathbb{F}_{p^n}$, with p^n elements. Moreover, every finite field is isomorphic to $\mathbb{F}_{p^n}$ for some p and n.*

We have thus classified all finite fields. However, our understanding of finite fields is merely theoretical. How do we add and multiply elements in $\mathbb{F}_{p^n}$? This is not clear from the definition. Solving the exercises should help the reader to get a better feeling for the subject.

At the end, let us say that finite fields are used in number theory and some other classical areas of mathematics. In modern times, they have also been found to be important in cryptography and coding theory, areas that lie at the border of mathematics and other sciences.

Exercises

7.111. Let K be an extension of a finite field F. Prove that $[K : F] = n$ if and only if $|K| = |F|^n$.

7.112. Prove that the product of all nonzero elements of a finite field is equal to -1. *Hint.* $X^{p^n-1} - 1 = \prod_{a \in \mathbb{F}_{p^n}^*} (X - a)$ (why?).

7.113. Let $f(X) \in \mathbb{Z}_p[X]$. Prove that $f(X^{p^n}) = f(X)^{p^n}$ for all $n \geq 0$.

7.114. Prove that the Frobenius endomorphism of a finite field is an automorphism.

7.115. Prove that the Frobenius endomorphism of the field of rational functions $\mathbb{Z}_p(X)$ is not surjective.
Hint. What is a possible degree of a polynomial lying in the image of φ?

7.116. Does there exist a ring homomorphism from $\mathbb{F}_4$ to $\mathbb{Z}_4$? Does there exist a ring homomorphism from $\mathbb{Z}_4$ to $\mathbb{F}_4$?

7.117. Let K be a finite field. Prove that the group $(K^*, \cdot)$ is cyclic.
Sketch of proof. Suppose it is not cyclic. The result of Exercise 5.158 then implies that there is a prime p such that $x^p = 1$ holds for more than p elements x in K^*. However, this is impossible since the number of roots of the polynomial $X^p - 1$ cannot exceed its degree p.

7.118. Prove that, for every $n \in \mathbb{N}$, there exists an irreducible polynomial of degree n over $\mathbb{Z}_p$.
Hint. Let a be a generator of the cyclic group $\mathbb{F}_{p^n}^*$ (see the preceding exercise). Then $\mathbb{F}_{p^n} = \mathbb{Z}_p(a)$. What is the degree of the minimal polynomial of a?

7.119. Let $q(X) \in \mathbb{Z}_p[X]$ be an irreducible polynomial of degree n. Prove that

$$\mathbb{F}_{p^n} = \mathbb{Z}_p[X]/(q(X))$$

and that $q(X)$ divides $X^{p^n} - X$.
Hint. The cosets $X^i + (q(X))$, $i = 0, 1, \ldots, n - 1$, are linearly independent over $\mathbb{Z}_p$, and each coset $f(X) + (q(X))$ is their linear combination.

7.120. Verify that the polynomial $X^2 + X + 1$ is irreducible over $\mathbb{Z}_2$. The result of the preceding exercise implies that

$$\mathbb{F}_4 = \mathbb{Z}_2[X]/(X^2 + X + 1).$$

Using this, write down the addition and multiplication tables for $\mathbb{F}_4$.

7.121. Using the result of Exercise 7.119, show that:

(a) $\mathbb{F}_8 = \mathbb{Z}_2[X]/(X^3 + X + 1)$.

(b) $\mathbb{F}_9 = \mathbb{Z}_3[X]/(X^2 + 1)$.
(c) $\mathbb{F}_{16} = \mathbb{Z}_2[X]/(X^4 + X + 1)$.
(d) $\mathbb{F}_{25} = \mathbb{Z}_5[X]/(X^2 + 3)$.
(e) $\mathbb{F}_{27} = \mathbb{Z}_3[X]/(X^3 - X - 1)$.

7.122. With the help of the previous exercises, write $X^4 - X$ and $X^8 - X$ as a product of irreducible polynomials in $\mathbb{Z}_2[X]$.

7.123. Let $d \mid n$. Prove that

$$L := \{x \in \mathbb{F}_{p^n} \mid x^{p^d} = x\}$$

is a subfield of $\mathbb{F}_{p^n}$, and that $|L| = p^d$.

Sketch of proof. Use the same argument as in the proof of Lemma 7.109 to show that L is a subfield. After that, show that $p^d - 1$ divides $p^n - 1$, and hence $X^{p^d-1} - 1$ divides $X^{p^n-1} - 1$ in $\mathbb{Z}_p[X]$. Consequently, $X^{p^d} - X$ divides $X^{p^n} - X$ in $\mathbb{Z}_p[X]$. As we know, $X^{p^n} - X$ has p^n distinct roots in $\mathbb{F}_{p^n}$. Hence conclude that $X^{p^d} - X$ has p^d distinct roots in $\mathbb{F}_{p^n}$, and so $|L| = p^d$.

7.124. Let L be a subfield of $\mathbb{F}_{p^n}$. Prove that there is a $d \in \mathbb{N}$ such that $|L| = p^d$, $d \mid n$, and $L = \{x \in \mathbb{F}_{p^n} \mid x^{p^d} = x\}$.

Comment. This and the preceding exercise show that for every divisor d of n, $\mathbb{F}_{p^n}$ contains exactly one subfield with p^d elements, and that these are actually the only subfields of $\mathbb{F}_{p^n}$. We may therefore regard $\mathbb{F}_{p^d}$ as a subfield of $\mathbb{F}_{p^n}$. For example, the subfields of $\mathbb{F}_{p^{16}}$ are

$$\mathbb{F}_p, \ \mathbb{F}_{p^2}, \ \mathbb{F}_{p^4}, \ \mathbb{F}_{p^8}, \ \mathbb{F}_{p^{16}}.$$

Each field in this sequence is a subfield of its successor. If we take $\mathbb{F}_{p^{12}}$, then the connection between its subfields

$$\mathbb{F}_p, \ \mathbb{F}_{p^2}, \ \mathbb{F}_{p^3}, \ \mathbb{F}_{p^4}, \ \mathbb{F}_{p^6}, \ \mathbb{F}_{p^{12}},$$

is more involved. For example, none of $\mathbb{F}_{p^2}$ and $\mathbb{F}_{p^3}$ is contained in another one, but they are both contained in $\mathbb{F}_{p^6}$. Furthermore, $\mathbb{F}_{p^2}$ is contained in $\mathbb{F}_{p^4}$, while $\mathbb{F}_{p^3}$ is not.

For any m and n, we can thus view $\mathbb{F}_{p^m}$ and $\mathbb{F}_{p^n}$ as subfields of $\mathbb{F}_{p^{mn}}$. This makes it possible to endow the set

$$\overline{\mathbb{Z}}_p := \bigcup_{n \in \mathbb{N}} \mathbb{F}_{p^n}$$

with addition and multiplication, which make it a field. It is not difficult to prove that $\overline{\mathbb{Z}}_p$ is an algebraic closure of $\mathbb{Z}_p$.

7.7 Fields of Characteristic Zero

Fields of characteristic zero seem intuitively closer than fields of prime characteristic. In this section, we will see they are also easier to deal with in some ways. We will prove three seemingly unrelated theorems that concern extensions of fields of characteristic zero. The first one will be used in the proof of the second and third, and these two will be used in the next section on Galois theory.

7.7.1 Roots of Irreducible Polynomials

Our first aim in this section is to show that irreducible polynomials over fields of characteristic 0 have only simple roots. To this end, we define the (formal) **derivative** of

$$f(X) = \lambda_0 + \lambda_1 X + \lambda_2 X^2 + \cdots + \lambda_n X^n \in F[X]$$

as

$$f'(X) = \lambda_1 + 2\lambda_2 X + \cdots + n\lambda_n X^{n-1} \in F[X].$$

We usually associate the notion of the derivative with tangents to curves. Therefore, it will perhaps surprise the reader that the formal derivative of a polynomial with coefficients in an abstract field is also important. As we will see, even the formula

$$\big(f(X)g(X)\big)' = f'(X)g(X) + f(X)g'(X),$$

with which the reader is familiar at least in the case where $F = \mathbb{R}$, is useful. If $f(X)$ and $g(X)$ are monomials, the formula can be checked immediately. The general case can be reduced to the monomial case with the help of the formula

$$\big(f(X) + g(X)\big)' = f'(X) + g'(X),$$

which follows at once from the definition of the derivative.

The formula

$$\deg(f'(X)) = \deg(f(X)) - 1, \quad \text{whenever } \deg(f(X)) \geq 1,$$

may also seem indisputable at first glance. However, it holds only if the characteristic of F is 0 (since then $\lambda_n \neq 0$ implies $n\lambda_n \neq 0$). If F has prime characteristic p, then, for example, $(X^p)' = 0$. The proof of the next theorem depends on the condition that the derivative of a non-constant polynomial is nonzero, and so does not work for prime characteristic.

Theorem 7.125. *Let F be a field of characteristic 0 and let $p(X) \in F[X]$ be an irreducible polynomial. Then every root of $p(X)$ (in any extension K of F) is simple.*

Proof. Suppose $a \in K$ is not a simple root. Then there exists a polynomial $k(X) \in K[X]$ such that $p(X) = (X - a)^2 k(X)$. The formula for the derivative of the product of polynomials then gives

$$p'(X) = 2(X - a)k(X) + (X - a)^2 k'(X).$$

Consequently, $p'(a) = 0$. As F has characteristic 0, this shows that a is a root of a nonzero polynomial in $F[X]$ of lower degree than $p(X)$. However, this is impossible since, being irreducible, $p(X)$ is, up to a multiple by a nonzero element in F, equal to the minimal polynomial of a over F (see Theorem 7.8). $\square$

7.7.2 The Primitive Element Theorem

Let F be any field. Theorem 7.39 states that $F(a_1, \ldots, a_n)$ is a finite extension of F, provided that each a_i is algebraic over F. The converse is also true: every finite extension K of F is of the form $F(a_1, \ldots, a_n)$ for some algebraic elements a_i. Indeed, we can, for example, take the elements of any basis of K over F (they are indeed algebraic by Lemma 7.32). If F has characteristic 0, then, perhaps surprisingly, the elements $a_1, \ldots, a_n$ can always be replaced by a single element. This is the content of the next theorem, but let us first look at a simple example.

Example 7.126. We claim that

$$\mathbb{Q}(\sqrt{2}, \sqrt{3}) = \mathbb{Q}(\sqrt{2} + \sqrt{3}). \tag{7.19}$$

Denote $\sqrt{2} + \sqrt{3}$ by a. The claim will be proved by showing that $\sqrt{2}, \sqrt{3} \in \mathbb{Q}(a)$ and $a \in \mathbb{Q}(\sqrt{2}, \sqrt{3})$. The latter is obvious, so we only have to verify the former. It is enough to show that $\sqrt{2} \in \mathbb{Q}(a)$, since $\sqrt{3} \in \mathbb{Q}(a)$ then follows from $\sqrt{3} = a - \sqrt{2}$. Squaring $a - \sqrt{2} = \sqrt{3}$ we obtain $a^2 - 2a\sqrt{2} + 2 = 3$, and hence

$$\sqrt{2} = \frac{1}{2}(a - a^{-1}).$$

This already shows that $\sqrt{2} \in \mathbb{Q}(a)$, since $\mathbb{Q}(a)$ is a field and therefore contains a^{-1}; on the other hand, we can easily derive that

$$\sqrt{2} = -\frac{9}{2}a + \frac{1}{2}a^3,$$

and in this way express $\sqrt{2}$ as an element of $\mathbb{Q}(a)$ in the standard way, as in Theorem 7.35. We also mention that in light of $[\mathbb{Q}(\sqrt{2}, \sqrt{3}) : \mathbb{Q}] = 4$ (see Example 7.40), (7.19) implies that $\sqrt{2} + \sqrt{3}$ is algebraic of degree 4. From the discussion at the end of Section 7.2 it now follows that its minimal polynomial is $X^4 - 10X^2 + 1$.

Recall that an extension K of F is said to be *simple* if there exists an $a \in K$ such that $K = F(a)$, and that in this case we call a a *primitive element* of K.

Theorem 7.127. **(Primitive Element Theorem)** *Every finite extension of a field with characteristic 0 is simple.*

Proof. Let K be a finite extension of a field F with characteristic 0. As explained above, there are $a_1, \ldots, a_n \in K$ such that $K = F(a_1, \ldots, a_n)$. We want to show that $K = F(a)$ for some $a \in K$. Writing K as

$$F_1(a_{n-1}, a_n), \text{ where } F_1 = F(a_1, \ldots, a_{n-2}),$$

we see by an obvious inductive argument that the general case follows from the case where $n = 2$.

Assume, therefore, that $K = F(b, c)$, i.e., K is obtained by adjoining two elements b and c to F. Since K is a finite extension of F, b and c are algebraic over F. Let $p(X)$ and $q(X)$ be the minimal polynomials for b and c over F, respectively, and let K_1 be an extension of K over which both $p(X)$ and $q(X)$ split (e.g., the splitting field for $p(X)q(X)$ over K). Further, let $b = b_1, b_2, \ldots, b_r$ be the roots of $p(X)$ in K_1, and let $c = c_1, c_2, \ldots, c_s$ be the roots of $q(X)$ in K_1. Of course, F is infinite since its characteristic is 0. Therefore, there exists a $\lambda \in F$ such that

$$\lambda \neq (b_j - b)(c - c_k)^{-1} \text{ for all } j \text{ and all } k \neq 1.$$

We will show that
$$a := b + \lambda c$$

satisfies $K = F(a)$. It is clear that $a \in F(b, c) = K$, and so $F(a) \subseteq K$. To prove the converse inclusion, it is enough to show that $F(a)$ contains c, since then it also contains $b = a - \lambda c$.

We begin the proof that $c \in F(a)$ by introducing the polynomial

$$f(X) := p(a - \lambda X) \in F(a)[X].$$

Since $f(c) = p(b) = 0$ and c is also a root of $q(X) \in F[X] \subseteq F(a)[X]$, the minimal polynomial $\widetilde{q}(X)$ of c over $F(a)$ divides both $f(X)$ and $q(X)$ (see Theorem 7.8). Therefore, every root of $\widetilde{q}(X)$ is also a root of both $f(X)$ and $q(X)$. Besides $c = c_1$, the only roots of $q(X)$ are $c_2, \ldots, c_s$. However, none of them is a root of $f(X)$ since

$$f(c_k) = p(a - \lambda c_k) = p(b + \lambda(c - c_k))$$

and, by the choice of λ,
$$b + \lambda(c - c_k) \neq b_j$$

for all j and all $k \neq 1$. Consequently, $c = c_1$ is the only root of $\widetilde{q}(X)$ in K_1. Moreover, by Theorem 7.125 it is a simple root, since $\widetilde{q}(X)$, being a minimal polynomial, is irreducible. Accordingly, $\widetilde{q}(X) = X - c$. By the very definition, $\widetilde{q}(X)$ is a polynomial in $F(a)[X]$. Therefore, $c \in F(a)$. $\qquad\square$

7.7.3   Counting Isomorphisms

We now return to the setting of Subsection 7.5.3 in which we established the uniqueness of splitting fields. Recall from Remark 7.84 that every isomorphism between fields $\varphi : F \to F'$ can be extended to an isomorphism between rings $F[X]$ and $F'[X]$, denoted $f(X) \mapsto f_\varphi(X)$. Theorem 7.86 states that if K is the splitting field for $f(X)$ over F and K' is the splitting field for $f_\varphi(X)$ over F', then φ can be extended to an isomorphism from K to K'. Under the assumption that F has characteristic 0, we will now refine this theorem by determining the exact number of such isomorphisms. The proof will be similar, but more subtle.

Theorem 7.128. *Let F be a field with characteristic 0, let $f(X) \in F[X]$ be a nonconstant polynomial, and let K be the splitting field for $f(X)$ over F. If $\varphi : F \to F'$ is an isomorphism of fields and K' is the splitting field for $f_\varphi(X)$ over F', then there are exactly $[K : F]$ isomorphisms from K to K' extending φ.*

Proof. We proceed by induction on $n := [K : F]$. If $n = 1$, then $f(X)$ splits over F and hence $f_\varphi(X)$ splits over F', so $K = F$ and $K' = F'$; therefore, the only extension of φ is φ itself. Let $n > 1$. Then $f(X)$ does not split over F, and so it is divisible by an irreducible polynomial $p(X) \in F[X]$ of degree $m > 1$. Pick and fix a root $a \in K$ of $p(X)$. We claim that every isomorphism $\widehat{\varphi}$ from K to K' extending φ sends a to a root of $p_\varphi(X)$. Indeed, if $p(X) = \sum_{i=0}^{m} \lambda_i X^i$, then

$$p_\varphi(\widehat{\varphi}(a)) = \sum_{i=0}^{m} \varphi(\lambda_i)\widehat{\varphi}(a)^i = \sum_{i=0}^{m} \widehat{\varphi}(\lambda_i)\widehat{\varphi}(a)^i = \widehat{\varphi}(p(a)) = \widehat{\varphi}(0) = 0.$$

Note that $p_\varphi(X)$ is irreducible over F'. Therefore, by Theorem 7.125, $p_\varphi(X)$ has exactly m distinct roots $a_1', \dots, a_m'$ in K'. In the next paragraph, we will prove that, for each $a' \in \{a_1', \dots, a_m'\}$, there are exactly $\frac{n}{m}$ isomorphisms from K to K' that extend φ and send a to a'. This then implies that the number of isomorphisms from K to K' extending φ is equal to

$$m \cdot \frac{n}{m} = n = [K : F],$$

as desired.

We start the proof by invoking Lemma 7.85 which tells us that there is exactly one isomorphism Φ from $F(a)$ to $F'(a')$ extending φ and sending a to a'. Now, K is the splitting field for $f(X)$ over $F(a)$ and K' is the splitting field for $f_\varphi(X)$ over $F'(a')$ (see Remark 7.82). Since

$$[F(a) : F] = m > 1$$

and hence

$$[K : F(a)] = \frac{[K : F]}{[F(a) : F]} = \frac{n}{m} < n,$$

the induction hypothesis implies that there are exactly $\frac{n}{m}$ isomorphisms from K to K' that agree with Φ on $F(a)$. Of course, these are exactly the isomorphisms from K to K' that extend φ and send a to a'. The proof is thereby completed.   □

As in Subsection 7.5.3, we are primarily interested in the case where $F = F'$ and $\varphi = \mathrm{id}_F$. Theorem 7.128 for this special case is an integral part of Galois theory which will be treated in the next section (specifically, see Theorem 7.145).

We conclude the section by mentioning a few important notions, which, however, will not be studied in this book. A polynomial in $F[X]$ that has only simple roots in every extension of F is called a **separable polynomial**. An algebraic extension K of F with the property that the minimal polynomial of every element in K is separable is called a **separable extension**. It turns out that the Primitive Element Theorem holds for every finite separable extension (regardless of characteristic). Finally, a field F with the property that every finite extension of F is separable is called a **perfect field**. Theorem 7.125 implies that all fields of characteristic 0 are perfect. It can be proved that so are all finite fields. The field of rational functions $\mathbb{Z}_p(X)$ is an example of a non-perfect field.

The assumption that our fields have characteristic zero will also be used in the next section. We could instead assume certain separability conditions. However, this would make the exposition more cumbersome.

Exercises

7.129. Determine all $\lambda \in \mathbb{Q}$ such that $\mathbb{Q}(\sqrt{2}, \sqrt{3}) = \mathbb{Q}(\sqrt{2} + \lambda\sqrt{3})$.
Hint. Either look at the proof of Theorem 7.127 or argue as in Example 7.126.

7.130. Prove that $[\mathbb{Q}(\sqrt{3}, i) : \mathbb{Q}] = 4$, and find an $a \in \mathbb{C}$ such that $\mathbb{Q}(a) = \mathbb{Q}(\sqrt{3}, i)$.

7.131. Determine $[\mathbb{Q}(\lambda\sqrt{3} + \mu\sqrt[3]{5}) : \mathbb{Q}]$ for all $\lambda, \mu \in \mathbb{Q}$.

7.132. Determine $[\mathbb{Q}(\sqrt{2} + \sqrt{3} + \sqrt{5}) : \mathbb{Q}]$.

7.133. Let F be a field of characteristic 0 and let K be an extension of F. Prove that $a \in K$ is a root of multiplicity k of $f(X) \in F[X]$ if and only if

$$f(a) = f'(a) = \cdots = f^{(k-1)}(a) = 0 \quad \text{and} \quad f^{(k)}(a) \neq 0.$$

Here, $f^{(i)}(X)$ is the ith derivative of $f(X)$.

7.134. Let F be any field and let $f(X)$ be a non-constant polynomial in $F[X]$. Prove that the polynomials $f(X)$ and $f'(X)$ are relatively prime if and only if every root of $f(X)$ (in any extension K of F) is simple.

7.135. Prove that the polynomial $f(Y) = Y^2 - X \in \mathbb{Z}_2(X)[Y]$ is irreducible over $\mathbb{Z}_2(X)$ and has a root of multiplicity 2 in its splitting field.

7.8 Galois Theory

We now have enough knowledge to get acquainted with one of the gems of mathematics, Galois theory.

7.8.1 Galois Theory Through Examples

We begin with an informal discussion, describing the main points of Galois theory and illustrating them by examples.

In short, Galois theory connects field theory and group theory. This connection is not only beautiful and interesting in its own right, but extremely useful, if not decisive, for solving some difficult problems in field theory. Indeed, Galois theory makes it possible to translate these problems into corresponding problems in group theory, where they become more approachable and powerful group-theoretic tools are available.

We first describe the framework in which we will work. Let F be a field with characteristic 0 and let K be its extension. We are interested in all extensions of F that are contained in K, i.e., in all fields L such that

$$F \subseteq L \subseteq K.$$

We call them the **intermediate fields**. Denote the set of all intermediate fields by $\mathfrak{F}$.

Now we turn to groups. Let F and K be as above. By $\mathrm{Aut}(K)$ we denote the set of all automorphisms of the field K. Thus, $\mathrm{Aut}(K)$ consists of all bijective maps $\sigma : K \to K$ that satisfy $\sigma(x + y) = \sigma(x) + \sigma(y)$ and $\sigma(xy) = \sigma(x)\sigma(y)$ for all $x, y \in K$. We write 1 for the identity automorphism, and $\rho\sigma$ for the composition $\rho \circ \sigma$ of automorphisms ρ and σ. Recall from Corollary 3.50 that $\mathrm{Aut}(K)$ is a group under the operation of composition. Our main object of study will be the subgroup G of $\mathrm{Aut}(K)$ that consists of all automorphisms σ leaving elements of F fixed, i.e., satisfying $\sigma(\lambda) = \lambda$ for all $\lambda \in F$. One immediately checks that G is indeed a subgroup. Later we will denote it by $\mathrm{Aut}(K/F)$ or $\mathrm{Gal}(K/F)$, but let us stick with G for now. Denote the set of all subgroups of G by $\mathfrak{G}$.

Galois theory describes the connection between the sets $\mathfrak{F}$ and $\mathfrak{G}$.

Example 7.136. Let $F = \mathbb{R}$ and $K = \mathbb{C}$. If L is an intermediate field different from $\mathbb{R}$, then L contains a number $z = \alpha + \beta i$ with $\beta \neq 0$, and hence it also contains $\beta^{-1}(z - \alpha) = i$. Therefore, $L = \mathbb{C}$, and so $\mathfrak{F} = \{\mathbb{R}, \mathbb{C}\}$. Let us describe $\mathfrak{G}$. Besides the identity 1, G obviously also contains the automorphism σ given by $\sigma(z) = \overline{z}$ where $\overline{z}$ is the complex conjugate of z. Now take an arbitrary ρ in G. Since ρ fixes elements of $\mathbb{R}$, we have $\rho(\lambda + \mu i) = \lambda + \mu\rho(i)$ for all $\lambda, \mu \in R$. From $i^2 = -1$ it follows that $\rho(i)^2 = -1$, and so $\rho(i) = i$ or $\rho(i) = -i$. Consequently, $\rho = 1$ or $\rho = \sigma$. Hence, $G = \{1, \sigma\}$, and so $\mathfrak{G} = \{\{1\}, \{1, \sigma\}\}$.

This example was too simple to make any general conjecture. Still, we notice that both $\mathfrak{F}$ and $\mathfrak{G}$ have two elements. This is no coincidence. One of the byproducts of Galois theory is that $\mathfrak{F}$ and $\mathfrak{G}$ always have the same number of elements, as long as K satisfies any of the following three conditions, which turn out to be equivalent (see Theorem 7.145):

- K is the splitting field for some polynomial $f(X) \in F[X]$.
- Any irreducible polynomial in $F[X]$ with a root in K splits over K.
- $|G| = [K : F]$.

When these conditions are met, we say that K is a *Galois extension* of F. Certainly, $\mathbb{C}$ is a Galois extension of $\mathbb{R}$ since it is the splitting field for $X^2 + 1$ over $\mathbb{R}$.

Example 7.137. The field $K = \mathbb{Q}(\sqrt{2})$ is a Galois extension of $F = \mathbb{Q}$ since it is the splitting field for $X^2 - 2$ over $\mathbb{Q}$. A straightforward modification of the discussion in Example 7.136 shows that $\mathfrak{F} = \{\mathbb{Q}, \mathbb{Q}(\sqrt{2})\}$, $G = \{1, \sigma\}$ where σ is given by $\sigma(\lambda + \mu\sqrt{2}) = \lambda - \mu\sqrt{2}$, and so $\mathfrak{G} = \{\{1\}, \{1, \sigma\}\}$.

Example 7.138. Let $F = \mathbb{Q}$ and $K = \mathbb{Q}(\sqrt[3]{2})$. If $\sigma \in G$, then $\sigma(\sqrt[3]{2})^3 = \sigma(2) = 2$. Now, K is contained in $\mathbb{R}$ and the only real solution of the equation $x^3 = 2$ is $x = \sqrt[3]{2}$. Therefore, σ fixes $\sqrt[3]{2}$, which readily implies that $\sigma = 1$. Hence, $G = \{1\}$ and so $\mathfrak{G}$ consists of $\{1\}$. On the other hand, $\mathfrak{F}$ contains two elements, F and K (and no other element, as can be checked). Thus, $\mathfrak{G}$ and $\mathfrak{F}$ have a different cardinality. This does not contradict what we said above since K is not a Galois extension of $\mathbb{Q}$. Indeed, $|G| \neq [K : F]$. Also, the polynomial $X^3 - 2$ is irreducible over $\mathbb{Q}$ and has a root in K, but does not split over K (its other two roots are complex numbers).

Assume now that K is a Galois extension of F. As we will explain in Remark 7.142, it is easy to see that G is a finite group and hence $\mathfrak{G}$ is a finite set. On the other hand, it does not seem obvious that $\mathfrak{F}$ is finite. Why are there not infinitely many intermediate fields? More specifically, if we choose an infinite sequence $a_1, a_2, \ldots$ of elements in K such that $Fa_1, Fa_2, \ldots$ are distinct 1-dimensional spaces, why it is not possible that the fields $F(a_1), F(a_2), \ldots$ are also distinct? The fact that Galois theory implies not only that $\mathfrak{F}$ is finite, but has the same cardinality as $\mathfrak{G}$, is therefore interesting and somewhat surprising. But this is not yet the essence. The theory tells us that there is a particularly nice bijective correspondence between $\mathfrak{G}$ and $\mathfrak{F}$, defined as follows: to each $H \in \mathfrak{G}$ we associate

$$K^H := \{x \in K \mid \sigma(x) = x \text{ for all } \sigma \in H\},$$

which is readily seen to be an intermediate field (i.e., $K^H \in \mathfrak{F}$). The Fundamental Theorem of Galois Theory (Theorem 7.148) in particular states the following:

- $H \mapsto K^H$ is a bijective map from $\mathfrak{G}$ to $\mathfrak{F}$.
- $H \subseteq H'$ if and only if $K^H \supseteq K^{H'}$.
- $|H| = [K : K^H]$.

Example 7.139. Let $F = \mathbb{Q}$ and $K = \mathbb{Q}(\sqrt{2}, \sqrt{3})$. Note that K is the splitting field for the polynomial $(X^2 - 2)(X^2 - 3)$ over $\mathbb{Q}$, and so it is a Galois extension of F. An automorphism in G is determined by its values on $\sqrt{2}$ and $\sqrt{3}$. The same argument as in the preceding examples shows that it can send $\sqrt{2}$ either to $\sqrt{2}$ or to $-\sqrt{2}$, and $\sqrt{3}$ either to $\sqrt{3}$ or to $-\sqrt{3}$. Thus, G consists of at most four automorphisms, determined by:

(a) $\sqrt{2} \mapsto \sqrt{2}, \sqrt{3} \mapsto \sqrt{3}$,
(b) $\sqrt{2} \mapsto \sqrt{2}, \sqrt{3} \mapsto -\sqrt{3}$,
(c) $\sqrt{2} \mapsto -\sqrt{2}, \sqrt{3} \mapsto \sqrt{3}$,
(d) $\sqrt{2} \mapsto -\sqrt{2}, \sqrt{3} \mapsto -\sqrt{3}$.

It is easy to check that each of these four possibilities indeed gives rise to an automorphism. On the other hand, this also follows from the theory. Indeed, we know that $[K : F] = 4$ (see Example 7.40), and since K is a Galois extension it follows that $|G| = 4$. Therefore, each of the four possibilities (a), (b), (c), (d) must induce an automorphism. The automorphism corresponding to (a) is the identity 1. Let σ denote the automorphism determined by (b). Explicitly, σ is defined by

$$\sigma\left(\lambda_0 + \lambda_1\sqrt{2} + \lambda_2\sqrt{3} + \lambda_3\sqrt{6}\right) = \lambda_0 + \lambda_1\sqrt{2} - \lambda_2\sqrt{3} - \lambda_3\sqrt{6}$$

for all $\lambda_i \in \mathbb{Q}$. Next, let ρ denote the automorphism determined by (c), i.e.,

$$\rho\left(\lambda_0 + \lambda_1\sqrt{2} + \lambda_2\sqrt{3} + \lambda_3\sqrt{6}\right) = \lambda_0 - \lambda_1\sqrt{2} + \lambda_2\sqrt{3} - \lambda_3\sqrt{6},$$

and note that the automorphism determined by (d) is $\rho\sigma$ ($= \sigma\rho$). Note that

$$G = \{1, \sigma, \rho, \rho\sigma\} \cong \mathbb{Z}_2 \oplus \mathbb{Z}_2.$$

The diagram representing the set of subgroups $\mathfrak{G}$ is given in Figure 7.6.

According to the theory, the only intermediate fields are

$$K^{\{1\}} = \mathbb{Q}(\sqrt{2}, \sqrt{3}), \quad K^{\{1,\sigma\}} = \mathbb{Q}(\sqrt{2}), \quad \text{etc.}$$

The diagram representing $\mathfrak{F}$, in which each field K^H occupies the same position as the group H in the diagram representing $\mathfrak{G}$, is given in Figure 7.7. Observe that if H is any of the three proper nontrivial subgroups of G, then $|H| = 2 = [K : K^H]$.

There is an important part of Galois theory that has not been mentioned so far. Under the usual assumption that K is a Galois extension of F, the following holds:

- H is a normal subgroup of G if and only if K^H is a Galois extension of F. In this case, G/H is isomorphic to the group of all automorphisms of K^H that leave elements of F fixed.

Example 7.140. Write ω for the third root of unity $e^{\frac{2\pi i}{3}}$, and consider the case where $F = \mathbb{Q}$ and $K = \mathbb{Q}(\sqrt[3]{2}, \omega)$. It is an easy exercise to show that K is the splitting field for the polynomial $X^3 - 2$ over F (so it is a Galois extension) and that $[K : F] = 6$. Consequently, $|G| = 6$. From $\omega^2 + \omega + 1 = 0$ it follows that an automorphism in G can send ω either to ω or to ω^2, since these are the only solutions of the equation

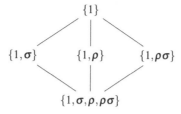

Fig. 7.6 Diagram of $\mathfrak{G}$ for $F = \mathbb{Q}$ and $K = \mathbb{Q}(\sqrt{2}, \sqrt{3})$

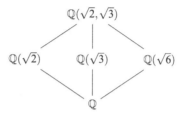

Fig. 7.7 Diagram of $\mathfrak{F}$ for $F = \mathbb{Q}$ and $K = \mathbb{Q}(\sqrt{2}, \sqrt{3})$

$x^2 + x + 1 = 0$. Similarly, it can send $\sqrt[3]{2}$ only to $\sqrt[3]{2}$, $\sqrt[3]{2}\omega$, or $\sqrt[3]{2}\omega^2$. Thus, the following conditions determine the six elements of G:

(a) $\omega \mapsto \omega$, $\sqrt[3]{2} \mapsto \sqrt[3]{2}$,
(b) $\omega \mapsto \omega$, $\sqrt[3]{2} \mapsto \sqrt[3]{2}\omega$,
(c) $\omega \mapsto \omega$, $\sqrt[3]{2} \mapsto \sqrt[3]{2}\omega^2$,
(d) $\omega \mapsto \omega^2$, $\sqrt[3]{2} \mapsto \sqrt[3]{2}$,
(e) $\omega \mapsto \omega^2$, $\sqrt[3]{2} \mapsto \sqrt[3]{2}\omega$,
(f) $\omega \mapsto \omega^2$, $\sqrt[3]{2} \mapsto \sqrt[3]{2}\omega^2$.

Denote the automorphism corresponding to (b) by σ. Using $\omega^2 = -\omega - 1$, we can write down σ explicitly as follows:

$$\sigma(\lambda_0 + \lambda_1 \sqrt[3]{2} + \lambda_2 \sqrt[3]{4} + \lambda_3 \omega + \lambda_4 \sqrt[3]{2}\omega + \lambda_5 \sqrt[3]{4}\omega)$$
$$= \lambda_0 - \lambda_4 \sqrt[3]{2} + (\lambda_5 - \lambda_2)\sqrt[3]{4} + \lambda_3 \omega + (\lambda_1 - \lambda_4)\sqrt[3]{2}\omega - \lambda_2 \sqrt[3]{4}\omega$$

for all $\lambda_i \in \mathbb{Q}$. Further, let ρ denote the automorphism corresponding to (d). The automorphisms corresponding to (c), (e), and (f) are then σ^2, $\rho\sigma^2$, and $\rho\sigma$, respectively. Note that $\sigma\rho = \rho\sigma^2$ and so $\rho\sigma \neq \sigma\rho$. Hence

$$G = \{1, \sigma, \sigma^2, \rho, \rho\sigma, \rho\sigma^2\} \cong S_3,$$

since S_3 is, up to isomorphism, the only non-Abelian group of order 6. The set of subgroups $\mathfrak{G}$ is represented by the diagram in Figure 7.8.

We can now list all intermediate fields K^H. For example, $K^{\{1,\rho\sigma\}} = \mathbb{Q}(\sqrt[3]{2}\omega)$, $K^{\{1,\sigma,\sigma^2\}} = \mathbb{Q}(\omega)$, etc. The diagram representing $\mathfrak{F}$ that corresponds to the above diagram representing $\mathfrak{G}$ is given in Figure 7.9.

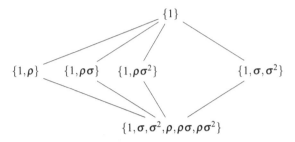

Fig. 7.8 Diagram of $\mathfrak{G}$ for $F = \mathbb{Q}$ and $\mathbb{Q}(\sqrt[3]{2}, \omega)$

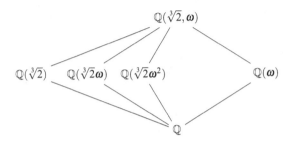

Fig. 7.9 Diagram of $\mathfrak{F}$ for $F = \mathbb{Q}$ and $\mathbb{Q}(\sqrt[3]{2}, \omega)$

If H is any of the three subgroups of order 2, then $|H| = 2 = [K : K^H]$. None of these subgroups is normal, and the corresponding fields $\mathbb{Q}(\sqrt[3]{2})$, $\mathbb{Q}(\sqrt[3]{2}\omega)$, and $\mathbb{Q}(\sqrt[3]{2}\omega^2)$ are not Galois extensions of $\mathbb{Q}$ (compare Example 7.138). Let now $H = \{1, \sigma, \sigma^2\}$. Then $K^H = \mathbb{Q}(\omega)$ and $|H| = 3 = [K : K^H]$. Further, K^H is the splitting field for the polynomial $X^2 + X + 1$ over $\mathbb{Q}$, so it is a Galois extension of $\mathbb{Q}$. One easily checks that the corresponding group of automorphisms has two elements. Note that H is a normal subgroup of G and that G/H is also a group with two elements.

7.8.2   The Fundamental Theorem of Galois Theory

We proceed to the formal treatment. To make this subsection self-contained, we repeat some definitions given above.

Let K be an extension field of a field F. By $\mathrm{Aut}(K)$ we denote the group of all automorphisms of K. Note that

$$\mathrm{Aut}(K/F) := \{\sigma \in \mathrm{Aut}(K) \,|\, \sigma(\lambda) = \lambda \text{ for all } \lambda \in F\}$$

is a subgroup of $\mathrm{Aut}(K)$. The following simple lemma was indirectly but repeatedly used in the above examples.

Lemma 7.141. *Let $\sigma \in \mathrm{Aut}(K/F)$. If $a \in K$ is a root of $f(X) \in F[X]$, then so is $\sigma(a)$.*

Proof. Let $f(X) = \sum_i \lambda_i X^i$ where $\lambda_i \in F$. By assumption, $\sum_i \lambda_i a^i = 0$. Applying σ we obtain $\sum_i \lambda_i \sigma(a)^i = 0$, so $\sigma(a)$ is a root of $f(X)$. □

Remark 7.142. Let K be a finite extension of F. We are interested only in the case of characteristic 0, and so, in view of the Primitive Element Theorem, we may assume that $K = F(a)$ for some algebraic element a. Let $p(X)$ be the minimal polynomial of a over F. Obviously, every $\sigma \in \mathrm{Aut}(K/F)$ is completely determined by its action on a. Since $\sigma(a)$ is a root of $p(X)$ by Lemma 7.141, $|\mathrm{Aut}(K/F)|$ cannot exceed the number of roots of $p(X)$ in K; in particular, $\mathrm{Aut}(K/F)$ is a finite group. On the other hand, Lemma 7.85, applied to the case where $F = F'$ and $\varphi = \mathrm{id}_F$, shows that for every root a' of $p(X)$ in K there exists a unique $\sigma \in \mathrm{Aut}(K/F)$ satisfying $\sigma(a) = a'$. Therefore,

$$|\mathrm{Aut}(K/F)| = \text{ the number of roots of } p(X) \text{ in } K.$$

Denote these roots by $a = a_1, a_2, \ldots, a_n$ (where, of course, $1 \leq n \leq \deg(p(X))$). By Lemma 7.141, each $\sigma \in \mathrm{Aut}(K/F)$ maps a_i into another root a_j. We can therefore consider the elements in $\mathrm{Aut}(K/F)$ as permutations of the set of roots of a polynomial (more formally, we can embed $\mathrm{Aut}(K/F)$ into the symmetric group S_n). This observation is important for understanding the philosophy of Galois theory, although we will not use it explicitly in this section. Incidentally, the concept of an automorphism did not yet exist in Galois' time. Galois himself considered only permutation groups.

In the next two lemmas, we assume that K is a finite extension of a field F with characteristic 0, and that H is a subgroup of $\mathrm{Aut}(K/F)$. As just noticed, $\mathrm{Aut}(K/F)$ is a finite group (the statements of the lemmas therefore make sense). We call

$$K^H := \{x \in K \mid \sigma(x) = x \text{ for all } \sigma \in H\}$$

the **fixed field** of H. Observe that K^H indeed is a field; more precisely, it is an *intermediate field*, i.e., a subfield of K containing F (incidentally, in the proof of Lemma 7.109 we have already encountered a field defined in this way).

Lemma 7.143. *Let $a \in K$ and let $a = a_1, a_2, \ldots, a_m$ be all the distinct elements of the set $\{\sigma(a) \mid \sigma \in H\}$. Then $p(X) = (X - a_1) \cdots (X - a_m)$ is the minimal polynomial of a over K^H.*

Proof. We must first show that the coefficients of $p(X)$ lie in K^H. Recall that every automorphism ρ of the field K induces the automorphism $f(X) \mapsto f_\rho(X)$ of the ring $K[X]$ (see Remark 7.84). Writing $p(X) = \sum_{i=0}^m \alpha_i X^i$ with $\alpha_i \in K$ (and $\alpha_m = 1$), we have $p_\rho(X) = \sum_{i=0}^m \rho(\alpha_i) X^i$. On the other hand, from the definition of $p(X)$ it follows that

$$p_\rho(X) = (X - \rho(a_1)) \cdots (X - \rho(a_m)).$$

Each a_i can be written as $\sigma_i(a)$ for some $\sigma_i \in H$. Hence, if $\rho \in H$ then

$$\rho(a_i) = (\rho\sigma_i)(a) \in \{\sigma(a) \mid \sigma \in H\} = \{a_1, \ldots, a_m\}.$$

Moreover, ρ is injective and so $\rho(a_i) \neq \rho(a_j)$ if $i \neq j$. Therefore,

$$\{\rho(a_1), \ldots, \rho(a_m)\} = \{a_1, \ldots, a_m\},$$

which yields $p_\rho(X) = p(X)$ for every $\rho \in H$. Consequently, $\rho(\alpha_i) = \alpha_i$ for every i and every $\rho \in H$, showing that $\alpha_i \in K^H$. That is, $p(X)$ is a polynomial in $K^H[X]$.

Now take any $f(X) \in K^H[X]$ having a as a root. By Lemma 7.141, $\sigma(a)$ is also a root of $f(X)$ for every $\sigma \in H$. This implies that $p(X)$ divides $f(X)$, and so $p(X)$ is the minimal polynomial of a over K^H. $\qquad\square$

Lemma 7.144. $|H| = [K : K^H]$ and $[K : F] = |H| \cdot [K^H : F]$.

Proof. By the Primitive Element Theorem, there is an $a \in K$ such that $K = F(a)$. Since $F \subseteq K^H \subseteq K$, we also have $K = K^H(a)$. Hence, $[K : K^H]$ is equal to the degree m of the minimal polynomial of a over K^H. By Lemma 7.143, m is the number of distinct elements in $\{\sigma(a) \mid \sigma \in H\}$. We claim that $\sigma(a) \neq \rho(a)$ whenever σ and ρ are different elements of H, so that $m = |H|$. Indeed, $\sigma \neq \rho$ means that ρ and σ do not coincide on $K = K^H(a)$, but since they coincide on K^H, $\sigma(a)$ and $\rho(a)$ must be different. We have thus proved that $|H| = [K : K^H]$. The second formula follows from Theorem 7.29. $\qquad\square$

In the following theorem, we consider the situation where H is the whole group $\mathrm{Aut}(K/F)$ and the corresponding fixed field coincides with F (this is not always the case, see Example 7.138).

Theorem 7.145. *Let K be a finite extension of a field F with characteristic 0. The following conditions are equivalent:*

(i) $|\mathrm{Aut}(K/F)| = [K : F]$.
(ii) $K^{\mathrm{Aut}(K/F)} = F$.
(iii) *Any irreducible polynomial in $F[X]$ with a root in K splits over K.*
(iv) *K is the splitting field for some irreducible polynomial in $F[X]$.*
(v) *K is the splitting field for some polynomial in $F[X]$.*

Proof. Write G for $\mathrm{Aut}(K/F)$.

(i) $\Longrightarrow$ (ii). The second formula of Lemma 7.144 shows that $|G| = [K : F]$ implies $[K^G : F] = 1$, i.e., $K^G = F$.

(ii) $\Longrightarrow$ (iii). Take an irreducible polynomial $p(X) \in F[X]$ that has a root a in K. We may assume that $p(X)$ is monic, so that it is the minimal polynomial of a over F. By Lemma 7.143, $K^G = F$ implies that $p(X)$ is a product of linear polynomials $X - a_i$ with $a_i \in K$. This means that $p(X)$ splits over K.

(iii) $\Longrightarrow$ (iv). By the Primitive Element Theorem, there is an $a \in K$ such that $K = F(a)$. By assumption, the minimal polynomial $p(X)$ of a over F splits over K. Hence, K is the splitting field for $p(X)$.

(iv) $\Longrightarrow$ (v). Trivial.

(v) $\Longrightarrow$ (i). Use Theorem 7.128 for the case where $F = F'$ and $\varphi = \mathrm{id}_F$. $\qquad\square$

Definition 7.146. A finite extension K of a field F with characteristic 0 is called a **Galois extension** if it satisfies one (and hence all) of the conditions (i)–(v) of Theorem 7.145. In this case, we call the group $\mathrm{Aut}(K/F)$ the **Galois group of K over F**, and denote it by $\mathrm{Gal}(K/F)$. If K is the splitting field for the polynomial $f(X) \in F[X]$ over F, then we call $\mathrm{Gal}(K/F)$ the **Galois group of $f(X)$ over F**.

Remark 7.147. If K is a Galois extension of F and L is an intermediate field, i.e., $F \subseteq L \subseteq K$, then K is also a Galois extension of L. Indeed, if K is the splitting field for $f(X)$ over F, then, as pointed out in Remark 7.82, it is also the splitting field for $f(X)$ over L. On the other hand, L is not necessarily a Galois extension of F.

We now have enough information to prove our central theorem.

Theorem 7.148. (Fundamental Theorem of Galois Theory) *Let K be a Galois extension of a field F with characteristic 0. Denote by $\mathfrak{F}$ the set of all intermediate fields between F and K, and by $\mathfrak{G}$ the set of all subgroups of $G := \mathrm{Gal}(K/F)$.*

(a) *The map*
$$\alpha : \mathfrak{G} \to \mathfrak{F}, \quad \alpha(H) = K^H,$$

is bijective and its inverse is the map

$$\beta : \mathfrak{F} \to \mathfrak{G}, \quad \beta(L) = \mathrm{Gal}(K/L).$$

(b) *If H corresponds to L—that is, $H = \mathrm{Gal}(K/L)$, or, equivalently, $L = K^H$—then*

$$|H| = [K : L] \quad and \quad [G : H] = [L : F].$$

(c) *If H and H' correspond to L and L', respectively, then $H \subseteq H'$ if and only if $L \supseteq L'$.*

(d) *If H corresponds to L, then H is a normal subgroup of G if and only if L is a Galois extension of F. In this case, $G/H \cong \mathrm{Gal}(L/F)$.*

Proof. (a) We have to prove that $\alpha \circ \beta = \mathrm{id}_{\mathfrak{F}}$ and $\beta \circ \alpha = \mathrm{id}_{\mathfrak{G}}$, i.e., $\alpha(\beta(L)) = L$ for every $L \in \mathfrak{F}$ and $\beta(\alpha(H)) = H$ for every $H \in \mathfrak{G}$; or, more explicitly,

$$K^{\mathrm{Gal}(K/L)} = L \quad and \quad \mathrm{Gal}(K/K^H) = H.$$

Since K is a Galois extension of L (Remark 7.147), the first equality follows from Theorem 7.145 (condition (ii)). Now take $H \in \mathfrak{G}$. It is clear that $H \subseteq \mathrm{Gal}(K/K^H)$. Therefore, in order to prove that these two sets are equal, it is enough to show that they have the same cardinality. This follows from Lemma 7.144, which tells us that $|H| = [K : K^H]$, and Theorem 7.145 (condition (i)), which tells us that $|\mathrm{Gal}(K/K^H)| = [K : K^H]$.

(b) As just mentioned, $|H| = [K : L]$ holds by Lemma 7.144. The second equality follows from this one. Indeed,

$$[G : H] = \frac{|G|}{|H|} = \frac{[K : F]}{[K : L]} = [L : F].$$

(c) We have $L = K^H$ and $L' = K^{H'}$. Thus, $x \in L$ (resp. $x \in L'$) if and only if $\rho(x) = x$ for all $\rho \in H$ (resp. $\rho \in H'$). Hence, $H \subseteq H'$ implies $L \supseteq L'$. Similarly, from $H = \text{Gal}(K/L)$ and $H' = \text{Gal}(K/L')$ we see that $L \supseteq L'$ implies $H \subseteq H'$.

(d) Since $H = \text{Gal}(K/L)$ consists of those elements $\rho \in G$ that leave elements of L fixed, and $L = K^H$ consists of those elements ℓ in K that are fixed by every $\rho \in H$, the following holds:

$$\begin{aligned} H \lhd G &\iff \sigma^{-1}\rho\sigma \in H \quad \text{for all } \rho \in H, \sigma \in G \\ &\iff (\sigma^{-1}\rho\sigma)(\ell) = \ell \quad \text{for all } \rho \in H, \sigma \in G, \ell \in L \\ &\iff \rho(\sigma(\ell)) = \sigma(\ell) \quad \text{for all } \rho \in H, \sigma \in G, \ell \in L \\ &\iff \sigma(\ell) \in L \quad \text{for all } \sigma \in G, \ell \in L. \end{aligned}$$

We have thus proved that

$$H \lhd G \iff \sigma(L) \subseteq L \text{ for every } \sigma \in G. \tag{7.20}$$

The condition that $\sigma(L) \subseteq L$ is actually equivalent to $\sigma(L) = L$. This is because L is a finite-dimensional vector space over F and σ is linear (as $\sigma(\lambda) = \lambda$ for every $\lambda \in F$), so L and $\sigma(L)$ are of the same (finite) dimension over F. The only subspace of L of the same dimension, however, is L itself.

Let us now prove the statement of the theorem. Assume that $H \lhd G$; equivalently, $\sigma(L) = L$ for every $\sigma \in G$. We may therefore regard $\sigma|_L$, the restriction of σ to L, as an element of $\text{Aut}(L/F)$. The map

$$\varphi : G \to \text{Aut}(L/F), \quad \varphi(\sigma) = \sigma|_L,$$

is clearly a group homomorphism. Its kernel consists of all $\sigma \in G$ that act as the identity on L, which are exactly the elements of $\text{Gal}(K/L) = H$. By the Isomorphism Theorem,

$$G/H \cong \text{im}\,\varphi \le \text{Aut}(L/F). \tag{7.21}$$

Hence, $|G/H| \le |\text{Aut}(L/F)|$. On the other hand, applying the second formula of Lemma 7.144 to the case where K is L, we see that $|\text{Aut}(L/F)|$ divides $[L : F]$, which is, by (b), equal to $[G : H] = |G/H|$. Consequently,

$$|\text{Aut}(L/F)| = [L : F] = |G/H|. \tag{7.22}$$

Theorem 7.145 (condition (i)) now shows that L is a Galois extension of F. Thus, we may write $\text{Gal}(L/F)$ for $\text{Aut}(L/F)$. From (7.21) and (7.22) we also infer that φ is surjective and so $G/H \cong \text{Gal}(L/F)$.

Finally, let L be a Galois extension of F. By Theorem 7.145 (condition (v)), L is the splitting field for a polynomial $f(X) \in F[X]$, so $L = F(a_1, \ldots, a_m)$ where $a_1, \ldots, a_m$ are the roots of $f(X)$. By Lemma 7.141, every $\sigma \in G$ sends each root a_i into another root a_j, showing that $\sigma(L) \subseteq L$. According to (7.20), H is a normal subgroup of G. □

Exercises

7.149. Let K be an extension of $\mathbb{Q}$. Prove that every $\sigma \in \mathrm{Aut}(K)$ satisfies $\sigma(\lambda) = \lambda$ for every $\lambda \in \mathbb{Q}$.

7.150. Prove that $\mathrm{Aut}(\mathbb{R}) = \{1\}$.

Sketch of proof. First show that every $\sigma \in \mathrm{Aut}(\mathbb{R})$ maps positive numbers to positive numbers (hint: $x = y^2$ implies $\sigma(x) = \sigma(y)^2$), hence derive that σ is a strictly increasing function, and then conclude that $\sigma = 1$ by using the preceding exercise.

7.151. Prove that the only continuous automorphisms of $\mathbb{C}$ are 1 and complex conjugation.

Comment. There actually exist discontinuous automorphisms of $\mathbb{C}$.

7.152. Let $[K : F] = 2$. Prove that K is a Galois extension of F.

Comment. We have developed the theory only for fields of characteristic 0, so we are tacitly assuming that this restriction holds for all fields in these exercises.

7.153. Let $F = \mathbb{Q}$ and $K = \mathbb{Q}(\sqrt[4]{2})$. Describe $\mathrm{Aut}(K/F)$. Is K a Galois extension of F?

7.154. Suppose that L is a Galois extension of F and K is a Galois extension of L. Is K necessarily a Galois extension of F?

7.155. Prove that the Galois group of a polynomial of degree n embeds into S_n and hence its order divides $n!$.

7.156. Let K be a Galois extension of F and let L be an intermediate subfield. Denote by H its corresponding subgroup. Prove that for every $\sigma \in G$, $\sigma(L)$ is an intermediate subfield whose corresponding subgroup is $\sigma H \sigma^{-1}$.

7.157. Let K be a finite extension of F. Prove that K is contained in a Galois extension of F, and hence deduce that there are only finitely many intermediate fields between F and K.

7.158. We say that K is a **biquadratic extension** of F if $[K : F] = 4$ and $K = F(a, b)$ where a, b are such that $a^2 \in F$ and $b^2 \in F$. Prove that K is a Galois extension of F and determine $\mathfrak{G}$ and $\mathfrak{F}$.

7.159. Determine the intermediate fields between $\mathbb{Q}$ and $\mathbb{Q}(\sqrt{2}, \sqrt{3})$ directly, without using Galois theory.

7.160. In Example 7.126, we mentioned that $X^4 - 10X^2 + 1$ is the minimal polynomial of $\sqrt{2} + \sqrt{3}$ over $\mathbb{Q}$. Show that this can also be deduced from Lemma 7.143.

7.161. Determine the minimal polynomial of $\sqrt{2} + i$ over $\mathbb{Q}$.

7.162. Explain why there exists an irreducible polynomial of degree 6 in $\mathbb{Q}[X]$ whose Galois group over $\mathbb{Q}$ is isomorphic to S_3.

Hint. Example 7.140.

7.163. Let $f(X) \in \mathbb{Q}[X]$ have degree 4. Suppose that the order of the Galois group of $f(X)$ over $\mathbb{Q}$ is 6. Prove that $f(X)$ has a root in $\mathbb{Q}$.

7.164. Let $n \geq 2$. Note that $\mathbb{Q}(e^{\frac{2\pi i}{n}})$ is the splitting field for the polynomial $X^n - 1$ over $\mathbb{Q}$. We call it the **cyclotomic field of nth roots of unity**. Prove that the Galois group $\mathrm{Gal}\big(\mathbb{Q}(e^{\frac{2\pi i}{n}})/\mathbb{Q}\big)$ is isomorphic to $\mathbb{Z}_n^*$, the group of invertible elements of the ring $\mathbb{Z}_n$.

Hint. You may use what was stated in Example 5.11 without proof.

7.165. Let p be a prime. What is the order of the Galois group of $X^p - 2$ over $\mathbb{Q}$? Determine its elements by their action on some set of generators.

7.166. Determine $\mathfrak{G}$ and $\mathfrak{F}$ and represent them by diagrams in the case where:

(a) $F = \mathbb{Q}(\sqrt{2})$ and $K = \mathbb{Q}(\sqrt{2}, \sqrt{3}, i)$.
(b) $F = \mathbb{Q}$ and $K = \mathbb{Q}(\sqrt{2}, \sqrt{3}, i)$.
(c) $F = \mathbb{Q}(i)$ and $K = \mathbb{Q}(\sqrt[4]{2}, i)$.
(d) $F = \mathbb{Q}$ and $K = \mathbb{Q}(\sqrt[4]{2}, i)$.

In each of these cases, K is a Galois extension of F. To which familiar group is $\mathrm{Gal}(K/F)$ isomorphic?

7.9   Solvability of Polynomial Equations by Radicals

In Section 7.1, we wrote that while there exist formulas for roots of polynomial equations of degree at most 4, there can be no such formula for polynomial equations of degree 5 or more. Armed with the knowledge of Galois theory, we are now in a position to prove the latter.

Let us be precise. By "formula" we mean an expression involving sums, differences, products, quotients, and nth roots of the coefficients of the polynomial. As we will soon see, this is connected with the following definition.

Definition 7.167. Let F be a field. A polynomial $f(X) \in F[X]$ is said to be **solvable by radicals over** F if there exist elements $a_1, \ldots, a_m$ in some extension of F such that:

(a) $f(X)$ splits over $F(a_1, \ldots, a_m)$.
(b) There exist $n_i \in \mathbb{N}$, $i = 1, \ldots, m$, such that $a_1^{n_1} \in F$ and $a_i^{n_i} \in F(a_1, \ldots, a_{i-1})$ for $i = 2, \ldots, m$.

Example 7.168. Let $f(X) = aX^2 + bX + c$, where a, b, c are complex numbers. Set $F = \mathbb{Q}(a, b, c)$ and $a_1 = \sqrt{b^2 - 4ac}$. From the formula for roots of a quadratic equation, we see that $f(X)$ splits over $F(a_1)$. Since $a_1^2 \in F$, $f(X)$ is solvable by radicals over F.

What does the condition that $f(X) \in F[X]$ is solvable by radicals over F really mean? Let us go slowly through the definition. The field F contains the coefficients of $f(X)$; for example, it may be the subfield generated by all its coefficients. The element a_1 has the property that $a_1^{n_1} \in F$, so we may write $a_1 = \sqrt[n_1]{\alpha}$ where $\alpha \in F$ (let us put aside the question of the precise meaning of the symbol $\sqrt[n_1]{\cdot}$ and use it just as a suggestive notation). Since a_1 is algebraic over F (of degree n_1 or less), every element in $F(a_1)$ is of the form $\sum_i \lambda_i \sqrt[n_1]{\alpha^i}$ with $\lambda_i \in F$. Likewise, we write $a_2 = \sqrt[n_2]{\beta}$ where $\beta \in F(a_1)$, so elements in $F(a_1, a_2)$ are of the form $\sum_j \gamma_j \sqrt[n_2]{\beta^j}$ with $\gamma_j \in F(a_1)$. Similarly, we describe $F(a_1, a_2, a_3)$, $F(a_1, a_2, a_3, a_4)$, etc. By assumption, $F(a_1, \ldots, a_m)$ contains all roots of $f(X)$.

The notion of solvability by radicals is thus a formalization of our intuitive understanding of expressing the roots of a polynomial in terms of its coefficients, using only the operations of addition, subtraction, multiplication, division, and extraction of roots. From the formulas for solutions of cubic and quartic equations, it can be inferred that every polynomial with complex coefficients of degree at most 4 is solvable by radicals over the field generated by its coefficients.

As we saw in Example 7.168, it is not difficult to show that some polynomials are solvable by radicals. But how to show that some are not? To answer this question, we will bring together three topics: solvability by radicals, Galois theory, and, surprisingly, solvable groups, which were considered in Section 6.4.

In what follows, we will restrict ourselves to subfields of $\mathbb{C}$. We could easily cover general fields of characteristic 0, but let us rather focus on a more concrete situation for simplicity of exposition. The next lemma considers the Galois group of the splitting field for the polynomial $X^n - \alpha$. As is evident from Example 7.140, this group may be non-Abelian. We will show, however, that it is always solvable.

Lemma 7.169. *Let F be a subfield of $\mathbb{C}$, let $\alpha \in F$, and let $n \in \mathbb{N}$. Then the Galois group of the polynomial $X^n - \alpha$ is solvable over F.*

Proof. Let K be the splitting field for $X^n - \alpha$ over F, and let $a \in K$ be one of its roots. The other roots are then $a\omega^i$, $i = 1, \ldots, n-1$, where $\omega = e^{\frac{2\pi i}{n}}$. If $\alpha = 0$, then $K = F$ and the theorem trivially holds. Thus, let $\alpha \neq 0$. Then $a \neq 0$ and therefore $\omega = a^{-1}(a\omega) \in K$, so $K = F(a, \omega)$.

Note that $F(\omega)$ is the splitting field for $X^n - 1$ over F. We may therefore speak about the Galois group $\mathrm{Gal}(F(\omega)/F)$. Let σ and ρ be two elements of this group. Since they necessarily send ω to some other root of $X^n - 1$, we have $\sigma(\omega) = \omega^i$ and $\rho(\omega) = \omega^j$ for some i, j. Hence,

$$(\rho\sigma)(\omega) = \rho(\sigma(\omega)) = \rho(\omega^i) = \rho(\omega)^i = (\omega^j)^i = \omega^{ji}.$$

Similarly we see that $(\sigma\rho)(\omega) = \omega^{ji}$. As the elements of $\mathrm{Gal}(F(\omega)/F)$ are completely determined by their action on ω, it follows that $\sigma\rho = \rho\sigma$. That is, the group $\mathrm{Gal}(F(\omega)/F)$ is Abelian (compare Exercise 7.164).

Since $F(\omega)$ is an intermediate field between F and K, we may also speak about the Galois group $\mathrm{Gal}(K/F(\omega))$. Elements in this group are completely determined by their action on a; if $\sigma' \in \mathrm{Gal}(K/F(\omega))$ sends a to $a\omega^k$ and $\rho' \in \mathrm{Gal}(K/F(\omega))$ sends a to $a\omega^\ell$, then both $\rho'\sigma'$ and $\sigma'\rho'$ send a to $a\omega^{k+\ell}$. Therefore, $\mathrm{Gal}(K/F(\omega))$ is an Abelian group too.

Theorem 7.148 (d) tells us that $H := \mathrm{Gal}(K/F(\omega))$ is a normal subgroup of $G := \mathrm{Gal}(K/F)$ and $G/H \cong \mathrm{Gal}(F(\omega)/F)$. Thus, G contains a normal subgroup H such that both H and G/H are Abelian. In particular, G is solvable. $\qquad\square$

We can now establish the connection between solvability by radicals and solvability of groups. In the proof, we will use Corollary 6.105, which states that if $H \lhd G$, then G is solvable if and only if both H and G/H are solvable.

Theorem 7.170. *Let F be a subfield of $\mathbb{C}$ and let $f(X) \in F[X]$. If $f(X) \in F[X]$ is solvable by radicals over F, then the Galois group of $f(X)$ over F is solvable.*

Proof. Let a_i, n_i, and m be as in Definition 7.167. In particular, $\alpha := a_1^{n_1} \in F$. Denote by K_f the splitting field for $f(X)$ over F, by K_α the splitting field for $X^{n_1} - \alpha$ over F, and by K the splitting field for $f(X)$ over K_α. Note that K is also the splitting field for $f(X)(X^{n_1} - \alpha)$ over F, so all K_α, K_f, and K are Galois extensions of F. We also remark that K_f and K_α are intermediate fields between F and K, $a_1 \in K_\alpha$ and so $F(a_1) \subseteq K_\alpha$, and $\mathrm{Gal}(K_\alpha/F)$ is a solvable group by Lemma 7.169.

Our task is to prove that $\mathrm{Gal}(K_f/F)$ is a solvable group. To this end, it suffices to show that the group $G := \mathrm{Gal}(K/F)$ is solvable. Indeed, from Theorem 7.148 (d) it follows that $H_f := \mathrm{Gal}(K/K_f)$ is a normal subgroup of G and $G/H_f \cong \mathrm{Gal}(K_f/F)$. Corollary 6.105 therefore shows that $\mathrm{Gal}(K_f/F)$ is solvable if G is.

We will prove that $\mathrm{Gal}(K_f/F)$ is solvable by induction on m.

If $m = 1$, then $f(X)$ splits over $F(a_1) \subseteq K_\alpha$ and so $K = K_\alpha$. Therefore, $G = \mathrm{Gal}(K/F)$ is solvable and the desired conclusion follows.

Let $m > 1$. From $a_1 \in K_\alpha$ it follows that $f(X)$ splits over $K_\alpha(a_2, \ldots, a_m)$, and so $f(X)$ is solvable by radicals over K_α. Since K is the splitting field for $f(X)$ over K_α, we can use the induction hypothesis to conclude that the group $H_\alpha := \mathrm{Gal}(K/K_\alpha)$ is solvable. Next, Theorem 7.148 (d) shows that H_α is a normal subgroup of G and $G/H_\alpha \cong \mathrm{Gal}(K_\alpha/F)$. Therefore, both H_α and G/H_α are solvable groups, and so G is solvable too by Corollary 6.105; the desired conclusion follows. $\qquad\square$

The converse of Theorem 7.170 is also true. That is, if the Galois group of $f(X)$ over F is solvable, then $f(X)$ is solvable by radicals over F. This is interesting and important, but not needed for our purposes. Theorem 7.170 is sufficient for our main goal which is settling the question of solvability of **quintic polynomials**, i.e., polynomials of degree 5.

Lemma 7.171. *An irreducible quintic polynomial $p(X) \in \mathbb{Q}[X]$ with exactly three real roots is not solvable by radicals over $\mathbb{Q}$.*

Proof. Since $p(X)$ is irreducible, it has five distinct roots (Theorem 7.125). Let a_1, a_2, a_3 be the real roots, and a_4, a_5 the nonreal complex roots. Since $p(X)$ has real coefficients, a_4 and a_5 are complex conjugates of each other.

Let $K := \mathbb{Q}(a_1, a_2, a_3, a_4, a_5)$ and $G := \operatorname{Gal}(K/\mathbb{Q})$. Thus, K is the splitting field for $p(X)$ over $\mathbb{Q}$, and G is the Galois group of $p(X)$ over $\mathbb{Q}$. Our goal is to show that G is not solvable. An automorphism $\sigma \in G$ is completely determined by its action on the roots a_i, and sends each a_i into another root a_j (Lemma 7.141). Note that the restriction of σ to the set $\{a_1, a_2, a_3, a_4, a_5\}$ is a permutation, and so we may consider G as a subgroup of the symmetric group S_5 (in formal terms, $\sigma \mapsto \sigma \mid \{a_1, a_2, a_3, a_4, a_5\}$ is an embedding of G into the group of permutations of a set with 5 elements).

The complex conjugation $z \mapsto \bar{z}$ is an element of G. Since it fixes a_1, a_2, a_3 and interchanges a_4 and a_5, it is, when regarded as a permutation, a transposition.

Let a be any of the roots a_i. Since $p(X)$ is, up to a scalar multiple, the minimal polynomial of a, $[\mathbb{Q}(a) : \mathbb{Q}]$ is equal to $\deg(p(X)) = 5$. Hence, $[K : \mathbb{Q}] = 5[K : \mathbb{Q}(a)]$. Since $[K : \mathbb{Q}] = |G|$ by Theorem 7.145, it follows that 5 divides the order of G. By Cauchy's Theorem, G contains an element of order 5. The only elements in S_5 of order 5, however, are the 5-cycles.

We now know that G contains a transposition τ and a 5-cycle σ. Without loss of generality we may assume that $\sigma = (1\,2\,3\,4\,5)$ and $\tau = (1\,k)$. Note that σ^{k-1} is a 5-cycle that sends 1 to k. Replacing σ by σ^{k-1} and renumbering, we may further assume that $\tau = (1\,2)$. Considering $\sigma^i \tau \sigma^{-i}$ we see that G contains the transpositions $(2\,3)$, $(3\,4)$, $(4\,5)$, and $(5\,1)$, and hence also $(1\,2)(2\,3)(1\,2) = (1\,3)$, $(1\,3)(3\,4)(1\,3) = (1\,4)$, etc. Thus, G contains all the transpositions of S_5, and is therefore equal to S_5.

Corollary 6.118 shows that S_5 is not solvable, and so $p(X)$ is not solvable by radicals over $\mathbb{Q}$ by Theorem 7.170. $\square$

It remains to show that such polynomials $p(X)$ exist. We will provide one concrete example (there is nothing very special about it, it is not difficult to find more examples).

Theorem 7.172. *There exists a quintic polynomial in $\mathbb{Q}[X]$ that is not solvable by radicals over $\mathbb{Q}$.*

Proof. Let $p(X) = X^5 - 3X^4 + 3$. By the Eisenstein Criterion, $p(X)$ is irreducible over $\mathbb{Q}$. Since the polynomial function $x \mapsto p(x)$ is continuous and $p(-1) = -1$, $p(0) = 3$, $p(2) = -13$, and $p(3) = 3$, the Intermediate Value Theorem shows that $p(X)$ has a real root a_1 between -1 and 0, a real root a_2 between 0 and 2, and a real root a_3 between 2 and 3. If $p(X)$ had more than three real roots, then its derivative $p'(X) = 5X^4 - 12X^3$ would have at least three roots by Rolle's Theorem. However, the equation $5x^4 = 12x^3$ has only two solutions in $\mathbb{R}$ (0 and $\frac{12}{5}$). Therefore, the other two roots a_4 and a_5 of $p(X)$ are not real. By Lemma 7.171, $p(X)$ is not solvable by radicals over $\mathbb{Q}$. $\square$

Theorem 7.172 is sometimes called the **Abel–Ruffini Theorem**. Abel and Ruffini actually showed that there can be no formula for solutions of a general quintic equation (see Section 7.1). Theorem 7.172 tells us more than just that.

Exercises

7.173. Use Cardano's formula (7.2) to show that the cubic polynomials $X^3 - pX - q$ are solvable by radicals.

7.174. Show that the (unproved) converse of Theorem 7.170 implies that every polynomial of degree at most 4 is solvable by radicals.

Hint. Example 6.98, Corollary 6.104.

7.175. Prove that every polynomial in $\mathbb{R}[X]$ is solvable by radicals over $\mathbb{R}$.

Comment. Polynomials such as $X^5 - 3X^4 + 3$ are thus solvable by radicals over $\mathbb{R}$, but not over $\mathbb{Q}$.

7.176. Give an example of an irreducible quintic polynomial in $\mathbb{Q}[X]$ that is solvable by radicals over $\mathbb{Q}$.

7.177. Show that the polynomial $X^5 - pX^4 + p$ is not solvable by radicals over $\mathbb{Q}$ for every prime p.

Hint. If $p \geq 3$, consider the evaluations at $-1, 0, p - 1$, and p.

7.178. Find $a, b \in \mathbb{Z}$ such that the polynomial $X^5 + aX + b$ is not solvable by radicals over $\mathbb{Q}$.

7.179. Does there exist an irreducible quintic polynomial in $\mathbb{Q}[X]$ such that its Galois group over $\mathbb{Q}$ embeds into S_4?

7.180. Use Theorem 7.172 to show that, for every $n \geq 5$, there exists a polynomial in $\mathbb{Q}[X]$ of degree n that is not solvable by radicals over $\mathbb{Q}$.

7.181. Generalize Lemma 7.171 by proving that an irreducible polynomial in $\mathbb{Q}[X]$ of prime degree $p \geq 5$ with exactly $p - 2$ real roots is not solvable by radicals over $\mathbb{Q}$.

7.182. Write ω for $e^{\frac{2\pi i}{5}}$. Then $a_i := \omega^{i-1}, i = 1, 2, 3, 4, 5$, are the roots of $X^5 - 1$. Determine all permutations of $\{a_1, a_2, a_3, a_4, a_5\}$ that correspond to the elements of the Galois group of $X^5 - 1$ over $\mathbb{Q}$.

7.183. Let a_i be as in the preceding exercise. Then $b_i := \sqrt[5]{2}a_i, i = 1, 2, 3, 4, 5$, are the roots of $X^5 - 2$. Provide an example of a permutation of $\{b_1, b_2, b_3, b_4, b_5\}$ that does not correspond to an element of the Galois group of $X^5 - 2$ over $\mathbb{Q}$.

7.10 The Fundamental Theorem of Algebra

This may look like a perfect title for the last section of a book on algebra. However, as we explained in Section 7.1, the adjective "fundamental" here should be considered with some reservation; the Fundamental Theorem of Algebra is indeed closely connected with the roots of algebra (hence the name), but is actually more an analytic than an algebraic result. Nevertheless, it does seem convenient to end the book with its proof, since this will give us an opportunity to see some "fundamentals of algebra" at work. In our proof we will use Galois theory and the Sylow theorems (Theorem 6.64) as the main tools. It should be pointed out that there are quite elementary analytic proofs, as well as other advanced proofs that use methods from different mathematical areas. The proof we are about to give is supposed to serve as an illustration of how abstract algebra can be used in "real life," that is, in solving problems that are not algebraic in nature.

We cannot produce a proof entirely without analysis. All we need, however, is the following elementary lemma.

Lemma 7.184. *A polynomial $f(X) \in \mathbb{R}[X]$ of odd degree has a real root.*

The lemma is geometrically obvious (think of the graph!), and also easily proven. Suppose the leading coefficient of $f(X)$ is positive. It is a simple exercise to show that $f(x) > 0$ for every large enough positive x (in fact, $\lim_{x \to \infty} f(x) = \infty$), and $f(x) < 0$ for every small enough negative x. Since the polynomial function $x \mapsto f(x)$ is continuous, there must exist an $a \in \mathbb{R}$ such that $f(a) = 0$.

Theorem 7.185. (Fundamental Theorem of Algebra) *The field of complex numbers is algebraically closed.*

Proof. Let $g(X) = \sum_{i=0}^{n} a_i X^i \in \mathbb{C}[X]$ be a non-constant polynomial. Define $\overline{g}(X) = \sum_{i=0}^{n} \overline{a}_i X^i$, where $\overline{a}_i$ is the complex conjugate of a_i, and observe that the polynomial $g(X)\overline{g}(X)$ has real coefficients. Since $g(\overline{z}) = \overline{\overline{g}(z)}$, $g(X)$ has a root in $\mathbb{C}$ if and only if $g(X)\overline{g}(X)$ has a root in $\mathbb{C}$. Therefore, it suffices to prove that every non-constant polynomial in $\mathbb{R}[X]$ has a root in $\mathbb{C}$.

Take a non-constant polynomial $f(X) \in \mathbb{R}[X]$ and denote by K its splitting field over $\mathbb{C}$. Our goal is to show that $K = \mathbb{C}$. Since K can also be described as the splitting field for $f(X)(X^2 + 1)$ over $\mathbb{R}$, it is a Galois extension of $\mathbb{R}$. Set $G := \mathrm{Gal}(K/\mathbb{R})$ and $G' := \mathrm{Gal}(K/\mathbb{C})$. As $[\mathbb{C} : \mathbb{R}] = 2$, it follows from Theorem 7.148 (b) that $[G : G'] = 2$, i.e., $|G| = 2|G'|$. Write $|G| = 2^k t$, where $k \geq 1$ and t is odd. By Theorem 6.64 (a), G has a Sylow 2-subgroup H (i.e., a subgroup of order 2^k). Setting $L := K^H$, we see from Theorem 7.148 (b) that

$$[L : \mathbb{R}] = [G : H] = t.$$

Therefore, $[L : \mathbb{R}(a)] \cdot [\mathbb{R}(a) : \mathbb{R}] = t$ for every $a \in L$, implying that $[\mathbb{R}(a) : \mathbb{R}]$ is odd; that is, the minimal polynomial of a over $\mathbb{R}$ has odd degree. However, among polynomials in $\mathbb{R}[X]$ of odd degree, only linear ones are irreducible in view of Lemma 7.184. Thus, $\mathbb{R}(a) = \mathbb{R}$ for every $a \in L$, so $L = \mathbb{R}$ and hence $t = 1$. Therefore, $|G| = 2^k$ and $|G'| = 2^{k-1}$.

Suppose $k > 1$. Using Theorem 6.64 (a) again, it follows that G' contains a subgroup H' of order 2^{k-2}. Hence, using Theorem 7.148 (b) again, we see that $L' := K^{H'}$ satisfies

$$[L' : \mathbb{C}] = [G' : H'] = 2.$$

Accordingly, if $b \in L' \setminus \mathbb{C}$, then b is algebraic of degree 2 over $\mathbb{C}$. If $\alpha, \beta \in \mathbb{C}$ are such that b is a root of $X^2 + \alpha X + \beta$, then $b + \frac{\alpha}{2}$ is a root of $X^2 - \gamma$ where $\gamma = \frac{\alpha^2}{4} - \beta$. However, it is an elementary property of complex numbers that an equation of the form $x^2 = \gamma$ with $\gamma \in \mathbb{C}$ has solutions in $\mathbb{C}$. Therefore, $b + \frac{\alpha}{2} \in \mathbb{C}$ and hence $b \in \mathbb{C}$, contrary to our assumption. Hence $k = 1$, i.e., $|G'| = 1$. Theorem 7.145 (condition (i)) now shows that $K = \mathbb{C}$, which means that all the roots of $f(X)$ lie in $\mathbb{C}$.   □

Exercises

7.186. Let $K \supsetneq \mathbb{R}$ be a finite extension of $\mathbb{R}$. Prove that $K \cong \mathbb{C}$.

7.187. Let $K \supsetneq \mathbb{C}$ be an extension of $\mathbb{C}$. Prove that the field of rational functions $\mathbb{C}(X)$ embeds into K.

Hint. Remark 7.37.

7.188. Let A be a finite-dimensional complex algebra. Prove that for every $a \in A$ there exist a scalar λ and a nonzero $x \in A$ such that $ax = \lambda x$. Is this also true if A is a real algebra?

Index

© Springer Nature Switzerland AG 2019
M. Brešar, *Undergraduate Algebra*, Springer Undergraduate Mathematics Series,
https://doi.org/10.1007/978-3-030-14053-3

Printed in the United States
By Bookmasters